PLANE AND SPHERICAL

TRIGONOMETRY AND TABLES

BY

G. A. WENTWORTH

AUTHOR OF A SERIES OF TEXT-BOOKS IN MATHEMATICS

SECOND REVISED EDITION

BOSTON, U.S.A.

GINN & COMPANY, PUBLISHERS

The Athenæum Press

1903

PREFACE

IN preparing this work the aim has been to furnish just so much of Trigonometry as is actually taught in our best schools and colleges. Consequently, all investigations that are important only for the special student have been omitted, except the development of functions in series. The principles have been unfolded with the utmost brevity consistent with simplicity and clearness, and interesting problems have been selected with a view to awaken a real love for the study. Much time and labor have been spent in devising the simplest proofs for the propositions, and in exhibiting the best methods of arranging the logarithmic work.

The author acknowledges his obligation to G. A. Hill, A.M., of Cambridge, Mass., to Dr. F. N. Cole, of New York, N.Y., to Professor S. F. Norris, of Baltimore, Md., and to Professor B. F. Yanney, of Alliance, Ohio.

G. A. WENTWORTH.

EXETER, N.H.,
January, 1903.

CONTENTS

[The numbers refer to the pages.]

PLANE TRIGONOMETRY

v

CHAPTER IV. THE OBLIQUE TRIANGLE:

CHAPTER V. MISCELLANEOUS EXAMPLES:

CHAPTER VI. CONSTRUCTION OF TABLES:

SPHERICAL TRIGONOMETRY

CHAPTER VII. THE RIGHT SPHERICAL TRIANGLE:

CHAPTER VIII. THE OBLIQUE SPHERICAL TRIANGLE:

PLANE TRIGONOMETRY

CHAPTER I

TRIGONOMETRIC FUNCTIONS OF ACUTE ANGLES

SECTION I

ANGULAR MEASURE

As lengths are measured in terms of various conventional units, the foot, the meter, etc., so different units for measuring angles are employed, or have been proposed.

In the **common** or **sexagesimal system** the circumference of a circle is divided into 360 equal parts. The angle at the centre subtended by each of these parts is taken as the unit angle and is called a **degree**. The degree is subdivided into 60 *minutes*, and the minute into 60 *seconds*. Degrees, minutes, and seconds are denoted by symbols. Thus, 6 degrees 5 minutes 7 seconds is written 6° 5′ 7″.

Note. The sexagesimal system was employed by the early Babylonian astronomers to conform with their year of 360 days.

In the **circular system** an arc of a circle is laid off equal in length to the radius. The angle at the centre subtended by this arc is taken as the unit angle and is called a **radian**.

The number of radians in 360° is equal to the number of times the length of the radius is contained in the length of the circumference. It is proved in Geometry that this number is

1

2π for all circles, π being equal to 3.1416, nearly. Therefore the radian is the same angle in all circles.

The circumference of a circle is 2π times the radius.

Hence, 2π radians $= 360°$, and π radians $= 180°$.

Therefore, $1 \text{ radian} = \dfrac{180°}{\pi} = 57°\,17'\,45''$,

and $1 \text{ degree} = \dfrac{\pi}{180} \text{ radian} = 0.017453 \text{ radian}.$

By the last two equations the measure of an angle can be changed from radians to degrees or from degrees to radians.

Thus, $2 \text{ radians} = 2 \times \dfrac{180°}{\pi} = 2 \times (57°\,17'\,45'') = 114°\,35'\,30''.$

NOTE. The circular system came into use early in the eighteenth century. It is found more convenient in the higher mathematics, where the radians are expressed simply as numbers. Thus, the angle π means π radians, and the angle 3 means 3 radians.

On the introduction of the metric system of weights and measures at the close of the eighteenth century, it was proposed to divide the right angle into 100 equal parts called *grades*, which were to be taken as units. The grade was subdivided into 100 *minutes* and the minute into 100 *seconds*. This *French* or *centesimal* system, however, never came into actual use.

EXERCISE I

[Assume $\pi = 3.1416.$]

1. Reduce the following angles to circular measure, expressing the results as fractions of π:

$60°$, $45°$, $150°$, $195°$, $11°\,15'$, $123°\,45'$, $37°\,30'$.

2. How many degrees are there in $\frac{2}{3}\pi$ radians? $\frac{3}{4}\pi$ radians? $\frac{5}{8}\pi$ radians? $\frac{14}{15}\pi$ radians? $\frac{7}{15}\pi$ radians?

3. What decimal part of a radian is $1°$? $1'$?

4. How many seconds in a radian?

5. Express in radians one of the interior angles of a regular octagon; of a regular dodecagon.

6. On the circumference of a circle of 50 feet radius an arc of 10 feet is laid off. How many degrees in the angle at the centre subtended by this arc?

7. The earth's equatorial radius is approximately 3963 miles. If two points on the equator are 1000 miles apart, what is their difference in longitude?

8. If the difference in longitude of two points on the equator is 1°, what is the distance between them in miles?

9. What is the radius of a circle, if an arc of 1 foot subtends an angle of 1° at the centre?

10. In how many hours is a point on the equator carried by the rotation of the earth on its axis through a distance equal to the earth's radius?

11. The minute hand of a clock is $3\frac{1}{2}$ feet long. How far does its extremity move in 25 minutes? (Take $\pi = \frac{22}{7}$.)

12. A wheel makes 15 revolutions a second. How long does it take to turn through 4 radians? (Take $\pi = \frac{22}{7}$.)

SECTION II

THE TRIGONOMETRIC FUNCTIONS

The sides and angles of a plane triangle are so related that any three given parts, provided at least one of them is a side, determine the shape and the size of the triangle.

Geometry shows how, from three such parts, to **construct** the triangle.

Trigonometry shows how to **compute** the unknown parts of a triangle from the numerical values of the given parts.

. Geometry shows in a general way that the sides and angles of a triangle are mutually dependent. Trigonometry begins by showing the exact nature of this dependence in the **right triangle**, and for this purpose employs the **ratios of the sides**.

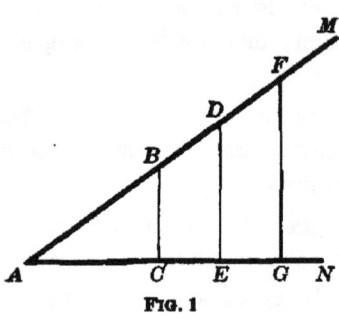

Fig. 1

Let MAN (Fig. 1) be an acute angle. If from any points B, D, F in one of its sides perpendiculars BC, DE, FG are let fall to the other side, then the right triangles ACB, AED, AGF thus formed have the angle A common, and are therefore mutually equiangular and similar. Hence, the ratios of their corresponding sides, pair by pair, are equal. That is,

$$\frac{AC}{AB} = \frac{AE}{AD} = \frac{AG}{AF}; \quad \frac{AC}{BC} = \frac{AE}{DE} = \frac{AG}{FG}; \quad \frac{BC}{AB} = \frac{DE}{AD} = \frac{FG}{AF}.$$

These ratios, therefore, remain unchanged so long as the angle A remains unchanged.

Hence, for every value of an acute angle A there are certain *numbers* that express the values of the **ratios of the sides** in all right triangles that have this acute angle A.

There are all together six different ratios :

I. The ratio of the opposite leg to the hypotenuse is called the **Sine of A**, and is written **sin A**.

II. The ratio of the adjacent leg to the hypotenuse is called the **Cosine of A**, and is written **cos A**.

III. The ratio of the opposite leg to the adjacent leg is called the **Tangent of A**, and is written **tan A**.

IV. The ratio of the adjacent leg to the opposite leg is called the **Cotangent of A**, and is written **cot A**.

V. The ratio of the hypotenuse to the adjacent leg is called the **Secant of A**, and is written **sec A**.

VI. The ratio of the hypotenuse to the opposite leg is called the **Cosecant of A**, and is written **csc A**.

These six ratios are called the **Trigonometric Functions** of the angle A.

To these six ratios are often added the two following functions, which also depend only on the angle A:

VII. The **Versed Sine of A** is $1 - \cos A$, and is written **vers A.**

VIII. The **Coversed Sine of A** is $1 - \sin A$, and is written **covers A.**

In the right triangle ACB (Fig. 2) let a, b, c denote the lengths of the sides opposite the acute angles A, B, and the right angle C, respectively, these lengths being all expressed in terms of a common unit. Then,

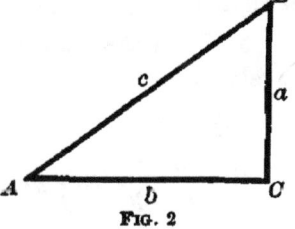

FIG. 2

$$\sin A = \frac{a}{c} = \frac{\text{opposite leg}}{\text{hypotenuse}}, \qquad \cos A = \frac{b}{c} = \frac{\text{adjacent leg}}{\text{hypotenuse}},$$

$$\tan A = \frac{a}{b} = \frac{\text{opposite leg}}{\text{adjacent leg}}, \qquad \cot A = \frac{b}{a} = \frac{\text{adjacent leg}}{\text{opposite leg}},$$

$$\sec A = \frac{c}{b} = \frac{\text{hypotenuse}}{\text{adjacent leg}}, \qquad \csc A = \frac{c}{a} = \frac{\text{hypotenuse}}{\text{opposite leg}},$$

$$\text{vers } A = 1 - \frac{b}{c} = \frac{c-b}{c}, \qquad \text{covers } A = 1 - \frac{a}{c} = \frac{c-a}{c}.$$

EXERCISE II

1. What are the functions of the other acute angle B of the triangle ACB (Fig. 2)?

2. Compare the functions of A and B, and show that

$\sin A = \cos B,$	$\sec A = \csc B,$
$\cos A = \sin B,$	$\csc A = \sec B,$
$\tan A = \cot B,$	$\text{vers } A = \text{covers } B,$
$\cot A = \tan B,$	$\text{covers } A = \text{vers } B.$

3. Find the values of the functions of A, if a, b, c, respectively, have the following values :

   (i) 3, 4, 5.   (iii) 8, 15, 17.   (v) 3.9, 8, 8.9.

   (ii) 5, 12, 13.   (iv) 9, 40, 41.   (vi) 1.19, 1.20, 1.69.

4. What condition must be fulfilled by the lengths of the three lines a, b, c (Fig. 2) in order to make them the sides of a right triangle ?   Is this condition fulfilled in Example 3 ?

5. Find the values of the functions of A, if a, b, c, respectively, have the following values:

   (i) $2\,mn$, $m^2 - n^2$, $m^2 + n^2$.   (iii) pqr, qrs, rsp.

   (ii) $\dfrac{2\,xy}{x-y}$, $x+y$, $\dfrac{x^2+y^2}{x-y}$.   (iv) $\dfrac{mn}{pq}$, $\dfrac{mv}{sq}$, $\dfrac{nr}{ps}$.

6. Prove that the values of a, b, c, in (i) and (ii), Example 5, satisfy the condition necessary to make them the sides of a right triangle.

7. What equations of condition must be satisfied by the values of a, b, c in (iii) and (iv), Example 5, in order that the values may represent the sides of a right triangle ?

Given $a^2 + b^2 = c^2$; find the functions of A and B when:

  8. $a = 24$, $b = 143$.   **11.** $a = \sqrt{p^2 + q^2}$, $b = \sqrt{2\,pq}$.

  9. $a = 0.264$, $c = 0.265$. **12.** $a = \sqrt{p^2 + pq}$, $c = p + q$.

 10. $b = 9.5$, $c = 19.3$.   **13.** $b = 2\sqrt{pq}$, $c = p + q$.

Given $a^2 + b^2 = c^2$; find the functions of A when:

 14. $a = 2\,b$.   **16.** $a + b = \frac{4}{5}\,c$.

 15. $a = \frac{2}{3}\,c$.   **17.** $a - b = \frac{1}{4}\,c$.

 18. Find a if $\sin A = \frac{3}{5}$, and $c = 20.5$.

 19. Find b if $\cos A = 0.44$, and $c = 3.5$.

 20. Find a if $\tan A = \frac{1}{3}$, and $b = 2\frac{4}{11}$.

 21. Find b if $\cot A = 4$, and $a = 17$.

22. Find c if sec $A = 2$, and $b = 20$.

23. Find c if csc $A = 6.45$, and $a = 35.6$.

Construct a right triangle, given :

24. $c = 6$, tan $A = \frac{2}{3}$. **26.** $b = 2$, sin $A = 0.6$.

25. $a = 3.5$, cos $A = \frac{1}{2}$. **27.** $b = 4$, csc $A = 4$.

28. In a right triangle $c = 2.5$ miles, sin $A = 0.6$, cos $A = 0.8$; compute the legs.

29. Construct with a protractor the angles 20°, 40°, and 70°; determine their functions by measuring the necessary lines, and compare the values obtained in this way with the more nearly correct values given in the following table :

	sin	cos	tan	cot	sec	csc
20°	0.342	0.940	0.364	2.747	1.064	2.924
40°	0.643	0.766	0.839	1.192	1.305	1.556
70°	0.940	0.342	2.747	0.364	2.924	1.064

30. Find, by means of the above table, the legs of a right triangle if $A = 20°$, $c = 1$; also if $A = 20°$, $c = 4$.

31. By dividing the length of a vertical rod by the length of its horizontal shadow, the tangent of the angle of elevation of the sun at the time of observation was found to be 0.82. How high is a tower, if the length of its horizontal shadow at the same time is 174.3 yards?

SECTION III

REPRESENTATION OF THE FUNCTIONS BY LINES

The functions of an angle, being ratios, are *numbers ;* but we may represent them by *lines* if we first choose a unit of length, and then construct right triangles, such that the denominators of the ratios shall be equal to this unit.

The most convenient way is the following:

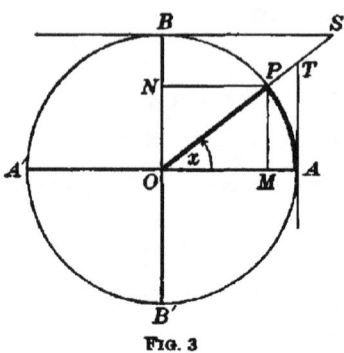

FIG. 3

About a point O (Fig. 3) as a centre, with a radius equal to one unit of length, describe a circle, and draw the horizontal diameter AA' and the diameter BB' perpendicular to AA'.

A circle with radius equal to 1 is called a *unit* circle.

Let AOP be an acute angle, and let its value (in degrees, etc.) be denoted by x. We may regard the angle x as generated by a line OP that revolves about O from the **initial position** OA to the **terminal position** OP.

Draw $PM \perp$ to OA, $PN \perp$ to OB.

In the rt. $\triangle OMP$ the hypotenuse $OP = 1$.

Therefore, $\sin x = \dfrac{MP}{OP} = MP$; $\cos x = \dfrac{OM}{OP} = OM$.

Through A and B draw tangents to the circle meeting OP produced in T and S, respectively; then, in the rt. $\triangle OAT$ and OBS, $OA = 1$, the leg $OB = 1$, and the $\angle OSB =$ the $\angle x$. Therefore,

$$\tan x = \frac{AT}{OA} = AT; \qquad\qquad \cot x = \frac{BS}{OB} = BS;$$

$$\sec x = \frac{OT}{OA} \doteq OT; \qquad\qquad \csc x = \frac{OS}{OB} = OS;$$

$$\operatorname{vers} x = 1 - OM = MA; \quad \operatorname{covers} x = 1 - ON = NB.$$

These eight *line* values of the functions are all expressed in terms of the radius of the circle as a unit; and it is clear that as the angle *varies in value* the line values of the functions will always remain equal *numerically* to the ratio values. Hence, in studying the changes in the functions as the angle is supposed to vary in value, we may employ the simpler line values instead of the ratio values.

1. Represent by lines the functions of an acute angle larger than that shown in Fig. 3.

If x is an acute angle, show that:

2. $\sin x$ is less than $\tan x$.

3. $\sec x$ is greater than $\tan x$.

4. $\csc x$ is greater than $\cot x$.

Construct the angle x, if:

5. $\tan x = 3$.   **7.** $\cos x = \frac{1}{2}$.   **9.** $\sin x = 2 \cos x$.

6. $\csc x = 2$.   **8.** $\sin x = \cos x$.   **10.** $4 \sin x = \tan x$.

11. Show that the sine of an angle is equal to one-half the chord of twice the angle.

12. Find x if $\sin x$ is equal to one-half the side of a regular inscribed decagon.

Given x and y, $x + y$ being less than 90°; construct:

13. The value of $\sin(x + y) - \sin x$.

14. The value of $\tan(x + y) - \sin(x + y) + \tan x - \sin x$.

Given an angle x; construct an angle y such that:

15. $\sin y = 2 \sin x$.   **17.** $\tan y = 3 \tan x$.

16. $\cos y = \frac{1}{2} \cos x$.   **18.** $\sec y = \csc x$.

19. Show by construction that $2 \sin A > \sin 2 A$.

20. Given two angles A and B, $A + B$ being less than 90°; show that $\sin(A + B) < \sin A + \sin B$.

21. Given $\sin x$ in a unit circle; find the length of a line corresponding in position to $\sin x$ in a circle whose radius is r.

22. In a right triangle, given the hypotenuse c, and also $\sin A = m$, $\cos A = n$; find the legs.

SECTION IV

CHANGES IN THE FUNCTIONS AS THE ANGLE CHANGES

If we suppose the $\angle AOP$, or x (Fig. 4), to increase gradually to 90° by the revolution of the moving radius OP about

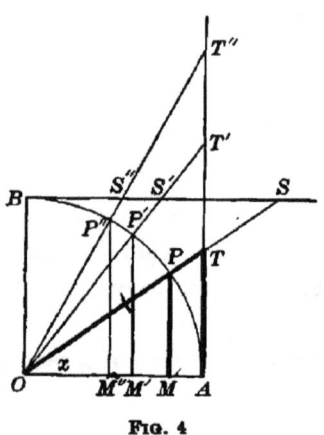

FIG. 4

O, the point P moves along the arc AB towards B, T moves along the tangent AT away from A, S moves along the tangent BS towards B, and M moves along the radius OA towards O.

Hence, the lines MP, AT, OT gradually increase in length, and the lines OM, BS, OS gradually decrease. That is,

As an acute angle increases to 90°, its sine, tangent, and secant also increase, while its cosine, cotangent, and cosecant decrease.

On the other hand, if we suppose x to decrease gradually, the reverse changes in its functions occur.

If we suppose x to decrease to 0°, OP coincides with OA and is parallel to BS. Therefore, MP and AT vanish, OM becomes equal to OA, while BS and OS are each infinitely long and are represented in value by the symbol ∞.

And if we suppose x to increase to 90°, OP coincides with OB and is parallel to AT. Therefore, MP and OS are each equal to OB, OM and BS vanish, while AT and OT are each infinite in length.

Hence, as the angle x increases from 0° to 90°,

sin x increases from 0 to 1,

cos x decreases from 1 to 0,

tan x increases from 0 to ∞,
cot x decreases from ∞ to 0,
sec x increases from 1 to ∞,
csc x decreases from ∞ to 1.

The values of the functions of 0° and of 90° are the *limiting* values of the functions of an acute angle. It is evident that for acute angles,

Sines and cosines are always less than 1;

Secants and cosecants are always greater than 1;

Tangents and cotangents have all values between 0 and ∞.

REMARK. We are now able to understand why the sine, cosine, etc., of an angle are called *functions* of the angle. By a *function* of any magnitude is meant another magnitude which remains constant so long as the first magnitude remains constant, but changes in value for every change in the value of the first magnitude. This, as we now see, is the relation in which the sine, cosine, etc., of an angle stand to the angle.

SECTION V

FUNCTIONS OF COMPLEMENTARY ANGLES

The general form of two complementary angles is A and $90° - A$.

In the rt. $\triangle ACB$ (Fig. 5),

$$A + B = 90°; \text{ hence } B = 90° - A.$$

Hence, putting $90° - A$ for B in the formulas on p. 5,

sin A = cos B = cos $(90° - A)$,

cos A = sin B = sin $(90° - A)$,

tan A = cot B = cot $(90° - A)$,

cot A = tan B = tan $(90° - A)$,

sec A = csc B = csc $(90° - A)$,

csc A = sec B = sec $(90° - A)$.

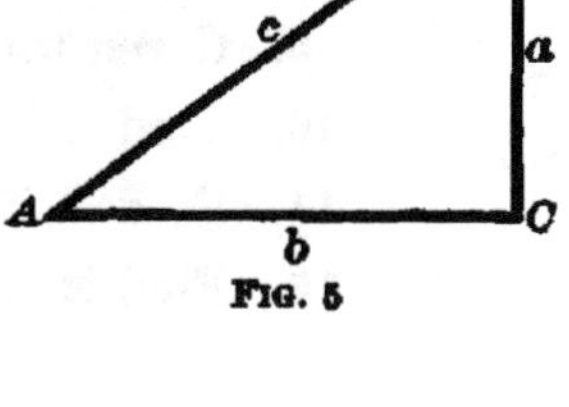

FIG. 5

Therefore,

Each function of an acute angle is equal to the co-named function of the complementary angle.

NOTE.   *Cosine, cotangent,* and *cosecant* are sometimes called *co-functions;* the words are simply abbreviated forms of *complement's sine, complement's tangent,* and *complement's secant.*

Hence, also,

Any function of an angle between 45° and 90° may be found by taking the co-named function of the complementary angle between 0° and 45°.

EXERCISE IV

1. Express as functions of the complementary angle:

   sin 30°.   tan 89°.   csc 18° 10'.   cot 82° 19'.
   cos 45°.   cot 15°.   cos 37° 24'.   csc 54° 46'.

2. Express as functions of an angle less than 45°:

   sin 60°.   tan 57°.   csc 69° 2'.   cot 89° 59'.
   cos 75°.   cot 84°.   cos 85° 39'.   csc 45° 1'.

3. Given $\tan 30° = \frac{1}{3}\sqrt{3}$; find cot 60°.

4. Given $\tan A = \cot A$; find A.

5. Given $\cos A = \sin 2A$; find A.

6. Given $\sin A = \cos 2A$; find A.

7. Given $\cos A = \sin (45° - \frac{1}{2}A)$; find A.

8. Given $\cot \frac{1}{2}A = \tan A$; find A.

9. Given $\tan (45° + A) = \cot A$; find A.

10. Find A if $\sin A = \cos 4A$.

11. Find A if $\cot A = \tan 8A$.

12. Find A if $\cot A = \tan nA$.

SECTION VI

RELATIONS OF THE FUNCTIONS OF AN ANGLE

Since (Fig. 6) $a^2 + b^2 = c^2$, therefore,

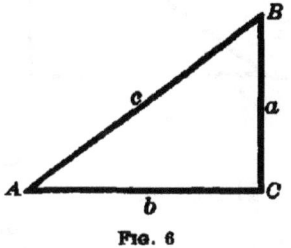

$$\frac{a^2}{c^2} + \frac{b^2}{c^2} = 1, \quad \text{or} \quad \left(\frac{a}{c}\right)^2 + \left(\frac{b}{c}\right)^2 = 1.$$

But $\frac{a}{c} = \sin A$, and $\frac{b}{c} = \cos A$;

therefore, $(\sin A)^2 + (\cos A)^2 = 1$;

or, as it is usually written for convenience,

Fig. 6

$$\sin^2 A + \cos^2 A = 1. \tag{1}$$

That is: *The sum of the squares of the sine and the cosine of an angle is equal to unity.*

Formula [1] enables us to find the cosine of an angle when the sine is known, and the sine when the cosine is known. The values of $\sin A$ and of $\cos A$ deduced from [1] are:

$$\sin A = \sqrt{1 - \cos^2 A}, \quad \cos A = \sqrt{1 - \sin^2 A}.$$

Since $$\frac{a}{c} \div \frac{b}{c} = \frac{a}{c} \times \frac{c}{b} = \frac{a}{b},$$

and since $\frac{a}{c} = \sin A$, $\frac{b}{c} = \cos A$, and $\frac{a}{b} = \tan A$,

therefore, $$\tan A = \frac{\sin A}{\cos A}. \tag{2}$$

That is: *The tangent of an angle is equal to the sine divided by the cosine.*

Formula [2] enables us to find the tangent of an angle when the sine and the cosine are known.

Now
$$\frac{a}{c} \times \frac{c}{a} = 1, \quad \frac{b}{c} \times \frac{c}{b} = 1, \quad \frac{a}{b} \times \frac{b}{a} = 1,$$

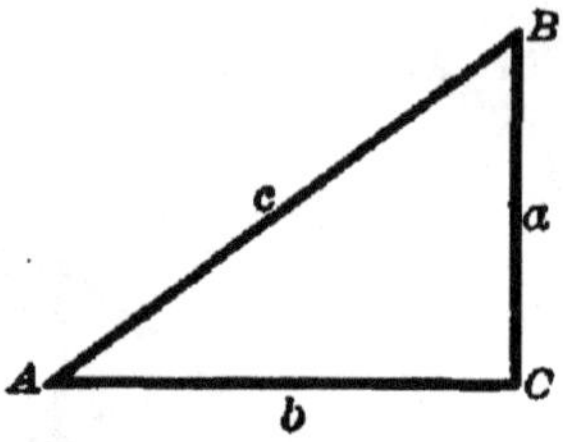

and

$$\frac{a}{c} = \sin A, \quad \frac{b}{c} = \cos A, \quad \frac{a}{b} = \tan A,$$

$$\frac{b}{a} = \cot A, \quad \frac{c}{b} = \sec A, \quad \frac{c}{a} = \csc A.$$

Fig. 7

Therefore,
$$\left. \begin{array}{l} \sin A \times \csc A = 1 \\ \cos A \times \sec A = 1 \\ \tan A \times \cot A = 1 \end{array} \right\}. \qquad [3]$$

That is: *The sine and the cosecant of an angle, the cosine and the secant, and the tangent and the cotangent, pair by pair, are reciprocals.*

The equations in [3] enable us to find an unknown function contained in any pair of these reciprocals when the other function in this pair is known.

EXERCISE V

1. Prove Formulas [1], [2], [3], using for the functions the line values in the unit circle given in Sect. III, page 8.

Prove that:

2. $1 + \tan^2 A = \sec^2 A.$ 　　　 3. $1 + \cot^2 A = \csc^2 A.$

NOTE. The equations in Examples 2 and 3 should be remembered.

4. $\cot A = \dfrac{\cos A}{\sin A}.$ 　　　 7. $\cos A \csc A = \cot A.$

5. $\sin A \sec A = \tan A.$ 　　　 8. $\tan A \cos A = \sin A.$

6. $\sin A \cot A = \cos A.$ 　　　 9. $\sin A \sec A \cot A = 1.$

10. $\cos A \csc A \tan A = 1.$

11. $(1 - \sin^2 A) \tan^2 A = \sin^2 A.$

12. $\sqrt{1 - \cos^2 A} \cot A = \cos A.$

13. $(1 + \tan^2 A) \sin^2 A = \tan^2 A.$

14. $(1 - \sin^2 A) \csc^2 A = \cot^2 A.$

15. $\tan^2 A \cos^2 A + \cos^2 A = 1.$

16. $(\sin^2 A - \cos^2 A)^2 = 1 - 4 \sin^2 A \cos^2 A.$

17. $(1 - \tan^2 A)^2 = \sec^4 A - 4 \tan^2 A.$

18. $\dfrac{\sin A}{\cos A} + \dfrac{\cos A}{\sin A} = \sec A \csc A.$

19. $\sin^4 A - \cos^4 A = \sin^2 A - \cos^2 A.$

20. $\sec A - \cos A = \sin A \tan A.$

21. $\csc A - \sin A = \cos A \cot A.$

22. $\dfrac{\cos A}{1 - \sin A} = \dfrac{1 + \sin A}{\cos A}.$

SECTION VII

APPLICATION OF FORMULAS [1], [2], [3]

Formulas [1], [2], and [3] enable us, when any one function of an angle is given, to find all the others. A given value of any one function, therefore, determines all the others.

EXAMPLE 1. Given $\sin A = \frac{2}{3}$; find the other functions.

By [1], p. 13, $\cos A = \sqrt{1 - \frac{4}{9}} = \sqrt{\frac{5}{9}} = \frac{1}{3} \sqrt{5}.$

By [2], p. 13, $\tan A = \frac{2}{3} \div \frac{1}{3} \sqrt{5} = \frac{2}{3} \times \dfrac{3}{\sqrt{5}} = \dfrac{2}{\sqrt{5}} = \frac{2}{5} \sqrt{5}.$

By [3], p. 14, $\cot A = \frac{1}{2} \sqrt{5}$, $\sec A = \frac{3}{5} \sqrt{5}$, $\csc A = \frac{3}{2}.$

EXAMPLE 2. Given $\tan A = 3$; find the other functions.

By [2], p. 13, $\qquad \dfrac{\sin A}{\cos A} = 3.$

And by [1], p. 13,
$$\sin^2 A + \cos^2 A = 1.$$

If we solve these equations (regarding $\sin A$ and $\cos A$ as two unknown quantities), we find
$$\sin A = \tfrac{3}{10}\sqrt{10}, \ \cos A = \tfrac{1}{10}\sqrt{10}.$$

Then, by [3], p. 14, $\quad \cot A = \tfrac{1}{3}, \ \sec A = \sqrt{10}, \ \csc A = \tfrac{1}{3}\sqrt{10}.$

EXAMPLE 3. Given $\sec A = m$; find the other functions.

By [3], p. 14, $\quad \cos A = \dfrac{1}{m}.$

By [1], p. 13, $\quad \sin A = \sqrt{1 - \dfrac{1}{m^2}} = \sqrt{\dfrac{m^2 - 1}{m^2}} = \dfrac{1}{m}\sqrt{m^2 - 1}.$

By [2], p. 13, $\quad \tan A = \sqrt{m^2 - 1}.$

By [3], p. 14,
$$\cot A = \dfrac{1}{m^2 - 1}\sqrt{m^2 - 1}; \quad \csc A = \dfrac{m}{m^2 - 1}\sqrt{m^2 - 1}.$$

EXERCISE VI

Find the values of the other functions, when :

1. $\sin A = \tfrac{12}{13}.$ 5. $\tan A = \tfrac{4}{3}.$ 9. $\csc A = \sqrt{2}.$

2. $\sin A = 0.8.$ 6. $\cot A = 1.$ 10. $\sin A = m.$

3. $\cos A = \tfrac{40}{41}.$ 7. $\cot A = 0.5.$ 11. $\sin A = \dfrac{2m}{1 + m^2}.$

4. $\cos A = 0.28.$ 8. $\sec A = 2.$ 12. $\cos A = \dfrac{2mn}{m^2 + n^2}.$

13. Given $\tan 45° = 1$; find the other functions of $45°$.

14. Given $\sin 30° = \tfrac{1}{2}$; find the other functions of $30°$.

15. Given csc $60° = \frac{2}{3}\sqrt{3}$; find the other functions of $60°$.

16. Given tan $15° = 2 - \sqrt{3}$; find the other functions of $15°$.

17. Given cot $22° 30' = \sqrt{2} + 1$; find the other functions of $22° 30'$.

18. Given sin $0° = 0$; find the other functions of $0°$.

19. Given sin $90° = 1$; find the other functions of $90°$.

20. Given tan $90° = \infty$; find the other functions of $90°$.

Express the values of all the other functions in terms of:

21. sin A.   **22.** cos A.   **23.** tan A.   **24.** cot A.

25. Given $2 \sin A = \cos A$; find sin A and cos A.

26. Given $4 \sin A = \tan A$; find sin A and tan A.

27. If sin A : cos A = 9 : 40, find sin A and cos A.

28. Transform the quantity $\tan^2 A + \cot^2 A - \sin^2 A - \cos^2 A$ into a form containing only cos A.

29. Prove that $\sin A + \cos A = (1 + \tan A) \cos A$.

30. Prove that $\tan A + \cot A = \sec A \times \csc A$.

SECTION VIII

FUNCTIONS OF 45°

Let ACB (Fig. 8) be an isosceles right triangle, in which the length of the hypotenuse AB is equal to 1; then AC is equal to BC, and the angle A is equal to $45°$. Since $\overline{AC}^2 + \overline{BC}^2 = 1$, therefore $2\overline{AC}^2 = 1$, and $AC = \sqrt{\frac{1}{2}} = \frac{1}{2}\sqrt{2}$.

Therefore, by Sect. II, p. 5,

$$\sin 45° = \cos 45° = \tfrac{1}{2}\sqrt{2};$$
$$\tan 45° = \cot 45° = 1;$$
$$\sec 45° = \csc 45° = \sqrt{2}.$$

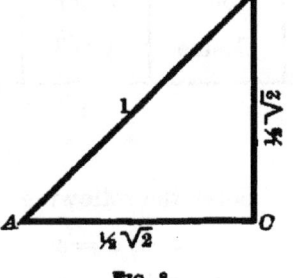

Fig. 8

SECTION IX

FUNCTIONS OF 30° AND 60°

Let ABC (Fig. 9) be an equilateral triangle, in which the length of each side is equal to 1; and let CD bisect the angle C. Then CD is perpendicular to AB and bisects AB.

Hence, $AD = \frac{1}{2}$, and $CD = \sqrt{1 - \frac{1}{4}} = \sqrt{\frac{3}{4}} = \frac{1}{2}\sqrt{3}$.

In the right triangle ADC, the angle $ACD = 30°$, and the angle $CAD = 60°$.   Whence, by Sect. II, p. 5,

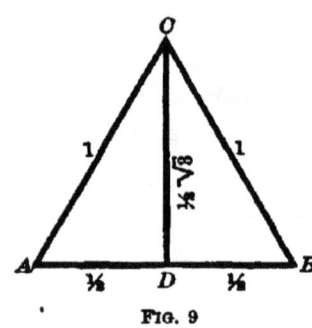

FIG. 9

$\sin 30° = \cos 60° = \frac{1}{2};$

$\cos 30° = \sin 60° = \frac{1}{2}\sqrt{3};$

$\tan 30° = \cot 60° = \dfrac{1}{\sqrt{3}} = \frac{1}{3}\sqrt{3};$

$\cot 30° = \tan 60° = \sqrt{3};$

$\sec 30° = \csc 60° = \dfrac{2}{\sqrt{3}} = \frac{2}{3}\sqrt{3};$

$\csc 30° = \sec 60° = 2.$

The results for sine and cosine of 30°, 45°, and 60° may be easily remembered by arranging them in the following form:

Angle	30°	45°	60°	$\frac{1}{2}\sqrt{1} = 0.5$
Sine	$\frac{1}{2}\sqrt{1}$	$\frac{1}{2}\sqrt{2}$	$\frac{1}{2}\sqrt{3}$	$\frac{1}{2}\sqrt{2} = 0.70711$
Cosine	$\frac{1}{2}\sqrt{3}$	$\frac{1}{2}\sqrt{2}$	$\frac{1}{2}\sqrt{1}$	$\frac{1}{2}\sqrt{3} = 0.86603$

EXERCISE VII

Solve the following equations:

1. $2 \cos x = \sec x.$

2. $4 \sin x = \csc x.$

3. $\tan x = 2 \sin x.$

4. $\sec x = \sqrt{2} \tan x.$

5. $\sin^2 x = 3 \cos^2 x$.

6. $2 \sin^2 x + \cos^2 x = \frac{3}{2}$.

7. $3 \tan^2 x - \sec^2 x = 1$.

8. $\tan x + \cot x = 2$.

9. $\sin^2 x - \cos x = \frac{1}{4}$.

10. $\tan^2 x - \sec x = 1$.

11. $\sin x + \sqrt{3} \cos x = 2$.

12. $\tan^2 x + \csc^2 x = 3$.

13. $2 \cos x + \sec x = 3$.

14. $\cos^2 x - \sin^2 x = \sin x$.

15. $2 \sin x + \cot x = 1 + 2 \cos x$.

16. $\sin^2 x + \tan^2 x = 3 \cos^2 x$.

17. $\tan x + 2 \cot x = \frac{5}{2} \csc x$.

NOTE. Wentworth & Hill's Five-place Logarithmic and Trigonometric Tables have full explanations, and directions for using them. Before proceeding to Chapter II the student should learn how to use these tables.

Table VI is to be used in solutions *without logarithms*. This four-place table contains the natural functions of angles at intervals of 1′. The decimal point must be inserted before each value given, except when it appears in the values of the table.

CHAPTER II

THE RIGHT TRIANGLE

SECTION X

THE GIVEN PARTS

In order to solve a right triangle, two parts besides the right angle must be given, one of them at least being a side.

The two given parts may be:

    I. An acute angle and the hypotenuse.

   II. An acute angle and the opposite leg.

  III. An acute angle and the adjacent leg.

  IV. The hypotenuse and a leg.

   V. The two legs.

SECTION XI

SOLUTION WITHOUT LOGARITHMS

The following examples illustrate the process of solution when logarithms are not employed.

CASE I

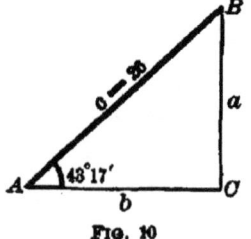

Fig. 10

Given $A = 43°\ 17'$, $c = 26$; find B, a, b.

   1. $B = 90° - A = 46°\ 43'$.

   2. $\dfrac{a}{c} = \sin A$; $\therefore a = c \sin A$.

   3. $\dfrac{b}{c} = \cos A$; $\therefore b = c \cos A$.

20

sin $A =$	0.6856	cos $A =$	0.7280
$c =$	26	$c =$	26
	$\overline{41136}$		$\overline{43680}$
	13712		14560
$a =$	$\overline{17.8256}$	$b =$	$\overline{18.9280}$

Case II

Given $A = 13°58'$, $a = 15.2$; find B, b, c.

1. $B = 90° - A = 76° \, 2'$.

2. $\dfrac{b}{a} = \cot A$; $\therefore b = a \cot A$.

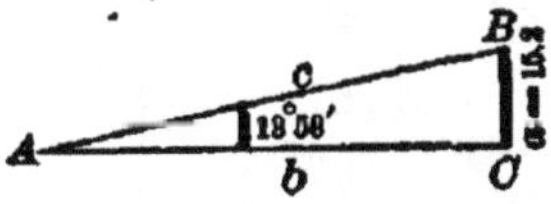

3. $\dfrac{a}{c} = \sin A$; $\therefore c = \dfrac{a}{\sin A}$.

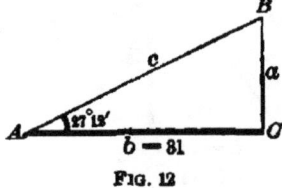

Fig. 11

$\cot A =$	4.0207	$a = 15.2$, $\sin A = 0.2414$.	
$a =$	15.2	$0.2414\,)\,15.200\,(\,62.9$	
	$\overline{80414}$	$14\,484$	
	201035	$\overline{7160}$	
	40207	4828	
$b =$	$\overline{61.11464}$ ·	$\overline{2332}$	
		$c = 62.9$	

Case III

Given $A = 27° \, 12'$, $b = 31$; find B, a, c.

1. $B = 90° - A = 62° \, 48'$.

2. $\dfrac{a}{b} = \tan A$; $\therefore a = b \tan A$.

3. $\dfrac{b}{c} = \cos A$; $\therefore c = \dfrac{b}{\cos A}$.

Fig. 12

$$\tan A = \quad 0.5139$$
$$b = \quad \underline{\quad 31}$$
$$\overline{5139}$$
$$\underline{15\,417}$$
$$a = 15.9309$$

$b = 31$, $\cos A = 0.8894$.

$$0.8894)\,31.000\,(34.9$$
$$\underline{26\,682}$$
$$\overline{4\,3180}$$
$$\underline{3\,5576}$$
$$c = 34.9 \quad \overline{7604}$$

CASE IV

Given $a = 47$, $c = 63$; find A, B, b.

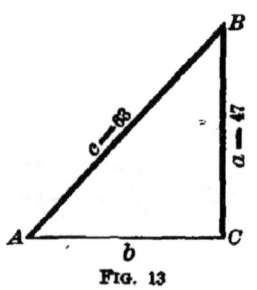

FIG. 13

1. $\sin A = \dfrac{a}{c}$.

2. $B = 90° - A$.

3. $b = \sqrt{c^2 - a^2} = \sqrt{(c+a)(c-a)}$.

$a = 47$, $c = 63$.

$$63)\,47.0\,(0.7460$$
$$\underline{44\,1}$$
$$\overline{2\,90}$$
$$\underline{2\,52}$$
$$\sin A = 0.7460 \quad \overline{380}$$
$$\therefore A = 48° \, 15' \quad \underline{378}$$
$$B = 41° \, 45' \quad 2$$

$$c + a = \quad 110$$
$$c - a = \quad \underline{16}$$
$$\overline{660}$$
$$\underline{110}$$
$$b^2 = \overline{1760}$$
$$b = \sqrt{1760}$$
$$= 41.95$$

CASE V

Given $a = 40$, $b = 27$; find A, B, c.

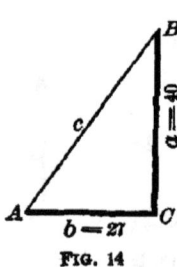

FIG. 14

1. $\tan A = \dfrac{a}{b}$.

2. $B = 90° - A$.

3. $c = \sqrt{a^2 + b^2}$.

$a = 40, \; b = 27.$

$\frac{40}{27} = 1.4815$

$\tan A = 1.4815$

$\therefore A = 55° \; 59'$

$B = 34° \; 1'$

$a^2 = 1600$

$b^2 = \;\; 729$

$c^2 = \overline{2329}$

$\therefore c = \sqrt{2329}$

$= 48.26$

SECTION XII

GENERAL METHOD OF SOLVING THE RIGHT TRIANGLE

From these five cases it appears that the general method of finding an unknown part in a right triangle is as follows:

Choose from the equation A + B = 90°, *and the equations that define the functions of the angles, an equation in which the required part only is unknown; solve this equation, if necessary, to find the value of the unknown part; then compute the value.*

NOTE. In Case IV, if the given sides (here a and c) are nearly alike in value, then A is near 90°, and its value cannot be accurately found from the tables, because the sines of large angles differ little in value (as is evident from Fig. 4). In this case it is better to find B first, by means of the formula given on page 59, namely,

$$\tan \tfrac{1}{2} B = \sqrt{\frac{c - a}{c + a}}.$$

EXAMPLE. Given $a = 49, \; c = 50$; find A, B, b.

$c - a = 1, \; c + a = 99.$

$\dfrac{c - a}{c + a} = 0.01010$

$\sqrt{\dfrac{c - a}{c + a}} = 0.1005$

$\tan \tfrac{1}{2} B = 0.1005$

$\therefore \tfrac{1}{2} B = 5° \; 44'$

$B = 11° \; 28'$

$A = 78° \; 32'$

$c - a = 1$

$c + a = 99$

$c^2 - a^2 = \overline{99}$

$b^2 = 99$

$b = \sqrt{99}$

$= 9.95$

EXERCISE VIII

' **1.** In Case II give another way of finding *c*, after *b* has been found.

2. In Case III give another way of finding *c*, after *a* has been found.

3. In Case IV give another way of finding *b*, after the angles have been found.

4. In Case V give another way of finding *c*, after the angles have been found.

5. Given *B* and *c*; find *A*, *a*, *b*.

6. Given *B* and *b*; find *A*, *a*, *c*.

7. Given *B* and *a*; find *A*, *b*, *c*.

8. Given *b* and *c*; find *A*, *B*, *a*.

Solve the following right triangles:

	GIVEN	REQUIRED
9	$a = 3$, $b = 4$.	$A = 36° 52'$, $B = 53° 8'$, $c = 5$.
10	$a = 7$, $c = 13$.	$A = 32° 35'$, $B = 57° 25'$, $b = 10.954$.
11	$a = 5.3$, $A = 12° 17'$.	$B = 77° 43'$, $b = 24.342$, $c = 24.918$.
12	$a = 10.4$, $B = 43° 18'$.	$A = 46° 42'$, $b = 9.800$, $c = 14.290$.
13	$c = 26$, $A = 37° 42'$.	$B = 52° 18'$, $a = 15.900$, $b = 20.572$.
14	$c = 140$, $B = 24° 12'$.	$A = 65° 48'$, $a = 127.694$, $b = 57.386$.
15	$b = 19$, $c = 23$.	$A = 34° 18'$, $B = 55° 42'$, $a = 12.961$.
16	$b = 98$, $c = 135.2$.	$A = 43° 33'$, $B = 46° 27'$, $a = 93.139$.
17	$b = 42.4$, $A = 32° 14'$.	$B = 57° 46'$, $a = 26.733$, $c = 50.124$.
18	$b = 200$, $B = 46° 11'$.	$A = 43° 49'$, $a = 191.900$, $c = 277.160$.
19	$a = 95$, $b = 37$.	$A = 68° 43'$, $B = 21° 17'$, $c = 101.951$.
20	$a = 6$, $c = 103$.	$A = 3° 21'$, $B = 86° 39'$, $b = 102.825$.
21	$a = 3.12$, $B = 5° 8'$.	$A = 84° 52'$, $b = 0.280$, $c = 3.133$.
22	$a = 17$, $c = 18$.	$A = 70° 48'$, $B = 19° 12'$, $b = 5.916$.
23	$c = 57$, $A = 38° 29'$.	$B = 51° 31'$, $a = 35.471$, $b = 44.620$.
24	$a + c = 18$, $b = 12$.	$A = 22° 37'$, $B = 67° 23'$, $a = 5$, $c = 13$.
25	$a + b = 9$, $c = 8$.	$A = 82° 18'$, $B = 7° 42'$, $\begin{cases} a = 7.928, \\ b = 1.072. \end{cases}$

SECTION XIII

SOLUTION BY LOGARITHMS

Case I

Given $A = 34° 28'$, $c = 18.75$; find B, a, b.

1. $B = 90° - A = 55° 32'$.

2. $\frac{a}{c} = \sin A$; $\therefore a = c \sin A$.

3. $\frac{b}{c} = \cos A$; $\therefore b = c \cos A$.

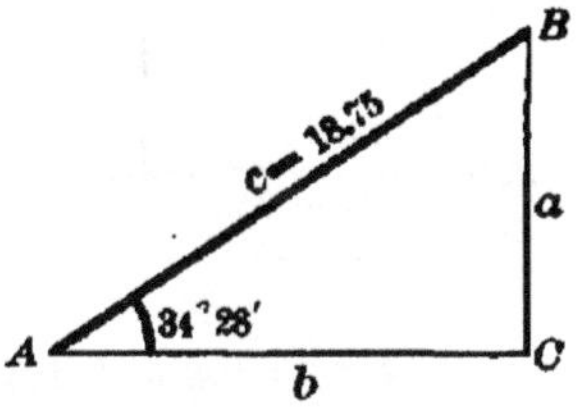

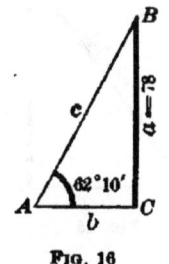

FIG. 15

$$\log a = \log c + \log \sin A$$

$$\begin{aligned} \log c &= 1.27300 \\ \log \sin A &= 9.75276 - 10 \\ \hline \log a &= 1.02576 \\ a &= 10.611 \end{aligned}$$

$$\log b = \log c + \log \cos A$$

$$\begin{aligned} \log c &= 1.27300 \\ \log \cos A &= 9.91617 - 10 \\ \hline \log b &= 1.18917 \\ b &= 15.459 \end{aligned}$$

Case II

Given $A = 62° 10'$, $a = 78$; find B, b, c.

1. $B = 90° - A = 27° 50'$.

2. $\frac{b}{a} = \cot A$; $\therefore b = a \cot A$.

3. $\frac{a}{c} = \sin A$.

$\therefore a = c \sin A$, and $c = \frac{a}{\sin A}$.

FIG. 16

$$\log b = \log a + \log \cot A$$

$$\begin{aligned} \log a &= 1.89209 \\ \log \cot A &= 9.72262 - 10 \\ \hline \log b &= 1.61471 \\ b &= 41.182 \end{aligned}$$

$$\log c = \log a + \operatorname{colog} \sin A$$

$$\begin{aligned} \log a &= 1.89209 \\ \operatorname{colog} \sin A &= 0.05340 \\ \hline \log c &= 1.94549 \\ c &= 88.204 \end{aligned}$$

Case III

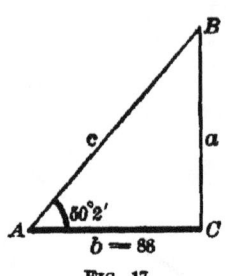

FIG. 17

Given $A = 50°\ 2'$, $b = 88$; find B, a, c

1. $B = 90° - A = 39°\ 58'$.

2. $\dfrac{a}{b} = \tan A'$; $\therefore a = b \tan A$.

3. $\dfrac{b}{c} = \cos A$.

   $\therefore b = c \cos A$, and $c = \dfrac{b}{\cos A}$.

$\log a = \log b + \log \tan A$

$\log b = \quad 1.94448$

$\log \tan A = \underline{10.07670 - 10}$

$\log a = \quad 2.02118$

$a = 105.00$

$\log c = \log b + \text{colog}\cos A$

$\log b = 1.94448$

$\text{colog}\cos A = \underline{0.19223}$

$\log c = 2.13671$

$c = 137.00$

Case IV

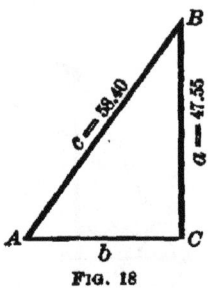

FIG. 18

Given $c = 58.40$, $a = 47.55$; find A, B, b.

1. $\sin A = \dfrac{a}{c}$.

2. $B = 90° - A$.

3. $\dfrac{b}{a} = \cot A$; $\therefore b = a \cot A$.

$\log \sin A = \log a + \text{colog } c$

$\log a = 1.67715$

$\text{colog } c = 8.23359 - 10$

$\log \sin A = \overline{9.91074 - 10}$

$A = 54°\ 31'$

$B = 35°\ 29'$

$\log b = \log a + \log \cot A$

$\log a = 1.67715$

$\log \cot A = 9.85300 - 10$

$\log b = \overline{1.53015}$

$b = 33.896$

CASE V

Given $a = 40$, $b = 27$; find A, B, c.

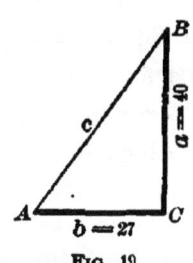

FIG. 19

1. $\tan A = \dfrac{a}{b}$.

2. $B = 90° - A$.

3. $\dfrac{a}{c} = \sin A$.

$\therefore a = c \sin A$, and $c = \dfrac{a}{\sin A}$.

$$\log \tan A = \log a + \operatorname{colog} b$$

$$\begin{aligned}
\log a &= \quad 1.60206 \\
\operatorname{colog} b &= \quad \underline{8.56864 - 10} \\
\log \tan A &= \underline{10.17070 - 10} \\
A &= 55° \ 59' \\
B &= 34° \ 1'
\end{aligned}$$

$$\log c = \log a + \operatorname{colog} \sin A$$

$$\begin{aligned}
\log a &= 1.60206 \\
\operatorname{colog} \sin A &= \underline{0.08151} \\
\log c &= \underline{1.68357} \\
c &= 48.258
\end{aligned}$$

NOTE. In Cases IV and V the unknown side may also be found from the equations

(for Case IV) $b = \sqrt{c^2 - a^2} = \sqrt{(c + a)(c - a)}$;

(for Case V) $c = \sqrt{a^2 + b^2}$.

These equations express the values of b and c directly in terms of the two given sides ; and if the values of the sides are simple numbers (*e.g.*, 5, 12, 13), it is often easier to find b or c in this way. But this value of c is not adapted to logarithms, and this value of b is not so readily found by logarithms as the value of b given under Case IV. See also p. 23.

SECTION XIV

AREA OF THE RIGHT TRIANGLE

The area of a triangle is equal to one-half the product of the base by the altitude; therefore, if a and b denote the legs of a right triangle, and F the area, $F = \frac{1}{2} ab$.

Hence, the area may be found when a and b are known.

For example: Find the area, having given:

<table>
<tr><td>Case I (Sect. XIII, p. 25).</td><td>Case IV (Sect. XIII, p. 26).</td></tr>
<tr><td>$A = 34° 28'$, $c = 18.75$.</td><td>$a = 47.55$, $c = 58.40$.</td></tr>
</table>

First find (as in Sect. XIII, p. 25) $\log a$ and $\log b$.

$\log F = \log a + \log b + \text{colog } 2$

$\log a = 1.02576$

$\log b = 1.18917$

$\text{colog } 2 = \underline{9.69897 - 10}$

$\log F = \overline{1.91390}$

$F = 82.016$

First find (as in Sect. XIII, p. 26) $\log a$ and $\log b$.

$\log F = \log a + \log b + \text{colog } 2$

$\log a = 1.67715$

$\log b = 1.53015$

$\text{colog } 2 = \underline{9.69897 - 10}$

$\log F = \overline{2.90627}$

$F = 805.88$

EXERCISE IX

Solve the following triangles by logarithms, finding the angles to the nearest minute:

	GIVEN		REQUIRED		
1	$a=6$,	$c=12$.	$A=30°$,	$B=60°$,	$b=10.392$.
2	$A=60°$,	$b=4$.	$B=30°$,	$c=8$,	$a=6.9282$.
3	$A=30°$,	$a=3$.	$B=60°$,	$c=6$,	$b=5.1961$.
4	$a=4$,	$b=4$.	$A=B=45°$,	$c=5.6568$.	
5	$a=2$,	$c=2.82843$.	$A=B=45°$,	$b=2$.	
6	$c=627$,	$A=23° 30'$.	$B=66° 30'$,	$a=250.02$,	$b=575.0$.
7	$c=2280$,	$A=28° 5'$.	$B=61° 55'$,	$a=1073.3$,	$b=2011.5$.
8	$c=72.15$,	$A=39° 34'$.	$B=50° 26'$,	$a=45.958$,	$b=55.620$.
9	$c=1$,	$A=36°$.	$B=54°$,	$a=0.58779$,	$b=0.80902$.
10	$c=200$,	$B=21° 47'$.	$A=68° 13'$,	$a=185.72$,	$b=74.22$.
11	$c=93.4$,	$B=76° 25'$.	$A=13° 35'$,	$a=21.936$,	$b=90.788$.
12	$a=637$,	$A= 4° 35'$.	$B=85° 25'$,	$b=7946$,	$c=7971.5$.
13	$a=48.532$,	$A=36° 44'$.	$B=53° 16'$,	$b=65.031$,	$c=81.144$.
14	$a=0.0008$,	$A=86°$.	$B=4°$,	$b=0.0000559$,	$c=0.000802$.
15	$b=50.937$,	$B=43° 48'$,	$A=46° 12'$,	$a=53.116$,	$c=73.59$.
16	$b=2$,	$B= 3° 38'$.	$A=86° 22'$,	$a=31.496$,	$c=31.559$.

	GIVEN		REQUIRED		
17	$a=992$,	$B=76°19'$.	$A=13°41'$,	$b=4074.5$,	$c=4198.5$.
18	$a=73$,	$B=68°52'$.	$A=21°8'$,	$b=188.86$,	$c=202.47$.
19	$a=2.189$,	$B=45°25'$.	$A=44°35'$,	$b=2.2211$,	$c=3.1185$.
20	$b=4$,	$A=37°56'$.	$B=52°4'$,	$a=3.1176$,	$c=5.0714$.
21	$c=8590$,	$a=4476$.	$A=31°24'$,	$B=58°36'$,	$b=7332.8$.
22	$c=86.53$,	$a=71.78$.	$A=56°3'$,	$B=33°57'$,	$b=48.324$.
23	$c=9.35$,	$a=8.49$.	$A=65°14'$,	$B=24°46'$,	$b=3.917$.
24	$c=2194$,	$b=1312.7$.	$A=53°15'$,	$B=36°45'$,	$a=1758$.
25	$c=30.69$,	$b=18.256$.	$A=53°30'$,	$B=36°30'$,	$a=24.67$.
26	$a=38.313$,	$b=19.522$.	$A=63°$,	$B=27°$,	$c=43$.
27	$a=1.2291$,	$b=14.950$.	$A=4°42'$,	$B=85°18'$,	$c=15$.
28	$a=415.38$,	$b=62.080$.	$A=81°30'$,	$B=8°30'$,	$c=420$.
29	$a=13.690$,	$b=16.926$.	$A=38°58'$,	$B=51°2'$,	$c=21.769$.
30	$c=91.92$,	$a=2.19$.	$A=1°22'$,	$B=88°38'$,	$b=91.894$.

Compute the unknown parts and also the area, having given:

31. $a = 5$, $\quad b = 6$.   36. $c = 68$, $\quad A = 69°54'$.

32. $a = 0.615$, $c = 70$.   37. $c = 27$, $\quad B = 44°4'$.

33. $b = \sqrt[3]{2}$, $\quad c = \sqrt{3}$.   38. $a = 47$, $\quad B = 48°49'$.

34. $a = 7$, $\quad A = 18°14'$.   39. $b = 9$, $\quad B = 34°44'$.

35. $b = 12$, $\quad A = 29°8'$.   40. $c = 8.462$, $B = 86°4'$.

41. Find the value of F in terms of c and A.

42. Find the value of F in terms of a and A.

43. Find the value of F in terms of b and A.

44. Find the value of F in terms of a and c.

45. Given $F = 58$, $a = 10$; solve the triangle.

46. Given $F = 18$, $b = 5$; solve the triangle.

47. Given $F = 12$, $A = 29°$; solve the triangle.

48. Given $F = 100$, $c = 22$; solve the triangle.

49. Find the angles of a right triangle if the hypotenuse is equal to three times one of the legs.

50. Find the legs of a right triangle if the hypotenuse is 6, and one angle is twice the other.

51. In a right triangle given c, and $A = nB$; find a and b.

52. In a right triangle the difference between the hypotenuse and the greater leg is equal to the difference between the two legs. Find the angles.

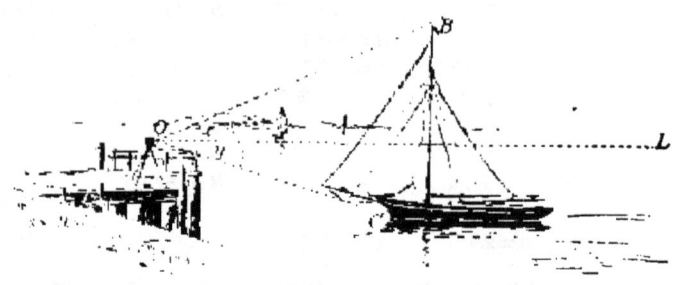

FIG. 20

The *angle of elevation* of an object, or the *angle of depression*, is the angle which a line from the eye to the object makes with a horizontal line in the same vertical plane.

Thus, if the observer is at O (Fig. 20), x is the angle of elevation of B, and y is the angle of depression of C.

53. At a horizontal distance of 120 feet from the foot of a steeple, the angle of elevation of the top was found to be 60° 30'. Find the height of the steeple.

54. From the top of a rock that rises vertically 326 feet out of the water, the angle of depression of a boat was found to be 24°. Find the distance of the boat from the foot of the rock.

55. How far is a monument, in a level plain, from the eye, if the height of the monument is 200 feet and the angle of elevation of the top 3° 30'?

56. A distance AB is measured 96 feet along the bank of a river from a point A opposite a tree C on the other bank. The angle ABC is 21° 14'. Find the breadth of the river.

57. What is the angle of elevation of an inclined plane if it rises 1 foot in a horizontal distance of 40 feet?

58. Find the angle of elevation of the sun when a tower 120 feet high casts a horizontal shadow 70 feet long.

59. How high is a tree that casts a horizontal shadow 80 feet in length when the angle of elevation of the sun is 50°?

60. A ship is sailing due northeast at a rate of 10 miles an hour. Find the rate at which she is moving due north, and also due east.

61. In front of a window 20 feet high is a flower-bed 6 feet wide. How long is a ladder that will just reach from the edge of the bed to the window?

62. A ladder 40 feet long may be so placed that it will reach a window 33 feet high on one side of the street, and by turning it over without moving its foot it will reach a window 21 feet high on the other side. Find the breadth of the street.

63. From the top of a hill the angles of depression of two successive milestones, on a straight level road leading to the hill, are observed to be 5° and 15°. Find the height of the hill.

64. A fort stands on a horizontal plain. The angle of elevation at a certain point on the plain is 30°, and at a point 100 feet nearer the fort it is 45°. How high is the fort?

65. From a certain point on the ground the angles of elevation of the belfry of a church and of the top of the steeple were found to be 40° and 51°, respectively. From a point 300 feet farther off, on a horizontal line, the angle of elevation of the top of the steeple is found to be 33° 45'. Find the distance from the belfry to the top of the steeple.

66. The angle of elevation of the top C of an inaccessible fort observed from a point A is 12°. At a point B, 219 feet from A and on a line AB perpendicular to AC, the angle ABC is 61° 45'. Find the height of the fort.

SECTION XV

THE ISOSCELES TRIANGLE

An isosceles triangle is divided by the perpendicular from the vertex to the base into two *equal right triangles.*

Therefore, an isosceles triangle is determined by any two parts that determine one of these right triangles.

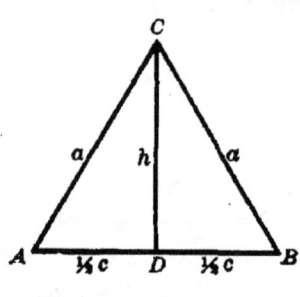

FIG. 21

Let the parts of an isosceles triangle CAB (Fig. 21), among which the altitude CD is to be included, be denoted as follows:

$a =$ one of the equal sides,

$c =$ the base,

$h =$ the altitude,

$A =$ one of the equal angles,

$C =$ the angle at the vertex.

For example: Given a and c; required A, C, h.

1. $\cos A = \dfrac{\frac{1}{2}c}{a} = \dfrac{c}{2\,a}$.

2. $C + 2\,A = 180°$; $\therefore C = 180° - 2\,A = 2\,(90° - A)$.

3. h may be found by any one of the equations:

$$h^2 + \frac{c^2}{4} = a^2;$$

whence $\qquad\qquad h = \sqrt{(a + \tfrac{1}{2}c)(a - \tfrac{1}{2}c)}.$

Also, $\qquad \dfrac{h}{a} = \sin A,\quad$ and $\quad \dfrac{h}{\frac{1}{2}c} = \tan A$;

whence $\qquad h = a \sin A,\quad$ and $\quad h = \frac{1}{2}c \tan A.$

When c and h are known, the area can be found by the formula

$$F = \tfrac{1}{2} ch.$$

Solve the following isosceles triangles, finding the angles to the nearest second:

1. Given a and A; find C, c, h.
2. Given a and C; find A, c, h.
3. Given c and A; find C, a, h.
4. Given c and C; find A, a, h.
5. Given h and A; find C, a, c.
6. Given h and C; find A, a, c.
7. Given a and h; find A, C, c.
8. Given c and h; find A, C, a.
9. Given $a = 14.3$, $c = 11$; find A, C, h.
10. Given $a = 0.295$, $A = 68°\ 10'$; find c, h, F.
11. Given $c = 2.352$, $C = 69°\ 49'$; find a, h, F.
12. Given $h = 7.4847$, $A = 76°\ 14'$; find a, c, F.
13. Given $a = 6.71$, $h = 6.6$; find A, C, c.
14. Given $c = 9$, $h = 20$; find A, C, a.
15. Given $c = 147$, $F = 2572.5$; find A, C, a, h.
16. Given $h = 16.8$, $F = 43.68$; find A, C, a, c.

17. Find the value of F in terms of a and c.
18. Find the value of F in terms of a and C.
19. Find the value of F in terms of a and A.
20. Find the value of F in terms of h and C.

21. A barn is 40×80 feet, the pitch of the roof is $45°$; find the length of the rafters and the area of the whole roof.

22. In a unit circle what is the length of the chord corresponding to the angle $45°$ at the centre?

23. If the radius of a circle is 30, and the length of a chord is 44, find the angle subtended at the centre.

24. Find the radius of a circle if a chord whose length is 5 subtends at the centre an angle of 133°.

25. What is the angle at the centre of a circle if the corresponding chord is equal to $\frac{3}{4}$ of the radius?

26. Find the area of a circular sector if the radius of the circle is 12, and the angle of the sector is 30°.

SECTION XVI

THE REGULAR POLYGON

Lines drawn from the centre of a regular polygon (Fig. 22) to the vertices are radii of the circumscribed circle; and lines drawn from the centre to the middle points of the sides are radii of the inscribed circle. These lines divide the polygon into *equal right triangles*. Therefore, a regular polygon is determined by a right triangle whose sides are the radius of the circumscribed circle, the radius of the inscribed circle, and half of one side of the polygon.

If the polygon has n sides, the angle of this right triangle at the centre of the polygon is equal to $\frac{1}{2}\left(\frac{360°}{n}\right)$, or $\frac{180°}{n}$; and the triangle may be solved when a side of the polygon or one of the radii is given.

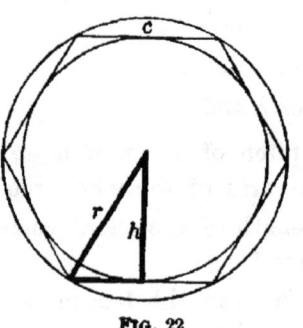

FIG. 22

Let

n = number of sides,
c = length of one side,
r = radius of circumscribed circle,
h = radius of inscribed circle,
p = the perimeter,
F = the area.

Then, by Geometry,

$$F = \tfrac{1}{2} hp.$$

EXERCISE XI

Find the remaining parts of a regular polygon, given:

1. $n = 10$, $c = 1$.   3. $n = 20$, $r = 20$.   5. $n = 11$, $F = 20$.
2. $n = 18$, $r = 1$.   4. $n = 8$, $h = 1$.   6. $n = 7$, $F = 7$.

Find the side of:

7. A regular decagon inscribed in a unit circle.

8. A regular decagon circumscribed about a unit circle.

9. If the side of an inscribed regular hexagon is 1, find the side of an inscribed regular dodecagon.

10. Given n and c, and let b denote the side of the inscribed regular polygon having $2\,n$ sides; find b in terms of n and c.

11. Compute the difference between the areas of a regular octagon and a regular nonagon if the perimeter of each is 16.

12. Compute the difference between the perimeters of a regular pentagon and a regular hexagon if the area of each is 12.

Find the area of:

13. The regular octagon formed by cutting away the corners of a square whose side is 1.

14. A regular pentagon if its diagonals are each equal to 12.

15. A regular polygon of 11 sides inscribed in a circle, if the area of an inscribed regular pentagon is 331.8.

16. A circle inscribed in an equilateral triangle whose perimeter is 20.

17. A regular polygon of 15 sides inscribed in a circle, if the area of a regular inscribed polygon of 16 sides is 100.

18. Find the perimeter of a regular dodecagon circumscribed about a circle the circumference of which is 1.

19. The area of a regular polygon of 25 sides is 40; find the area of the ring comprised between the circumferences of the inscribed and circumscribed circles.

CHAPTER III

GONIOMETRY

SECTION XVII

DEFINITION OF GONIOMETRY

To prepare the way for the solution of the oblique triangle, we now proceed to extend the definitions of the trigonometric functions to angles of all magnitudes, and to deduce certain useful relations of the functions of different angles.

That branch of Trigonometry which treats of trigonometric functions in general, and of their relations, is called **Goniometry**.

SECTION XVIII

POSITIVE AND NEGATIVE QUANTITIES

In measurements it is convenient to mark the distinction between two magnitudes that are measured in *opposite directions*, by calling one of them **positive** and the other **negative**.

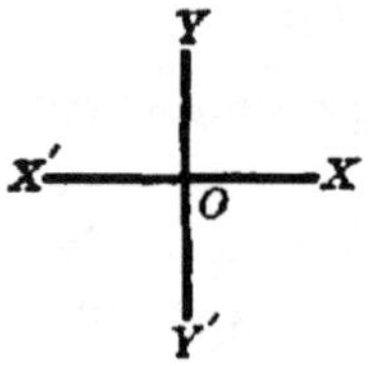

Thus, if *OX* (Fig. 23) is considered to be positive, then *OX'* is considered to be negative; and if *OY* is considered to be positive, then *OY'* is considered to be negative.

Fig. 23

When this distinction is applied to angles, an angle is considered to be *positive*, if the rotating line that describes it moves counter-clockwise, that is, in the direction opposite

36

to the hands of a clock, and to be *negative*, if the rotating line moves clockwise, that is, in the same direction as the hands of a clock.

Arcs corresponding to positive angles are considered *positive*, and arcs corresponding to negative angles are considered *negative*.

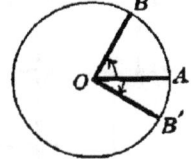

FIG. 24

Thus, the angle AOB (Fig. 24) described by a line rotating about O from OA to OB is positive, and the arc AB is positive; the angle AOB' described by the line rotating about O from OA to OB' is negative, and the arc AB' is negative.

SECTION XIX

CO-ORDINATES OF A POINT IN A PLANE

Let XX' (Fig. 25) be a horizontal line and let YY' be a line perpendicular to XX' at the point O. Then the plane determined by the lines XX' and YY' is divided into four quadrants which are numbered I, II, III, IV.

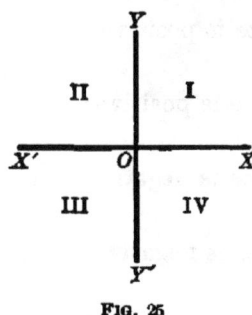

FIG. 25

Any point in the plane is determined by its *distance* and *direction* from each of the perpendiculars XX' and YY'. Its distance from YY', measured on XX', is called the **abscissa** of the point; its distance from XX', measured on YY', is called the **ordinate** of the point.

The abscissa and the ordinate of a point are called the **co-ordinates** of the point; and the lines XX' and YY' are called the **axes of co-ordinates**. XX' is called the **axis of abscissas** or the **axis of x**; YY' is called the **axis of ordinates** or the **axis of y**; and the point O is called the **origin**.

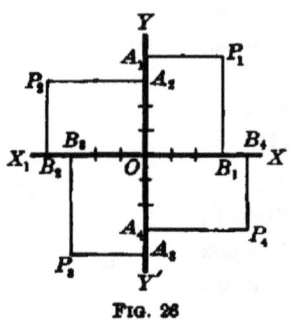

FIG. 26

In Fig. 26 the co-ordinates P_1, P_2, P_3, P_4 are as follows:

The abscissa of P_1 is OB_1,
 and the ordinate of P_1 is OA_1;
the abscissa of P_2 is OB_2,
 and the ordinate of P_2 is OA_2;
the abscissa of P_3 is OB_3,
 and the ordinate of P_3 is OA_3;
the abscissa of P_4 is OB_4,
 and the ordinate of P_4 is OA_4.

Abscissas to the *right* of YY' are **positive**.
Abscissas to the *left* of YY' are **negative**.
Ordinates *above* XX' are **positive**.
Ordinates *below* XX' are **negative**.

Therefore,

in Quadrant I,
 the abscissa is positive, the ordinate is positive;
in Quadrant II,
 the abscissa is negative, the ordinate is positive;
in Quadrant III,
 the abscissa is negative, the ordinate is negative;
in Quadrant IV,
 the abscissa is positive, the ordinate is negative.

SECTION XX

ANGLES OF ANY MAGNITUDE

If the line OP (Figs. 27–30) is revolved about O from OX as its **initial position** counter-clockwise, as shown by the curved arrows, the line during one revolution will form with OX all angles from 0° to 360°.

Any particular angle is said to be an angle of that quadrant in which its **terminal side** lies.

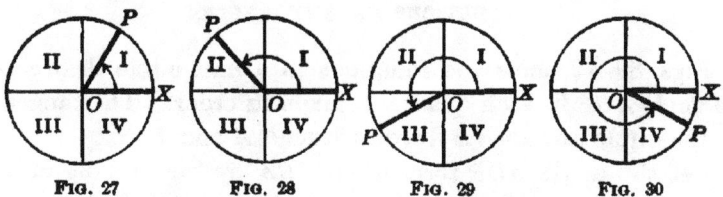

Fig. 27   Fig. 28   Fig. 29   Fig. 30

Angles between   0° and   90° are angles of Quadrant I.
Angles between   90° and 180° are angles of Quadrant II.
Angles between 180° and 270° are angles of Quadrant III.
Angles between 270° and 360° are angles of Quadrant IV.

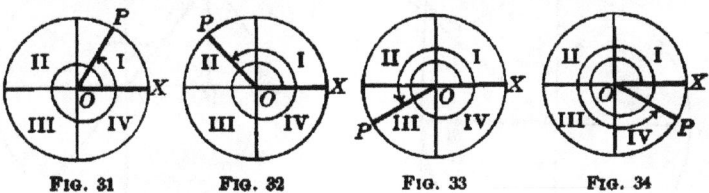

Fig. 31   Fig. 32   Fig. 33   Fig. 34

If the revolving line makes another revolution (Figs. 31–34), it will describe all angles from 360° to 720°; and so on.

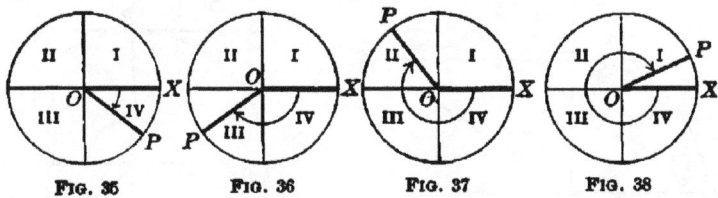

Fig. 35   Fig. 36   Fig. 37   Fig. 38

If the line *OP* is revolved from *OX* clockwise (Figs. 35–38), it will describe all *negative* angles.

Thus we arrive at the conception of an angle of any magnitude, positive or negative.

SECTION XXI

FUNCTIONS OF ANY ANGLE

Figs. 39–42 show the functions in a unit circle drawn for an angle AOP in each quadrant, taken in order. The tangents to the circle are *always* drawn through A and B.

Let the angle AOP formed with OA by the moving radius OP be denoted by x; then, in each quadrant,

$$\sin x = MP, \qquad \tan x = AT, \qquad \sec x = OT,$$
$$\cos x = OM, \qquad \cot x = BS, \qquad \csc x = OS.$$

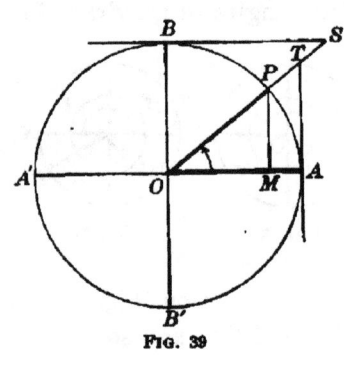

Fig. 39

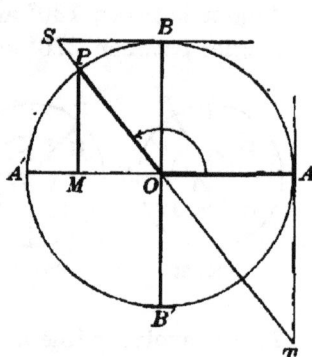

Fig. 40

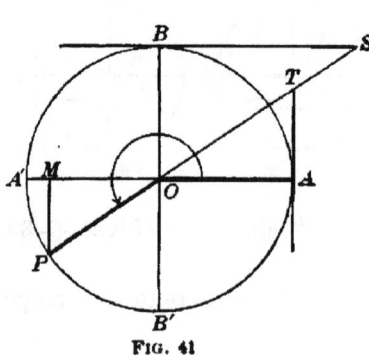

Fig. 41

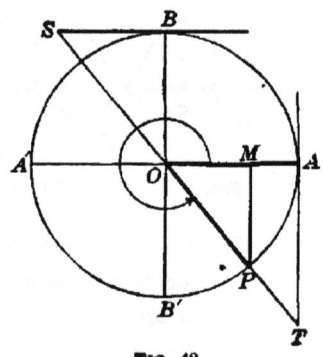

Fig. 42

If the terminal line of any angle x extends through the vertex indefinitely both ways, and if the circumference of a unit circle cuts the terminal line at P, the axis of abscissas at A, and the axis of ordinates at B, then

sin $x = the\ ordinate\ of$ P ;
cos $x = the\ abscissa\ of$ P ;
tan $x = the\ tangent\ from$ A *to meet the terminal line* ;
cot $x = the\ tangent\ from$ B *to meet the terminal line* ;
sec $x = the\ segment\ of\ the\ terminal\ line\ between\ the\ vertex
\qquad\ and\ the\ tangent$;
csc $x = the\ segment\ of\ the\ terminal\ line\ between\ the\ vertex
\qquad\ and\ the\ cotangent.$

Sines and tangents extending from the axis of abscissas *upwards* are positive; *downwards*, negative.

Cosines and cotangents extending from the axis of ordinates *towards the right* are positive; *towards the left*, negative.

The signs of the secant and cosecant are determined by the signs of the cosine and sine, respectively. Therefore, secants and cosecants extending from the centre, *in the direction of the terminal line*, are considered positive; *in the opposite direction*, negative. Hence,

QUADRANT	I	II	III	IV
sin and csc	+	+	−	−
cos and sec	+	−	−	+
tan and cot	+	−	+	−

In Quadrant I *all* the functions are positive.
In Quadrant II the *sine and cosecant only* are positive.
In Quadrant III the *tangent and cotangent only* are positive.
In Quadrant IV the *cosine and secant only* are positive.

The signs of all the functions of any quadrant are known when the signs of the sine and cosine are known.

If the sine and cosine have like signs, the tangent and cotangent are positive; if unlike signs, negative. The sine and cosecant have like signs; the cosine and secant have like signs.

SECTION XXII

FUNCTIONS OF A VARIABLE ANGLE

Let the angle AOP (Fig. 43) increase continuously from 0° to 360°. The values of its functions change as follows:

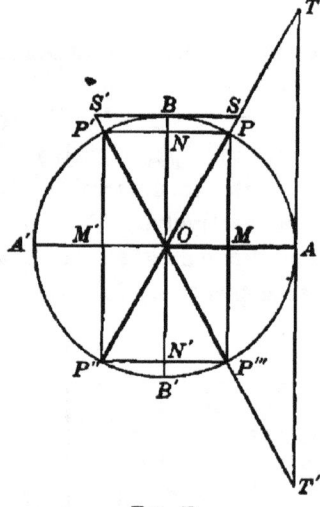

Fig. 43

1. *The Sine.* In the first quadrant the sine MP increases from 0 to 1; in the second it remains positive, and decreases from 1 to 0; in the third it is negative, and increases in absolute value from 0 to 1; in the fourth it is negative, and decreases in absolute value from 1 to 0.

2. *The Cosine.* In the first quadrant the cosine OM decreases from 1 to 0; in the second it becomes negative, and increases in absolute value from 0 to 1; in the third it is negative, and decreases in absolute value from 1 to 0; in the fourth it is positive, and increases from 0 to 1.

3. *The Tangent.* In the first quadrant the tangent AT increases from 0 to ∞; in the second it becomes negative, and decreases in absolute value from ∞ to 0; in the third it is positive, and increases from 0 to ∞; in the fourth it is negative, and decreases in absolute value from ∞ to 0.

4. *The Cotangent.* In the first quadrant, the cotangent *BS* decreases from ∞ to 0; in the second it is negative, and increases in absolute value from 0 to ∞; in the third and fourth quadrants, it has the same sign, and undergoes the same changes as in the first and second quadrants, respectively.

5. *The Secant.* In the first quadrant, the secant *OT* increases from 1 to ∞; in the second it is negative, and decreases in absolute value from ∞ to 1; in the third it is negative, and increases in absolute value from 1 to ∞; in the fourth it is positive, and decreases from ∞ to 1.

6. *The Cosecant.* In the first quadrant, the cosecant *OS* decreases from ∞ to 1; in the second it is positive, and increases from 1 to ∞; in the third it is negative, and decreases in absolute value from ∞ to 1; in the fourth it is negative, and increases in absolute value from 1 to ∞.

The limiting values of the functions are as follows:

	0°	90°	180°	270°	360°
Sine	± 0	+ 1	± 0	− 1	± 0
Cosine	+ 1	± 0	− 1	± 0	+ 1
Tangent	± 0	± ∞	± 0	± ∞	± 0
Cotangent	± ∞	± 0	± ∞	± 0	± ∞
Secant	+ 1	± ∞	− 1	± ∞	+ 1
Cosecant	± ∞	+ 1	± ∞	− 1	± ∞

Sines and cosines vary in value from + 1 to − 1; tangents and cotangents, from + ∞ to − ∞; secants and cosecants, from + ∞ to + 1, and from − 1 to − ∞.

In the table given above the double sign ± is placed before 0 and ∞. From the preceding investigation it appears that the functions *always change sign in passing through 0 and ∞*; and the sign + or − prefixed to 0 or ∞ simply shows the direction from which the value is reached.

SECTION XXIII

FUNCTIONS OF ANGLES LARGER THAN 360°

The functions of $360° + x$ are the same in sign and in absolute value as those of x; for the moving radius has the same position in both cases. If n is a positive integer,

The functions of $(n \times 360° + x)$ *are the same as those of* x.

For example: The functions of $2200°(6 \times 360° + 40°)$ are equal to the functions of $40°$.

SECTION XXIV

EXTENSION OF FORMULAS

The Formulas [1], [2], [3] established for *acute* angles on pp. 13, 14 hold true for *all* angles. Thus, in each quadrant

$$\overline{MP}^2 + \overline{OM}^2 = \overline{OP}^2.$$

Therefore,

$$\sin^2 x + \cos^2 x = 1. \quad [1]$$

We have in each quadrant from the similar triangles OMP, OAT, OBS the proportions

$AT : OA = MP : OM,$

or $\tan x : 1 = \sin x : \cos x$;

$MP : OP = OB : OS,$

or $\sin x : 1 = 1 : \csc x$;

$OM : OP = OA : OT,$

or $\cos x : 1 = 1 : \sec x$;

$AT : OA = OB : BS,$

or $\tan x : 1 = 1 : \cot x.$

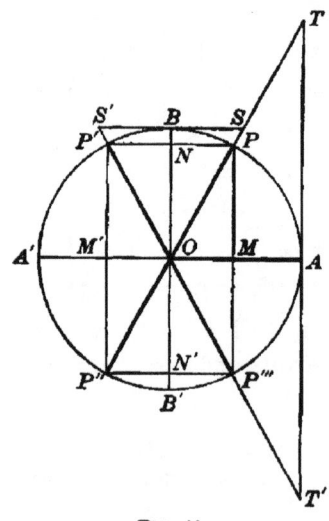

Fig. 44

That is, $$\tan x = \frac{\sin x}{\cos x}. \qquad [2]$$

$$\left.\begin{array}{l} \sin x \times \csc x = 1 \\ \cos x \times \sec x = 1 \\ \tan x \times \cot x = 1 \end{array}\right\} . \qquad [3]$$

Formulas [1]–[3] enable us, from a given value of one function, to find the *absolute* values of the other five functions, and also the sign of the reciprocal function. But in order to determine the proper signs to be placed before the other four functions, we must know the quadrant to which the angle in question belongs, or the sign of any *one* of these four functions; for, by Sect. XXI, p. 40, it will be seen that *the signs of any two functions that are not reciprocals determine the quadrant to which the angle belongs.*

EXAMPLE. Given $\sin x = + \frac{4}{5}$, and $\tan x$ negative; find the values of the other functions.

Since $\sin x$ is positive, x is an angle in Quadrant I or II; but, since $\tan x$ is negative, Quadrant I is inadmissible.

By [1], $\cos x = \pm \sqrt{1 - \frac{16}{25}} = \pm \frac{3}{5}$.

Since the angle is in Quadrant II, the minus sign must be taken, and we have

$$\cos x = - \tfrac{3}{5}.$$

By [2] and [3],

$\tan x = - \frac{4}{3}, \quad \cot x = - \frac{3}{4}, \quad \sec x = - \frac{5}{3}, \quad \csc x = \frac{5}{4}.$

EXERCISE XII

1. Construct the functions of an angle in Quadrant II. What are their signs?

2. Construct the functions of an angle in Quadrant III. What are their signs?

3. Construct the functions of an angle in Quadrant IV. What are their signs?

4. What are the signs of the functions of the following angles: 340°, 239°, 145°, 400°, 700°, 1200°, 3800°?

5. How many angles less than 360° have the value of the sine equal to $+\frac{4}{5}$, and in what quadrants do they lie?

6. How many values less than 720° can the angle x have if $\cos x = +\frac{2}{3}$, and in what quadrants do they lie?

7. If we take into account only angles less than 180°, how many values can x have if $\sin x = \frac{4}{7}$? if $\cos x = \frac{1}{2}$? if $\cos x = -\frac{4}{5}$? if $\tan x = \frac{2}{3}$? if $\cot x = -7$?

8. Within what limits must the angle x lie if $\cos x = -\frac{2}{3}$? if $\cot x = 4$? if $\sec x = 80$? if $\csc x = -3$? (If $x < 360°$.)

9. In what quadrant does an angle lie if sine and cosine are both negative? if cosine and tangent are both negative? if cotangent is positive and sine negative?

10. Between 0° and 3600° how many angles are there whose sines have the absolute value $\frac{2}{3}$? Of these sines how many are positive and how many negative?

11. In finding $\cos x$ by means of the equation $\cos x = \pm\sqrt{1 - \sin^2 x}$, when must we choose the positive sign and when the negative sign?

12. Given $\cos x = -\sqrt{\frac{1}{4}}$; find the other functions when x is an angle in Quadrant II.

13. Given $\tan x = \sqrt{3}$; find the other functions when x is an angle in Quadrant III.

14. Given $\sec x = +7$, and $\tan x$ negative; find the other functions of x.

15. Given $\cot x = -3$; find all the possible values of the other functions.

16. What functions of an angle of a triangle may be negative? In what case are they negative?

17. Why may cot 360° be considered equal either to $+ \infty$ or to $- \infty$?

18. Obtain by means of Formulas [1]–[3] the other functions of the angles, given:

    (i) $\tan 90° = \infty$.   (iii) $\cot 270° = 0$.

    (ii) $\cos 180° = -1$.   (iv) $\csc 360° = -\infty$.

19. Find the values of $\sin 450°$, $\tan 540°$, $\cos 630°$, $\cot 720°$, $\sin 810°$, $\csc 900°$.

Compute the values of the following expressions:

20. $a \sin 0° + b \cos 90° - c \tan 180°$.

21. $a \cos 90° - b \tan 180° + c \cot 90°$.

22. $a \sin 90° - b \cos 360° + (a - b) \cos 180°$.

23. $(a^2 - b^2) \cos 360° - 4\,ab \sin 270°$.

SECTION XXV

REDUCTION OF FUNCTIONS TO THE FIRST QUADRANT

In a unit circle (Fig. 45) draw two diameters PR and QS equally inclined to the horizontal diameter AA', or so that the angles AOP, $A'OQ$, $A'OR$, and AOS shall be equal. From the points P, Q, R, S let fall perpendiculars to AA'; the four right triangles thus formed, with a common vertex at O, are equal; because they have equal hypotenuses (radii of the circle) and equal acute angles at O. Therefore, the perpendiculars PM, QN, RN, SM are equal, and are the

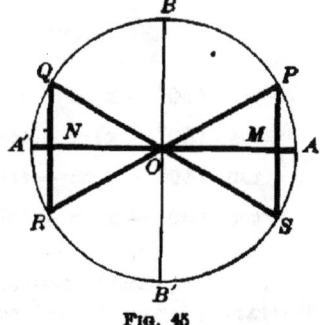

Fig. 45

sines of the angles AOP, AOQ, AOR, AOS, respectively.

Therefore, *in absolute value,*

$$\sin AOP = \sin AOQ = \sin AOR = \sin AOS.$$

And from Sect. XXIV, p. 44, it follows that *in absolute value* the cosines of these angles are also equal; and likewise the tangents, the cotangents, the secants, and the cosecants.*

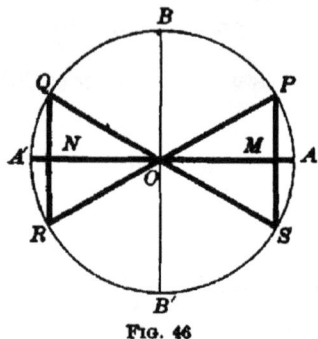

Fig. 46

Hence, *For every acute angle there is an angle in each of the higher quadrants whose functions, in absolute value, are equal to those of this acute angle.*

Let $\angle AOP = x$, $\angle POB = y$; then $x + y = 90°$, and the functions of x are equal to the co-named functions of y (Sect. V, p. 11); and

$$\angle AOQ \text{ (in Quadrant II)} = 180° - x = \ \ 90° + y,$$
$$\angle AOR \text{ (in Quadrant III)} = 180° + x = 270° - y,$$
$$\angle AOS \text{ (in Quadrant IV)} = 360° - x = 270° + y.$$

Hence, prefixing the proper sign (Sect. XXI, p. 40), we have:

ANGLE IN QUADRANT II

$\sin (180° - x) = \ \ \ \ \sin x.$	$\sin (90° + y) = \ \ \ \ \cos y.$
$\cos (180° - x) = - \cos x.$	$\cos (90° + y) = - \sin y.$
$\tan (180° - x) = - \tan x.$	$\tan (90° + y) = - \cot y.$
$\cot (180° - x) = - \cot x.$	$\cot (90° + y) = - \tan y.$

* In future, secants, cosecants, versed sines, and coversed sines will be disregarded. Secants and cosecants may be found by Formula [3], versed sines and coversed sines by VII and VIII, p. 5, if wanted, but they are seldom used in computations.

ANGLE IN QUADRANT III

$$\sin (180° + x) = - \sin x. \qquad \sin (270° - y) = - \cos y.$$
$$\cos (180° + x) = - \cos x. \qquad \cos (270° - y) = - \sin y.$$
$$\tan (180° + x) = \tan x. \qquad \tan (270° - y) = \cot y.$$
$$\cot (180° + x) = \cot x. \qquad \cot (270° - y) = \tan y.$$

ANGLE IN QUADRANT IV

$$\sin (360° - x) = - \sin x. \qquad \sin (270° + y) = - \cos y.$$
$$\cos (360° - x) = \cos x. \qquad \cos (270° + y) = \sin y.$$
$$\tan (360° - x) = - \tan x. \qquad \tan (270° + y) = - \cot y.$$
$$\cot (360° - x) = - \cot x. \qquad \cot (270° + y) = - \tan y.$$

REMARK. The tangents and cotangents may be found directly from the figure, or by Formula [2].

It is evident, from these formulas,

1. *The functions of all angles can be reduced to the functions of angles in the first quadrant, and therefore to functions of angles not greater than 45°* (Sect. V, p. 11).

2. *If an acute angle is added to or subtracted from 180° or 360°, the functions of the resulting angle are equal in absolute value to the* **like-named** *functions of the acute angle; but if an acute angle is added to or subtracted from 90° or 270°, the functions of the resulting angle are equal in absolute value to the* **co-named** *functions of the acute angle.*

3. *A given value of a sine or cosecant determines two supplementary angles, one acute, the other obtuse; a given value of any other function determines only one angle: acute if the value is positive, obtuse if the value is negative.* [See functions of (180° − x).]

SECTION XXVI

FUNCTIONS OF ANGLES THAT DIFFER BY 90°

The general form of two angles whose difference is 90° is x and $90° + x$, and they must lie in adjoining quadrants. The relations between their functions were found in Sect. XXV, p. 48, but only for the case when x is acute. These relations, however, may be shown to hold true for all values of x.

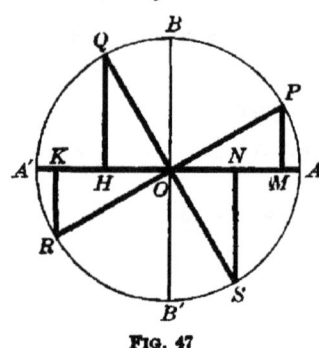

FIG. 47

In a unit circle (Fig. 47) draw two diameters PR and QS perpendicular to each other, and let fall to AA' the perpendiculars PM, QH, RK, and SN. The right triangles OMP, QHO, OKR, and SNO are equal, because they have equal hypotenuses and equal acute angles POM, OQH, ROK, and OSN.

Therefore, $OM = QH = OK = NS$,

and $PM = OH = RK = ON$.

Hence, taking into account the algebraic sign,

$\sin AOQ = \ \ \cos AOP$; $\sin AOS = \ \ \cos AOR$;

$\cos AOQ = - \sin AOP$; $\cos AOS = - \sin AOR$;

$\sin AOR = \ \ \cos AOQ$; $\sin (360° + AOP) = \ \ \cos AOS$;

$\cos AOR = - \sin AOQ$; $\cos (360° + AOP) = - \sin AOS$.

In all these equations, if x denotes the angle on the right-hand side, the angle on the left-hand side is $90° + x$.

Therefore, if x is an angle in any one of the four quadrants,

$\sin (90° + x) = \ \ \cos x$, $\tan (90° + x) = - \cot x$.

$\cos (90° + x) = - \sin x$, $\cot (90° + x) = - \tan x$.

In like manner, it can be shown that all the formulas of Sect. XXV, p. 48, hold true, whatever the values of x and y.

Hence, *In every case the algebraic sign of the function of the resulting angle is the same as when* x *and* y *are both acute.*

SECTION XXVII

FUNCTIONS OF A NEGATIVE ANGLE

If the angle x is generated by the radius moving from the initial position OA to the terminal position OS, it will have the sign $-$, and its terminal side will be identical with its position for the angle $360° - x$. Therefore, the functions of the angle $- x$ are the same as those of the angle $360° - x$; or (Sect. XXV, p. 49),

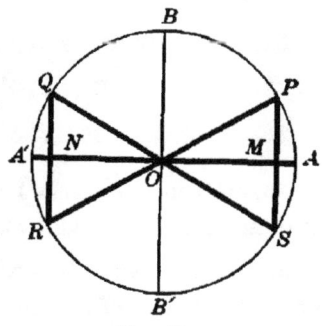

$$\sin (- x) = - \sin x,$$
$$\cos (- x) = \cos x,$$
$$\tan (- x) = - \tan x,$$
$$\cot (- x) = - \cot x.$$

Fig. 48

EXERCISE XIII

1. Express sin 250° in terms of the functions of an acute angle less than 45°.

SOLUTION. $\sin 250° = \sin (270° - 20°) = - \cos 20°.$

Express the following functions in terms of the functions of angles less than 45°:

2. sin 172°.	5. cot 91°.	8. sin 204°.
3. cos 100°.	6. sec 110°.	9. cos 359°.
4. tan 125°.	7. csc 157°.	10. tan 300°.

11. cot 264°.	14. sin 163° 49'.	17. cot 139° 17'.
12. sec 244°.	15. cos 195° 33'.	18. sec 299° 45'.
13. csc 271°.	16. tan 269° 15'.	19. csc 92° 25'.

Express all the functions of the following negative angles in terms of the functions of positive angles less than 45°:

20. − 75°.	22. − 200°.	24. − 52° 37'.
21. − 127°.	23. − 345°.	25. − 196° 54'.

26. Find the functions of 120°.

Hint. 120° = 180° − 60°, or 90° + 30° ; then apply Sect. XXV, p. 48.

Find the functions of the following angles :

27. 135°.	29. 210°.	31. 240°.	33. − 30°.
28. 150°.	30. 225°.	32. 300°.	34. − 225°.

35. Given $\sin x = − \frac{1}{2} \sqrt{2}$, and $\cos x$ negative; find the other functions of x, and the value of x.

36. Given $\cot x = − \sqrt{3}$, and x in Quadrant II; find the other functions of x, and the value of x.

37. Find the functions of 3540°.

38. What angles less than 360° have a sine equal to $− \frac{1}{2}$? a tangent equal to $− \sqrt{3}$?

39. Which of the angles mentioned in Examples 27–34 have a cosine equal to $− \frac{1}{2} \sqrt{2}$? a cotangent equal to $− \sqrt{3}$?

40. What values of x between 0° and 720° will satisfy the equation $\sin x = + \frac{1}{2}$?

41. Find the other angle between 0° and 360° for which the corresponding function (sign included) has the same value as sin 12°, cos 26°, tan 45°, cot 72°, sin 191°, cos 120°, tan 244°, cot 357°.

42. Given $\tan 238° = 1.6$; find $\sin 122°$.

43. Given $\cos 333° = 0.89$; find $\tan 117°$.

Simplify the following expressions:

44. $a \cos(90° - x) + b \cos(90° + x)$.

45. $m \cos(90° - x)\sin(90° - x)$.

46. $(a - b)\tan(90° - x) + (a + b)\cot(90° + x)$.

47. $a^2 + b^2 - 2\,ab \cos(180° - x)$.

48. $\sin(90° + x)\sin(180° + x) + \cos(90° + x)\cos(180° - x)$.

49. $\cos(180° + x)\cos(270° - y) - \sin(180° + x)\sin(270° - y)$.

50. $\tan x + \tan(-y) - \tan(180° - y)$.

51. For what values of x is the expression $\sin x + \cos x$ positive, and for what values negative?

52. Answer the questions of Example 51 for $\sin x - \cos x$.

53. Find the functions of $x - 90°$ in functions of x.

54. Find the functions of $x - 180°$ in functions of x.

SECTION XXVIII

FUNCTIONS OF THE SUM OF TWO ANGLES

In a unit circle (Fig. 49) let the angle $AOB = x$, the angle $BOC = y$; then the angle $AOC = x + y$.

In order to express $\sin(x + y)$ and $\cos(x + y)$ in terms of the sines and cosines of x and y, draw $CF \perp OA$, $CD \perp OB$, $DE \perp OA$, $DG \perp CF$; then $CD = \sin y$, $OD = \cos y$, and the angle $DCG = x$. Also,

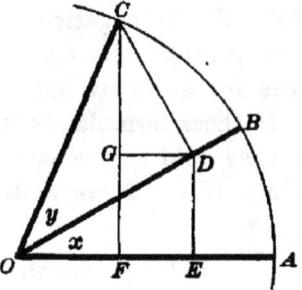

Fig. 49

$$\sin(x + y) = CF = DE + CG.$$

Now $\dfrac{DE}{OD} = \sin x$; hence, $DE = \sin x \times OD = \sin x \cos y$.

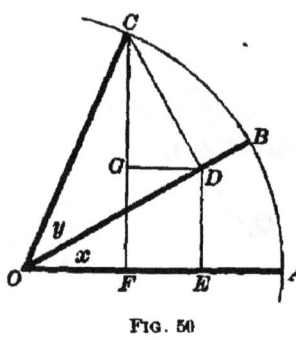

FIG. 50

And $\dfrac{CG}{CD} = \cos x$; hence,

$$CG = \cos x \times CD = \cos x \sin y.$$

Therefore,

$$\sin(x+y) = \sin x \cos y + \cos x \sin y. \quad [4]$$

Again,

$$\cos (x + y) = OF = OE - DG.$$

$\dfrac{OE}{OD} = \cos x$; hence,

$$OE = \cos x \times OD = \cos x \cos y.$$

$\dfrac{DG}{CD} = \sin x$; hence, $DG = \sin x \times CD = \sin x \sin y.$

Therefore, $\qquad \cos (x + y) = \cos x \cos y - \sin x \sin y. \quad [5]$

In this proof x and y, and also the sum $x + y$, are assumed to be acute angles. If the sum $x + y$ of the acute angles x and y is obtuse, as in Fig. 51, the proof remains, word for word, the same as above, the only difference being that the sign of OF will be negative, as DG is

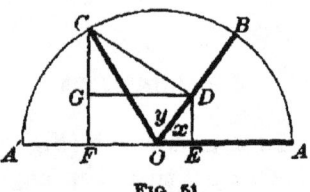

FIG. 51

now greater than OE. The above formulas, therefore, hold true for all acute angles x and y.

If these formulas hold true for any two acute angles x and y, they hold true when one of the angles is increased by 90°. Thus, if for x we write $x' = 90° + x$, then, by Sect. XXV, p. 48,

$$\sin (x' + y) = \sin (90° + x + y) = \cos (x + y),$$
$$\cos (x' + y) = \cos (90° + x + y) = - \sin (x + y).$$

Hence, by [5], $\sin (x' + y) = \cos x \cos y - \sin x \sin y,$

    by [4], $\cos (x' + y) = - \sin x \cos y - \cos x \sin y.$

Now, by Sect. XXV, p. 48,

$$\cos x = \sin (90° + x) = \sin x',$$
$$\sin x = - \cos (90° + x) = - \cos x'.$$

Substitute these values of $\cos x$ and $\sin x$, then

$$\sin (x' + y) = \sin x' \cos y + \cos x' \sin y,$$
$$\cos (x' + y) = \cos x' \cos y - \sin x' \sin y.$$

It follows that Formulas [4] and [5] hold true if either angle is repeatedly increased by 90°; therefore they apply to all angles whatever.

By Sect. XXIV, p. 45, Formula [2],

$$\tan (x + y) = \frac{\sin (x + y)}{\cos (x + y)} = \frac{\sin x \cos y + \cos x \sin y}{\cos x \cos y - \sin x \sin y}.$$

If we divide each term of the numerator and denominator of the last fraction by $\cos x \cos y$, we have

$$\tan (x + y) = \frac{\dfrac{\sin x}{\cos x} + \dfrac{\sin y}{\cos y}}{1 - \dfrac{\sin x \sin y}{\cos x \cos y}}.$$

That is, $\qquad \tan (x + y) = \dfrac{\tan x + \tan y}{1 - \tan x \tan y}.$ [6]

Also, $\cot (x + y) = \dfrac{\cos (x + y)}{\sin (x + y)} = \dfrac{\cos x \cos y - \sin x \sin y}{\sin x \cos y - \cos x \sin y}.$

Divide each term of the numerator and denominator by $\sin x \sin y$, remembering that $\dfrac{\cos x}{\sin x} = \cot x$ and $\dfrac{\cos y}{\sin y} = \cot y$; we have

$$\cot (x + y) = \frac{\cot x \cot y - 1}{\cot y + \cot x}. \qquad [7]$$

SECTION XXIX

FUNCTIONS OF THE DIFFERENCE OF TWO ANGLES

In a unit circle (Fig. 52) let the angle $AOB = x$, $COB = y$; then the angle $AOC = x - y$.

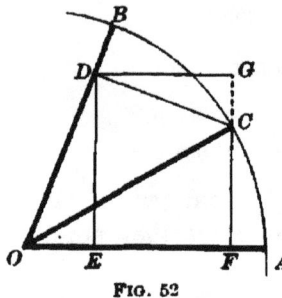

FIG. 52

In order to express $\sin(x - y)$ and $\cos(x - y)$ in terms of the sines and cosines of x and y, draw $CF \perp OA$, $CD \perp OB$, $DE \perp OA$, $DG \perp FC$ prolonged; then $CD = \sin y$, $OD = \cos y$, and the angle $DCG = x$.

Now $\sin(x - y) = CF = DE - CG$.

$\dfrac{DE}{OD} = \sin x$; hence, $DE = \sin x \times OD = \sin x \cos y$.

$\dfrac{CG}{CD} = \cos x$; hence, $CG = \cos x \times CD = \cos x \sin y$.

Therefore,   **$\sin(x - y) = \sin x \cos y - \cos x \sin y$.**   [8]

Again, $\cos(x - y) = OF = OE + DG$.

$\dfrac{OE}{OD} = \cos x$; hence, $OE = \cos x \times OD = \cos x \cos y$.

$\dfrac{DG}{CD} = \sin x$; hence, $DG = \sin x \times CD = \sin x \sin y$.

Therefore,   **$\cos(x - y) = \cos x \cos y + \sin x \sin y$.**   [9]

In this proof, both x and y are assumed to be acute angles; but, whatever be the values of x and y, the same method of proof will always lead to Formulas [8] and [9], when due regard is paid to the algebraic signs.

The general application of these formulas may be at once shown by deducing them from the general formulas established in Sect. XXVIII, p. 54, as follows·

It is obvious that $(x - y) + y = x$. If we apply Formulas [4] and [5] to $(x - y) + y$, then

$$\sin\{(x - y) + y\} \text{ or } \sin x = \sin (x - y) \cos y + \cos (x - y) \sin y,$$
$$\cos\{(x - y) + y\} \text{ or } \cos x = \cos (x - y) \cos y - \sin (x - y) \sin y.$$

Multiply the first equation by $\cos y$, the second by $\sin y$,

$$\sin x \cos y = \quad \sin (x - y) \cos^2 y + \cos (x - y) \sin y \cos y,$$
$$\cos x \sin y = - \sin (x - y) \sin^2 y + \cos (x - y) \sin y \cos y\,;$$

whence, by subtraction,

$$\sin x \cos y - \cos x \sin y = \sin (x - y) (\sin^2 y + \cos^2 y).$$

But $\quad \sin^2 y + \cos^2 y = 1$ (Sect. XXIV, p. 44).

Therefore, by substitution and transposition,

$$\sin (x - y) = \sin x \cos y - \cos x \sin y.$$

Again, if we multiply the first equation by $\sin y$, the second equation by $\cos y$, and add the results, we obtain, by reducing,

$$\cos (x - y) = \cos x \cos y + \sin x \sin y.$$

Therefore, Formulas [8] and [9], like [4] and [5], from which they have been derived, are universally true.

From [8] and [9], by proceeding as in Sect. XXVIII, p. 55, we obtain

$$\tan (x - y) = \frac{\tan x - \tan y}{1 + \tan x \tan y}. \qquad [10]$$

$$\cot (x - y) = \frac{\cot x \cot y + 1}{\cot y - \cot x}. \qquad [11]$$

Formulas [4]–[11] may be combined as follows:

$$\sin (x \pm y) = \sin x \cos y \pm \cos x \sin y,$$
$$\cos (x \pm y) = \cos x \cos y \mp \sin x \sin y,$$
$$\tan (x \pm y) = \frac{\tan x \pm \tan y}{1 \mp \tan x \tan y},$$
$$\cot (x \pm y) = \frac{\cot x \cot y \mp 1}{\cot y \pm \cot x}.$$

SECTION XXX

FUNCTIONS OF TWICE AN ANGLE

If $y = x$, Formulas [4]–[7] become

$$\sin 2x = 2 \sin x \cos x. \qquad [12]$$

$$\cos 2x = \cos^2 x - \sin^2 x. \qquad [13]$$

$$\tan 2x = \frac{2 \tan x}{1 - \tan^2 x}. \qquad [14]$$

$$\cot 2x = \frac{\cot^2 x - 1}{2 \cot x}. \qquad [15]$$

By these formulas the functions of twice an angle may be found when the functions of the angle are given.

SECTION XXXI

FUNCTIONS OF HALF AN ANGLE

Formula [1] is $\qquad \cos^2 x + \sin^2 x = 1.$
Formula [13] is $\qquad \cos^2 x - \sin^2 x = \cos 2x.$
Subtract, $\qquad\qquad\quad \overline{\qquad 2 \sin^2 x = 1 - \cos 2x.}$
Add, $\qquad 2 \cos^2 x \qquad\quad = 1 + \cos 2x.$

Whence,

$$\sin x = \pm \sqrt{\frac{1 - \cos 2x}{2}}, \quad \cos x = \pm \sqrt{\frac{1 + \cos 2x}{2}}.$$

These values, if z is put for $2x$, and hence $\tfrac{1}{2}z$ for x, become

$$\sin \tfrac{1}{2}z = \pm \sqrt{\frac{1 - \cos z}{2}}. \qquad [16]$$

$$\cos \tfrac{1}{2}z = \pm \sqrt{\frac{1 + \cos z}{2}}. \qquad [17]$$

Hence, by division (Sect. XXIV, p. 45),

$$\tan \tfrac{1}{2} z = \pm \sqrt{\frac{1 - \cos z}{1 + \cos z}}. \qquad [18]$$

$$\cot \tfrac{1}{2} z = \pm \sqrt{\frac{1 + \cos z}{1 - \cos z}}. \qquad [19]$$

By these formulas the functions of half an angle may be computed when the cosine of the entire angle is given.

The proper sign to be placed before the root in each case depends on the quadrant in which the angle $\tfrac{1}{2} z$ lies (Sect. XXII, p. 42).

Let the student show from Formula [18] that

$$\tan \tfrac{1}{2} B = \sqrt{\frac{c - a}{c + a}}. \quad \text{(See p. 23, Note.)}$$

SECTION XXXII

SUMS AND DIFFERENCES OF FUNCTIONS

From [4], [5], [8], and [9], by addition and subtraction,

$$\sin (x + y) + \sin (x - y) = \quad 2 \sin x \cos y,$$
$$\sin (x + y) - \sin (x - y) = \quad 2 \cos x \sin y,$$
$$\cos (x + y) + \cos (x - y) = \quad 2 \cos x \cos y,$$
$$\cos (x + y) - \cos (x - y) = - 2 \sin x \sin y;$$

or, by making $x + y = A$, and $x - y = B$, and, therefore, $x = \tfrac{1}{2}(A + B)$, and $y = \tfrac{1}{2}(A - B)$,

$$\sin A + \sin B = \quad 2 \sin \tfrac{1}{2}(A + B) \cos \tfrac{1}{2}(A - B). \qquad [20]$$
$$\sin A - \sin B = \quad 2 \cos \tfrac{1}{2}(A + B) \sin \tfrac{1}{2}(A - B). \qquad [21]$$
$$\cos A + \cos B = \quad 2 \cos \tfrac{1}{2}(A + B) \cos \tfrac{1}{2}(A - B). \qquad [22]$$
$$\cos A - \cos B = - 2 \sin \tfrac{1}{2}(A + B) \sin \tfrac{1}{2}(A - B). \qquad [23]$$

From [20] and [21], by division, we obtain

$$\frac{\sin A + \sin B}{\sin A - \sin B} = \tan \tfrac{1}{2}(A + B)\cot \tfrac{1}{2}(A - B);$$

or, since $\quad\quad \cot \tfrac{1}{2}(A - B) = \dfrac{1}{\tan \tfrac{1}{2}(A - B)},$

$$\frac{\sin A + \sin B}{\sin A - \sin B} = \frac{\tan \tfrac{1}{2}(A + B)}{\tan \tfrac{1}{2}(A - B)}. \quad\quad [24]$$

EXERCISE XIV

1. Find the value of $\sin (x + y)$ and $\cos (x + y)$ when $\sin x = \tfrac{3}{5}$, $\cos x = \tfrac{4}{5}$, $\sin y = \tfrac{5}{13}$, $\cos y = \tfrac{12}{13}$.

2. Find $\sin (90° - y)$ and $\cos (90° - y)$ by making $x = 90°$ in Formulas [8] and [9].

Find, by Formulas [4]–[11], the first four functions of:

3. $90° + y$.	8. $360° - y$.	13. $- y$.
4. $180° - y$.	9. $360° + y$.	14. $45° - y$.
5. $180° + y$.	10. $x - 90°$.	15. $45° + y$.
6. $270° - y$.	11. $x - 180°$.	16. $30° + y$.
7. $270° + y$.	12. $x - 270°$.	17. $60° - y$.

18. Find $\sin 3x$ in terms of $\sin x$.

19. Find $\cos 3x$ in terms of $\cos x$.

20. Given $\tan \tfrac{1}{2} x = 1$; find $\cos x$.

21. Given $\cot \tfrac{1}{2} x = \sqrt{3}$; find $\sin x$.

22. Given $\sin x = 0.2$; find $\sin \tfrac{1}{2} x$ and $\cos \tfrac{1}{2} x$.

23. Given $\cos x = 0.5$; find $\cos 2x$ and $\tan 2x$.

24. Given $\tan 45° = 1$; find the functions of $22° 30'$.

25. Given $\sin 30° = 0.5$; find the functions of $15°$.

26. Prove that $\tan 18° = \dfrac{\sin 33° + \sin 3°}{\cos 33° + \cos 3°}$.

Prove the following formulas:

27. $\sin 2x = \dfrac{2 \tan x}{1 + \tan^2 x}.$

29. $\tan \tfrac{1}{2} x = \dfrac{\sin x}{1 + \cos x}.$

28. $\cos 2x = \dfrac{1 - \tan^2 x}{1 + \tan^2 x}.$

30. $\cot \tfrac{1}{2} x = \dfrac{\sin x}{1 - \cos x}.$

31. $\sin \tfrac{1}{2} x \pm \cos \tfrac{1}{2} x = \sqrt{1 \pm \sin x}.$

32. $\dfrac{\tan x \pm \tan y}{\cot x \pm \cot y} = \pm \tan x \tan y.$

33. $\tan (45° - x) = \dfrac{1 - \tan x}{1 + \tan x}.$

If A, B, C are the angles of a triangle, prove that:

34. $\sin A + \sin B + \sin C = 4 \cos \tfrac{1}{2} A \cos \tfrac{1}{2} B \cos \tfrac{1}{2} C.$

35. $\cos A + \cos B + \cos C = 1 + 4 \sin \tfrac{1}{2} A \sin \tfrac{1}{2} B \sin \tfrac{1}{2} C.$

36. $\tan A + \tan B + \tan C = \tan A \times \tan B \times \tan C.$

37. $\cot \tfrac{1}{2} A + \cot \tfrac{1}{2} B + \cot \tfrac{1}{2} C = \cot \tfrac{1}{2} A \times \cot \tfrac{1}{2} B \times \cot \tfrac{1}{2} C.$

Change to a form more convenient for logarithmic computation:

38. $\cot x + \tan x.$

43. $1 + \tan x \tan y.$

39. $\cot x - \tan x.$

44. $1 - \tan x \tan y.$

40. $\cot x + \tan y.$

45. $\cot x \cot y + 1.$

41. $\cot x - \tan y.$

46. $\cot x \cot y - 1.$

42. $\dfrac{1 - \cos 2x}{1 + \cos 2x}.$

47. $\dfrac{\tan x + \tan y}{\cot x + \cot y}$

SECTION XXXIII

ANTI-TRIGONOMETRIC FUNCTIONS

If y is any trigonometric function of an angle x, then x is said to be the corresponding *anti-trigonometric* function of y.

Thus, if $y = \sin x$, x is the *anti-sine* or *inverse sine* of y.

The anti-trigonometric functions of y are written

$$\sin^{-1}y, \qquad \tan^{-1}y, \qquad \sec^{-1}y, \qquad \text{vers}^{-1}y,$$
$$\cos^{-1}y, \qquad \cot^{-1}y, \qquad \csc^{-1}y, \qquad \text{covers}^{-1}y.$$

These are read, the angle whose sine is y, and so on.

For example, $\sin 30° = \frac{1}{2}$; hence, $30° = \sin^{-1}\frac{1}{2}$. Similarly, $90° = \cos^{-1}0 = \sin^{-1}1$, and $45° = \tan^{-1}1 = \sin^{-1}\frac{1}{2}\sqrt{2}$, etc.

The symbol $^{-1}$ must not be confused with the exponent -1. Thus, $\sin^{-1}x$ is a very different expression from $\dfrac{1}{\sin x}$, which would be written $(\sin x)^{-1}$. On the continent of Europe mathematical writers employ the notation *arc sin*, *arc cos*, etc., for $\sin^{-1}$, $\cos^{-1}$, etc.

There is an important difference between the trigonometric and the anti-trigonometric functions. When an angle is given, its functions are all completely determined; but when one of the functions is given, the angle may have any one of an indefinite number of values. Thus, if $\sin y = \frac{1}{2}$, y may be 30°, or 150°, or either of these increased or diminished by any integral multiple of 360° or 2π, but cannot take any other values. Accordingly, $\sin^{-1}\frac{1}{2} = 30° \pm 2\,n\pi$, or $150° \pm 2\,n\pi$, where n is any positive integer. Similarly, $\tan^{-1}1 = 45° \pm 2\,n\pi$ or $225° \pm 2\,n\pi$; *i.e.*, $\tan^{-1}1 = 45° \pm n\pi$.

Since one of the angles whose sine is x and one of the angles whose cosine is x together make 90°, and since similar relations hold for the tangent and cotangent, for the secant and cosecant, and for the versed sine and coversed sine, we have

$$\sin^{-1}x + \cos^{-1}x = \frac{\pi}{2}, \qquad \sec^{-1}x + \csc^{-1}x = \frac{\pi}{2},$$

$$\tan^{-1}x + \cot^{-1}x = \frac{\pi}{2}, \quad \text{vers}^{-1}x + \text{covers}^{-1}x = \frac{\pi}{2},$$

where it must be understood that each equation is true only for a particular choice of the various possible values of the functions. Thus, if x is positive, and if the angles are always taken in the first quadrant, the equations are correct.

EXERCISE XV

1. Find all the values of the following functions:

$\sin^{-1}\frac{1}{2}\sqrt{3}$,　$\tan^{-1}\frac{1}{3}\sqrt{3}$,　$\operatorname{vers}^{-1}\frac{1}{2}$,　$\cos^{-1}(-\frac{1}{2}\sqrt{2})$,
$\csc^{-1}\sqrt{2}$,　$\tan^{-1}\infty$,　$\sec^{-1}2$,　$\cos^{-1}(-\frac{1}{2}\sqrt{3})$.

2. Prove that

$\sin^{-1}(-x) = -\sin^{-1}x$;　$\cos^{-1}(-x) = \pi - \cos^{-1}x$.

3. If $\sin^{-1}x + \sin^{-1}y = \pi$, prove that $x = y$.

4. If $y = \sin^{-1}\frac{1}{3}$, find $\tan y$.

5. Prove that $\cos(\sin^{-1}x) = \sqrt{1-x^2}$.

6. Prove that $\cos(2\sin^{-1}x) = 1 - 2x^2$.

7. Prove that $\tan(\tan^{-1}x + \tan^{-1}y) = \dfrac{x+y}{1-xy}$.

8. If $x = \sqrt{\frac{1}{2}}$, find all the values of $\sin^{-1}x + \cos^{-1}x$.

9. Prove that $\tan^{-1}\left(\dfrac{x}{\sqrt{1-x^2}}\right) = \sin^{-1}x$.

10. Find the value of $\sin(\tan^{-1}\frac{5}{12})$.

11. Find the value of $\cot(2\sin^{-1}\frac{2}{3})$.

12. Find the value of $\sin(\tan^{-1}\frac{1}{2} + \tan^{-1}\frac{1}{3})$.

13. If $\sin^{-1}x = 2\cos^{-1}x$, find x.

14. Prove that $\tan(2\tan^{-1}x) = \dfrac{2x}{1-x^2}$.

15. Prove that $\sin(2\tan^{-1}x) = \dfrac{2x}{1+x^2}$.

CHAPTER IV

THE OBLIQUE TRIANGLE

SECTION XXXIV

LAW OF SINES

Let A, B, C denote the angles of a triangle ABC (Figs. 53 and 54), and a, b, c, respectively, the lengths of the opposite sides.

Draw $CD \perp AB$, and meeting AB (Fig. 53) or AB produced (Fig. 54) at D. Let $CD = h$.

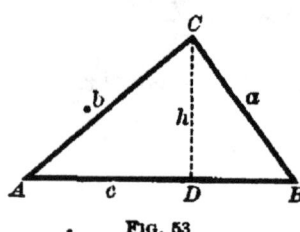

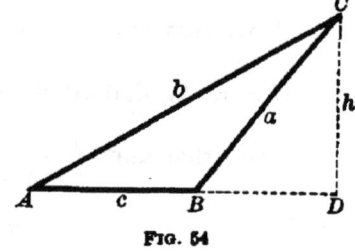

FIG. 53   FIG. 54

In either figure, $\dfrac{h}{b} = \sin A.$

In Fig. 53, $\dfrac{h}{a} = \sin B.$

In Fig. 54, $\dfrac{h}{a} = \sin (180° - B) = \sin B.$

Therefore, whether h lies within or without the triangle, we obtain, by division,

$$\frac{a}{b} = \frac{\sin A}{\sin B}. \qquad [25]$$

64

By drawing perpendiculars from the vertices A and B to the opposite sides, we may obtain, in the same way,

$$\frac{b}{c} = \frac{\sin B}{\sin C}, \qquad \frac{a}{c} = \frac{\sin A}{\sin C}.$$

Hence the Law of Sines:

The sides of a triangle are proportional to the sines of the opposite angles.

If we regard these three equations as proportions, and take them by alternation, it is evident that they may be written in the symmetrical form

$$\frac{a}{\sin A} = \frac{b}{\sin B} = \frac{c}{\sin C}.$$

Each of these equal ratios has a simple geometrical meaning which will appear if the Law of Sines is proved as follows:

Circumscribe a circle about the triangle ABC (Fig. 55), and draw the radii OB, OC. Let R denote the radius. Draw $OM \perp BC$. By Geometry, the angle $BOC = 2A$; hence, the angle $BOM = A$, then

$$BM = R \sin BOM = R \sin A.$$

$$\therefore BC \text{ or } a = 2R \sin A.$$

In like manner,

$b = 2R \sin B$, and $c = 2R \sin C$.

Whence we obtain

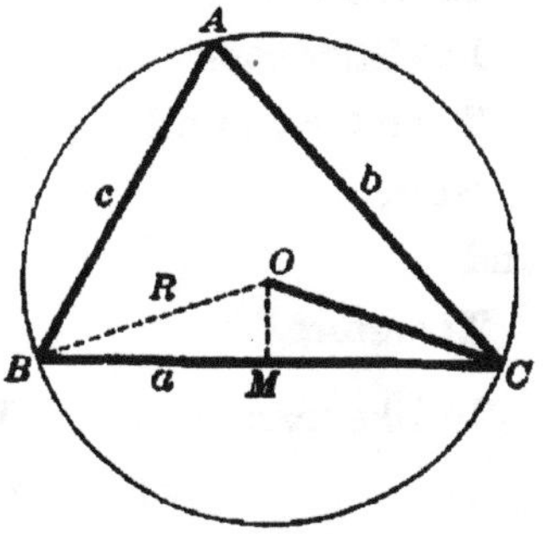

Fig. 55

$$2R = \frac{a}{\sin A} = \frac{b}{\sin B} = \frac{c}{\sin C}.$$

That is: *The ratio of any side of a triangle to the sine of the opposite angle is numerically equal to the diameter of the circumscribed circle.*

SECTION XXXV

LAW OF COSINES

This law gives the value of one side of a triangle in terms of the other two sides and the angle included between them.

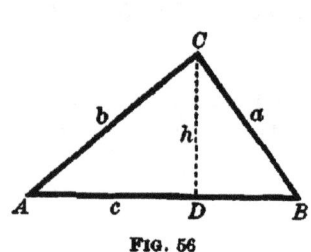

FIG. 56

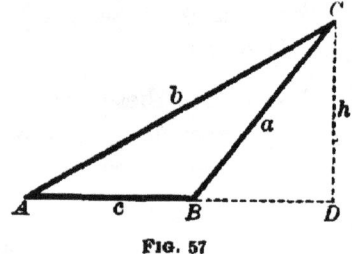

FIG. 57

In Figs. 56 and 57, $a^2 = h^2 + \overline{BD}^2$.

In Fig. 56, $BD = c - AD$.

In Fig. 57, $BD = AD - c$;

In either case, $\overline{BD}^2 = \overline{AD}^2 - 2c \times AD + c^2$.

Therefore, in all cases, $a^2 = h^2 + \overline{AD}^2 + c^2 - 2c \times AD$.

Now, $h^2 + \overline{AD}^2 = b^2$,

and $AD = b \cos A$.

Therefore, $\mathbf{a^2 = b^2 + c^2 - 2\,bc\cos A}$.   [26]

In like manner it may be proved that

$$b^2 = a^2 + c^2 - 2\,ac\cos B,$$
$$c^2 = a^2 + b^2 - 2\,ab\cos C.$$

The three formulas have precisely the same form, and the Law of Cosines may be stated as follows:

The square of any side of a triangle is equal to the sum of the squares of the other two sides diminished by twice their product into the cosine of the included angle.

SECTION XXXVI

LAW OF TANGENTS

By Sect. XXXIV, p. 64, $a : b = \sin A : \sin B$;
whence, by the Theory of Proportion,

$$\frac{a-b}{a+b} = \frac{\sin A - \sin B}{\sin A + \sin B}.$$

But by [24], p. 60,

$$\frac{\sin A - \sin B}{\sin A + \sin B} = \frac{\tan \frac{1}{2}(A-B)}{\tan \frac{1}{2}(A+B)}.$$

Therefore, $\dfrac{a-b}{a+b} = \dfrac{\tan \frac{1}{2}(A-B)}{\tan \frac{1}{2}(A+B)}.$ [27]

By merely changing the letters,

$$\frac{a-c}{a+c} = \frac{\tan \frac{1}{2}(A-C)}{\tan \frac{1}{2}(A+C)}, \qquad \frac{b-c}{b+c} = \frac{\tan \frac{1}{2}(B-C)}{\tan \frac{1}{2}(B+C)}.$$

Hence the Law of Tangents:

The difference of two sides of a triangle is to their sum as the tangent of half the difference of the opposite angles is to the tangent of half their sum.

NOTE. If in [27] $b > a$, then $B > A$. The formula is still true, but to avoid negative numbers the formula in this case should be written

$$\frac{b-a}{b+a} = \frac{\tan \frac{1}{2}(B-A)}{\tan \frac{1}{2}(B+A)}.$$

EXERCISE XVI

1. What do the formulas of Sect. XXXIV, p. 64, become when one of the angles is a right angle?

2. Prove by means of the Law of Sines that the bisector of an angle of a triangle divides the opposite side into parts proportional to the adjacent sides.

3. What does Formula [26] become when $A = 90°$? when $A = 0°$? when $A = 180°$? What does the triangle become in each of these cases?

NOTE. The case when $A = 90°$ explains why the theorem of Sect. XXXV, p. 66, is sometimes called the *Generalized Theorem of Pythagoras.*

4. Prove (Figs. 56 and 57) that whether the angle B is acute or obtuse $c = a \cos B + b \cos A$. What are the two symmetrical formulas obtained by changing the letters? What does the formula become when $B = 90°$?

5. From the three following equations (found in the last example) prove the theorem of Sect. XXXV, p. 66:

$$c = a \cos B + b \cos A,$$
$$b = a \cos C + c \cos A,$$
$$a = b \cos C + c \cos B.$$

HINT. Multiply the first equation by c, the second by b, the third by a; then from the first subtract the *sum* of the second and third.

6. In Formula [27] what is the maximum value of $\frac{1}{2}(A-B)$?

7. Find the form to which Formula [27] reduces, and describe the nature of the triangle, when

(i) $C = 90°$;   (ii) $A - B = 90°$, and $B = C$.

SECTION XXXVII

THE GIVEN PARTS

The formulas established in Sects. XXXIV–XXXVI, pp. 64–67, together with the equation $A + B + C = 180°$, are sufficient for solving every case of an oblique triangle. The three parts that determine an oblique triangle may be:

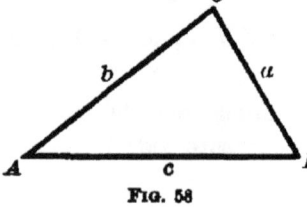

FIG. 58

I. One side and two angles;

II. Two sides and the angle opposite one of these sides;

III. Two sides and the included angle ;

IV. The three sides.

SECTION XXXVIII

SOLUTION OF AN OBLIQUE TRIANGLE

CASE I

Given one side a *and two angles* A *and* B; *find the remaining parts* C, b, *and* c.

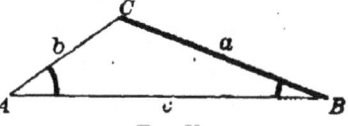

FIG. 59

1. $C = 180° - (A + B).$

2. $\dfrac{b}{a} = \dfrac{\sin B}{\sin A};$ $\quad \therefore b = \dfrac{a \sin B}{\sin A} = \dfrac{a}{\sin A} \times \sin B.$

3. $\dfrac{c}{a} = \dfrac{\sin C}{\sin A};$ $\quad \therefore c = \dfrac{a \sin C}{\sin A} = \dfrac{a}{\sin A} \times \sin C.$

EXAMPLE. $a = 24.31,\ A = 45° 18',\ B = 22° 11'.$

The work may be arranged as follows:

$a = 24.31$	$\log a = 1.38578$	$= 1.38578$
$A = 45° 18'$	colog $\sin A = 0.14825$	$= 0.14825$
$B = 22° 11'$	$\log \sin B = \overline{9.57700}$	$\log \sin C = \overline{9.96556}$
$A + B = \overline{67° 29'}$	$\log b = \overline{1.11103}$	$\log c = \overline{1.49959}$
$C = 112° 31'$	$b = 12.913$	$c = 31.593$

NOTE. When -10 is omitted after a logarithm or cologarithm, it must be remembered that the log or colog is 10 too large.

EXERCISE XVII

1. Given $a = 500,$ $\quad A = 10° 12',$ $\quad B = 46° 36';$
   find $C = 123° 12',$ $\quad b = 2051.5,$ $\quad c = 2362.6.$

2. Given $a = 795,$ $\quad A = 79° 59',$ $\quad B = 44° 41';$
   find $C = 55° 20',$ $\quad b = 567.69,$ $\quad c = 663.99.$

3. Given $a = 804$, $A = 99°\ 55'$, $B = 45°\ 1'$;
    find $C = 35°\ 4'$, $b = 577.31$, $c = 468.93$.

4. Given $a = 820$, $A = 12°\ 49'$, $B = 141°\ 59'$;
    find $C = 25°\ 12'$, $b = 2276.6$, $c = 1573.9$.

5. Given $c = 1005$, $A = 78°\ 19'$, $B = 54°\ 27'$;
    find $C = 47°\ 14'$, $a = 1340.6$, $b = 1113.8$.

6. Given $b = 13.57$, $B = 13°\ 57'$, $C = 57°\ 13'$;
    find $A = 108°\ 50'$, $a = 53.276$, $c = 47.324$.

7. Given $a = 6412$, $A = 70°\ 55'$, $C = 52°\ 9'$;
    find $B = 56°\ 56'$, $b = 5685.9$, $c = 5357.5$.

8. Given $b = 999$, $A = 37°\ 58'$, $C = 65°\ 2'$;
    find $B = 77°$, $a = 630.77$, $c = 929.48$.

9. In order to determine the distance of a hostile fort A from a place B, a line BC and the angles ABC and BCA were measured and found to be 322.55 yards, 60° 34', and 56° 10', respectively. Find the distance AB.

10. The angles B and C of a triangle ABC are 50° 30' and 122° 9', respectively, and BC is 9 miles. Find AB and AC.

11. Two observers 5 miles apart on a plain, and facing each other, find that the angles of elevation of a balloon in the same vertical plane with themselves are 55° and 58°, respectively. Find the distance from the balloon to each observer, and also the height of the balloon above the plain.

12. In a parallelogram given a diagonal d and the angles x and y which this diagonal makes with the sides; find the sides. Find the sides if $d = 11.237$, $x = 19°\ 1'$, and $y = 42°\ 54'$.

13. A lighthouse was observed from a ship to bear N. 34° E.; after the ship sailed due south 3 miles it bore N. 23° E. Find the distance from the lighthouse to the ship in each position.

NOTE. The phrase *to bear N. 34° E.* means that the line of sight to the lighthouse is in the northeast quarter of the horizon and makes, with a line due north, an angle of 34°.

14. In a trapezoid given the parallel sides a and b, and the angles x and y at the ends of one of the parallel sides; find the non-parallel sides. Compute the results when $a = 15$, $b = 7$, $x = 70°$, $y = 40°$.

Solve the following examples without using logarithms:

15. Given $b = 7.07107$, $A = 30°$, $C = 105°$; find a and c.

16. Given $c = 9.562$, $A = 45°$, $B = 60°$; find a and b.

17. The base of a triangle is 600 feet and the angles at the base are 30° and 120°. Find the other sides and the altitude.

18. Two angles of a triangle are, the one 20°, the other 40°. Find the ratio of the opposite sides.

19. The angles of a triangle are as $5 : 10 : 21$, and the side opposite the smallest angle is 3. Find the other sides.

20. Given one side of a triangle equal to 27, the adjacent angles equal each to 30°; find the radius of the circumscribed circle. (See Sect. XXXIV, p. 65.)

SECTION XXXIX

CASE II

Given two sides a *and* b *and the angle* A *opposite the side* a; *find the remaining parts* B, C, c.

This case, like the preceding case, is solved by means of the Law of Sines.

Since $\dfrac{\sin B}{\sin A} = \dfrac{b}{a}$, therefore $\sin B = \dfrac{b \sin A}{a}$;

$$C = 180° - (A + B).$$

And since $\dfrac{c}{a} = \dfrac{\sin C}{\sin A}$, therefore $c = \dfrac{a \sin C}{\sin A}$.

When an angle is determined by its sine it admits of two values which are supplements of each other (Sect. XXV, p. 48); hence, either value of B may be taken unless excluded by the conditions of the problem.

If $a > b$, then by Geometry $A > B$, and B must be acute whatever be the value of A; for a triangle can have only one obtuse angle. Hence, there is *one, and only one, triangle* that will satisfy the given conditions.

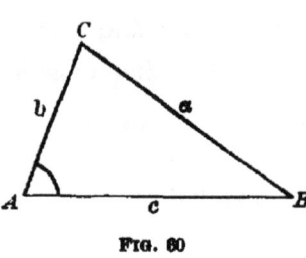

FIG. 60

If $a = b$, then by Geometry $A = B$; both A and B must be acute, and *the required triangle is isosceles.*

If $a < b$, then by Geometry $A < B$, and A must be acute in order that the triangle may be possible. If A is acute, it is evident from **Fig. 61**, where $\angle BAC = A$, $AC = b$, $CB = CB' = a$, that *the two triangles* ACB *and* ACB' *will satisfy the given conditions*, provided a is greater than the perpendicular CP; that is, provided a is greater

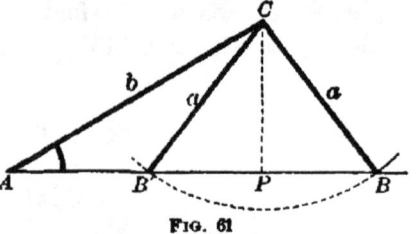

FIG. 61

than $b \sin A$ (Sect. XI, p. 20). The angles ABC and $AB'C$ are supplementary (since $\angle ABC = \angle BB'C$); they are, in fact, the supplementary angles obtained from the formula

$$\sin B = \frac{b \sin A}{a}.$$

If, however, $a = b \sin A = CP$ (Fig. 61), then $\sin B = 1$, $B = 90°$, and *the triangle required is a right triangle.*

If $a < b \sin A$, that is, $< CP$, then $\sin B > 1$, and *the triangle is impossible.*

These results, for convenience, may be thus stated:

Two solutions; if A is acute and the value of a lies between b and $b \sin A$.

No solution; if A is acute and $a < b \sin A$;

or   if A is obtuse and $a < b$, or $a = b$.

One solution; in all other cases.

The number of solutions can often be determined by inspection. In case of doubt, find the value of $b \sin A$.

Or we may proceed to compute $\log \sin B$. If $\log \sin B = 0$, the triangle required is a right triangle. If $\log \sin B > 0$, the triangle is impossible. If $\log \sin B < 0$, there is *one solution* when $a > b$; there are *two solutions* when $a < b$.

When there are two solutions, let B', C', c', denote the unknown parts of the second triangle; then,

$$B' = 180° - B, \quad C' = 180° - (A + B') = B - A,$$

$$c' = \frac{a \sin C'}{\sin A}.$$

EXAMPLE 1. Given $a = 16$, $b = 20$, $A = 106°$; find the remaining parts.

In this case $a < b$ and $A > 90°$; therefore, the triangle is impossible.

EXAMPLE 2. Given $a = 36$, $b = 80$, $A = 30°$; find the remaining parts.

Here we have $b \sin A = 80 \times \frac{1}{2} = 40$; so that $a < b \sin A$ and the triangle is impossible.

EXAMPLE 3. Given $a = 72,630$, $b = 117,480$, $A = 80° \, 0' \, 50''$; find B, C, c.

$a = 72,630$	$\text{colog } a = 5.13888$	Here $\log \sin B > 0$.
$b = 117,480$	$\log b = 5.06997$	$\therefore$ *no solution.*
$A = 80° \, 0' \, 50''$	$\log \sin A = 9.99337$	
	$\log \sin B = 0.20222$	

EXAMPLE 4. Given $a = 13.2$, $b = 15.7$, $A = 57° 13' 15''$; find B, C, c.

$a = 13.2$	colog $a = 8.87943$	$c = b \cos A$
$b = 15.7$	$\log b = 1.19590$	$\log b = 1.19590$
$A = 57° 13' 15''$	$\log \sin A = 9.92467$	$\log \cos A = 9.73352$
Here $\log \sin B = 0$.	$\log \sin B = 0.00000$	$\log c = 0.92942$
∴ a *right* triangle.	$B = 90°$	$c = 8.5$
	∴ $C = 32° 46' 45''$	

EXAMPLE 5. Given $a = 767$, $b = 242$, $A = 36° 53' 2''$; find B, C, c.

$a = 767$	colog $a = 7.11520$	$\log a = 2.88480$
$b = 242$	$\log b = 2.38382$	$\log \sin C = 9.86970$
$A = 36° 53' 2''$	$\log \sin A = 9.77830$	colog $\sin A = 0.22170$
Here $a > b$,	$\log \sin B = 9.27732$	$\log c = 2.97620$
and $\log \sin B < 0$.	$B = 10° 54' 58''$	$c = 946.68$
∴ *one solution.*	∴ $C = 132° 12' 0''$	

EXAMPLE 6. Given $a = 177.01$, $b = 216.45$, $A = 35° 36' 20''$; find the other parts.

$a = 177.01$	colog $a = 7.75200$	$\log a = 2.24800$	2.24800
$b = 216.45$	$\log b = 2.33536$	$\log \sin C = 9.99462$	9.23035
$A = 35° 36' 20''$	$\log \sin A = 9.76507$	colog $\sin A = 0.23493$	0.23493
Here $a < b$,	$\log \sin B = 9.85243$	$\log c = 2.47755$	1.71328
and $\log \sin B < 0$.	$B = 45° 23' 28''$	$c = 300.29$ or 51.675	
∴ *two solutions.*	or $134° 36' 32''$		
	∴ $C = 99° 0' 12''$		
	or $9° 47' 8''$		

EXERCISE XVIII

1. Find the number of solutions of the following:

    (i) $a = 80$, $b = 100$, $A = 30°$.

    (ii) $a = 50$, $b = 100$, $A = 30°$.

    (iii) $a = 40$, · $b = 100$, $A = 30°$.

    (iv) $a = 13.4$, $b = 11.46$, $A = 77° 20'$.

    (v) $a = 70$, $b = 75$, $A = 60°$.

    (vi) $a = 134.16$, $b = 84.54$, $B = 52° 9' 11''$.

    (vii) $a = 200$, $b = 100$, $A = 30°$.

2. Given $a = 840$, $b = 485$, $A = 21° 31'$;

    find $B = 12° 13' 34''$, $C = 146° 15' 26''$, $c = 1272.1$.

3. Given $a = 9.399$, $b = 9.197$, $A = 120° 35'$;

    find $B = 57° 23' 40''$, $C = 2° 1' 20''$, $c = 0.38525$.

4. Given $a = 91.06$, $b = 77.04$, $A = 51° 9' 6''$;

    find $B = 41° 13'$, $C = 87° 37' 54''$, $c = 116.82$.

5. Given $a = 55.55$, $b = 66.66$, $B = 77° 44' 40''$;

    find $A = 54° 31' 13''$, $C = 47° 44' 7''$, $c = 50.481$.

6. Given $a = 309$, $b = 360$, $A = 21° 14' 25''$;

    find $B = 24° 57' 54''$, $C = 133° 47' 41''$, $c = 615.67$,

    $B' = 155° 2' 6''$, $C' = 3° 43' 29''$, $c' = 55.41$.

7. Given $a = 8.716$, $b = 9.787$, $A = 38° 14' 12''$;

    find $B = 44° 1' 28''$, $C = 97° 44' 20''$, $c = 13.954$,

    $B' = 135° 58' 32''$, $C' = 5° 47' 16''$, $c' = 1.4202$.

8. Given $a = 4.4$, $b = 5.21$, $A = 57° 37' 17''$;

    find $B = 90°$, $C = 32° 22' 43''$, $c = 2.7901$.

9. Given $a = 34$, $b = 22$, $B = 30° 20'$;

    find $A = 51° 18' 27''$, $C = 98° 21' 33''$, $c = 43.098$,

    $A' = 128° 41' 33''$, $C' = 20° 58' 27''$, $c' = 15.593$.

10. Given $b = 19$, $c = 18$, $C = 15° 49'$;

    find $B = 16° 43' 13''$, $A = 147° 27' 47''$, $a = 35.519$,

    $B' = 163° 16' 47''$, $A = 0° 54' 13''$, $a' = 1.0415$.

11. Given $a = 75$, $b = 29$, $B = 16° 15' 36''$; find the difference between the areas of the two corresponding triangles.

12. Given in a parallelogram the side a, a diagonal d, and the angle A made by the two diagonals; find the other diagonal. Special case: $a = 35$, $d = 63$, $A = 21° 36' 30''$.

SECTION XL

Case III

Given two sides a *and* b *and the included angle* C; *find the remaining parts* A, B, *and* c.

SOLUTION I. The angles A and B may both be found by means of Formula [27], Sect. XXXVI, p. 67, which may be written

$$\tan \tfrac{1}{2}(A - B) = \frac{a - b}{a + b} \times \tan \tfrac{1}{2}(A + B).$$

Since $\tfrac{1}{2}(A + B) = \tfrac{1}{2}(180° - C)$, the value of $\tfrac{1}{2}(A + B)$ is known, so that this equation enables us to find the value of $\tfrac{1}{2}(A - B)$. We then have

$$\tfrac{1}{2}(A + B) + \tfrac{1}{2}(A - B) = A$$

and $$\tfrac{1}{2}(A + B) - \tfrac{1}{2}(A - B) = B.$$

After A and B are known, the side c may be found by the Law of Sines, which gives its value in two ways, as follows:

$$c = \frac{a \sin C}{\sin A}, \quad \text{or} \quad c = \frac{b \sin C}{\sin B}.$$

SOLUTION II. The third side c may be found directly from the equation (Sect. XXXV, p. 66)

$$c = \sqrt{a^2 + b^2 - 2\,ab \cos C};$$

and then, by the Law of Sines, the following equations for computing the values of the angles A and B are obtained:

$$\sin A = a \times \frac{\sin C}{c}, \quad \sin B = b \times \frac{\sin C}{c}.$$

SOLUTION III. If, in the triangle ABC (Fig. 62), BD is drawn perpendicular to the side AC, then

$$\tan A = \frac{BD}{AD} = \frac{BD}{AC - DC}.$$

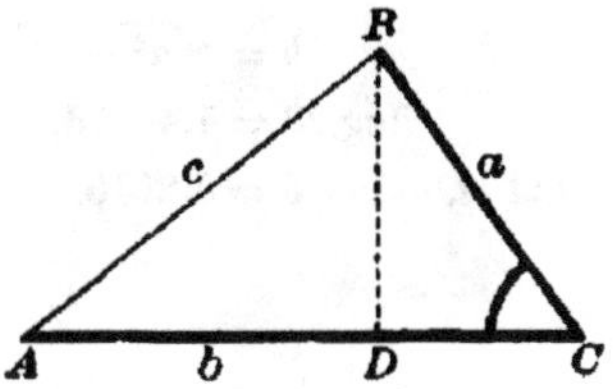

Now $$BD = a \sin C$$

and $$DC = a \cos C.$$

$$\therefore \tan A = \frac{a \sin C}{b - a \cos C}.$$

FIG. 62

By merely changing the letters,

$$\tan B = \frac{b \sin C}{a - b \cos C}.$$

It is not necessary, however, to use both formulas. When one angle, as A, has been found, the other, B, may be found from the relation $A + B + C = 180°$.

When the angles are known, the third side is found by the Law of Sines, as in Solution I.

NOTE. When all three unknown parts are required, Solution I is the most convenient in practice. When only the third side, c, is desired, Solution II may be used to advantage, provided the values of a^2 and b^2 can be obtained readily without the aid of logarithms. But Solutions II and III are not adapted to logarithmic work.

EXAMPLE 1. Given $a = 748$, $b = 375$, $C = 63° 35' 30''$; find A, B, and c.

$a + b = 1123$		
$a - b = 373$	$\log (a - b) = 2.57171$	$\log b = 2.57403$
$A + B = 116° 24' 30''$	$\operatorname{colog} (a + b) = 6.94962$	$\log \sin C = 9.95214$
$\frac{1}{2}(A + B) = 58° 12' 15''$	$\log \tan \frac{1}{2}(A + B) = 0.20766$	$\operatorname{colog} \sin B = 0.30073$
$\frac{1}{2}(A - B) = 28° 10' 54''$	$\log \tan \frac{1}{2}(A - B) = 9.72899$	$\log c = 2.82690$
$A = 86° 23' 9'$	$\frac{1}{2}(A - B) = 28° 10' 54''$	$c = 671.27$
$B = 30° 1' 21''$		

NOTE. In the above example we use the angle B in finding the side c rather than the angle A, because A is near 90°, and therefore the use of its sine should be avoided. See Note, p. 23.

EXAMPLE 2. Given $a = 4$, $c = 6$, $B = 60°$; find the third side b.

Here Solution II may be used to advantage. We have

$$b = \sqrt{a^2 + c^2 - 2\,ac \cos B} = \sqrt{16 + 36 - 24} = \sqrt{28};$$

$$\log 28 = 1.44716, \qquad \log \sqrt{28} = 0.72358, \qquad \sqrt{28} = 5.2915;$$

that is, $\quad b = 5.2915.$

EXERCISE XIX

1. Given $a = 77.99$, $b = 83.39$, $C = 72° 15'$;
   find $A = 51° 15'$, $B = 56° 30'$, $c = 95.24$.

2. Given $b = 872.5$, $c = 632.7$, $A = 80°$;
   find $B = 60° 45' 2''$, $C = 39° 14' 58''$, $a = 984.83$.

3. Given $a = 17$, $b = 12$, $C = 59° 17'$;
   find $A = 77° 12' 53''$, $B = 43° 30' 7''$, $c = 14.987$.

4. Given $b = \sqrt{5}$, $c = \sqrt{3}$, $A = 35° 53'$;
   find $B = 93° 28' 36''$, $C = 50° 38' 24''$, $a = 1.3131$.

5. Given $a = 0.917$, $b = 0.312$, $C = 33° 7' 9''$;
   find $A = 132° 18' 27''$, $B = 14° 34' 24''$, $c = 0.6775$.

6. Given $a = 13.715$, $c = 11.214$, $B = 15° 22' 36''$;
   find $A = 118° 55' 49''$, $C = 45° 41' 35''$, $b = 4.1554$.

7. Given $b = 3000.9$, $c = 1587.2$, $A = 86° 4' 4''$;
   find $B = 65° 13' 51''$, $C = 28° 42' 5''$, $a = 3297.2$.

8. Given $a = 4527$, $b = 3465$, $C = 66° 6' 27''$;
   find $A = 68° 29' 15''$, $B = 45° 24' 18''$, $c = 4449$.

9. Given $a = 55.14$, $b = 33.09$, $C = 30° 24'$;
   find $A = 117° 24' 32''$, $B = 32° 11' 28''$, $c = 31.431$.

10. Given $a = 47.99$, $b = 33.14$, $C = 175° 19' 10''$;
    find $A = 2° 46' 8''$, $B = 1° 54' 42''$, $c = 81.066$.

11. If two sides of a triangle are each equal to 6, and the included angle is 60°, find the third side.

12. If two sides of a triangle are each equal to 6, and the included angle is 120°, find the third side.

13. Apply Solution I to the case in which a is equal to b; that is, the case in which the triangle is isosceles.

14. If two sides of a triangle are 10 and 11, and the included angle is 50°, find the third side.

15. If two sides of a triangle are 43.301 and 25, and the included angle is 30°, find the third side.

16. In order to find the distance between two objects, A and B, separated by a swamp, a station C was chosen, and the distances $CA = 3825$ yards, $CB = 3475.6$ yards, together with the angle $ACB = 62°\ 31'$, were measured. Find the distance from A to B.

17. Two inaccessible objects, A and B, are each viewed from two stations, C and D, on the same side of AB and 562 yards apart. The angle ACB is $62°\ 12'$, BCD $41°\ 8'$, ADB $60°\ 49'$, and ADC $34°\ 51'$; required the distance AB.

18. Two trains start at the same time from the same station and move along straight tracks that form an angle of 30°, one train at the rate of 30 miles an hour, the other at the rate of 40 miles an hour. How far apart are the trains at the end of half an hour?

19. In a parallelogram given the two diagonals 5 and 6, and the angle that they form $49°\ 18'$; find the sides.

20. In a triangle one angle is $139°\ 54'$, and the sides forming the angle have the ratio $5:9$. Find the other two angles.

21. In order to find the distance between two objects, A and B, separated by a pond, a station C was chosen, and the distances $CA = 426$ yards, $CB = 322.4$ yards, together with the angle $ACB = 68°\ 42'$, were measured. Find the distance from A to B.

SECTION XLI

CASE IV

Given the three sides a, b, c; *find the angles* A, B, C.

The angles may be found directly from the formulas estab-lished in Sect. XXXV, p. 66.   Thus, from the formula

$$a^2 = b^2 + c^2 - 2\,bc\cos A,$$

$$\cos A = \frac{b^2 + c^2 - a^2}{2\,bc}.$$

From this equation formulas adapted to logarithmic work are deduced as follows :

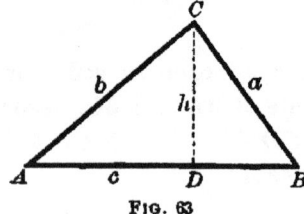

FIG. 63

For the sake of brevity, let

$$a + b + c = 2\,s;$$

then $b + c - a = 2\,(s - a),$

$$a - b + c = 2\,(s - b),$$

and $a + b - c = 2\,(s - c).$

Then the value of $1 - \cos A$ is

$$1 - \frac{b^2 + c^2 - a^2}{2\,bc} = \frac{2\,bc - b^2 - c^2 + a^2}{2\,bc} = \frac{a^2 - (b - c)^2}{2\,bc}$$

$$= \frac{(a + b - c)\,(a - b + c)}{2\,bc}$$

$$= \frac{2\,(s - b)\,(s - c)}{bc},$$

and the value of $1 + \cos A$ is

$$1 + \frac{b^2 + c^2 - a^2}{2\,bc} = \frac{2\,bc + b^2 + c^2 - a^2}{2\,bc} = \frac{(b + c)^2 - a^2}{2\,bc}$$

$$= \frac{(b + c + a)\,(b + c - a)}{2\,bc} = \frac{2\,s\,(s - a)}{bc}.$$

But from Formulas [16] and [17], p. 58, it follows that

$$1 - \cos A = 2 \sin^2 \tfrac{1}{2} A, \text{ and } 1 + \cos A = 2 \cos^2 \tfrac{1}{2} A.$$

$$\therefore 2 \sin^2 \tfrac{1}{2} A = \frac{2(s-b)(s-c)}{bc}, \text{ and } 2 \cos^2 \tfrac{1}{2} A = \frac{2s(s-a)}{bc},$$

whence

$$\sin \tfrac{1}{2} A = \sqrt{\frac{(s-b)(s-c)}{bc}}, \qquad [28]$$

$$\cos \tfrac{1}{2} A = \sqrt{\frac{s(s-a)}{bc}}, \qquad [29]$$

and by [2],

$$\tan \tfrac{1}{2} A = \sqrt{\frac{(s-b)(s-c)}{s(s-a)}}. \qquad [30]$$

By merely changing the letters,

$$\sin \tfrac{1}{2} B = \sqrt{\frac{(s-a)(s-c)}{ac}}, \quad \sin \tfrac{1}{2} C = \sqrt{\frac{(s-a)(s-b)}{ab}}.$$

$$\cos \tfrac{1}{2} B = \sqrt{\frac{s(s-b)}{ac}}, \qquad \cos \tfrac{1}{2} C = \sqrt{\frac{s(s-c)}{ab}}.$$

$$\tan \tfrac{1}{2} B = \sqrt{\frac{(s-a)(s-c)}{s(s-b)}}, \quad \tan \tfrac{1}{2} C = \sqrt{\frac{(s-a)(s-b)}{s(s-c)}}.$$

There is then a choice of three different formulas for finding the value of each angle. If half the angle is very near 0°, the formula for the cosine will not give a very accurate result, because the cosines of angles near 0° differ little in value; and the same holds true of the formula for the sine when half the angle is very near 90°. Hence, in the first case the formula for the sine, in the second that for the cosine, should be used.

But, in general, the formulas for the tangent are to be preferred.

It is not necessary to compute by the formulas more than two angles; for the third may then be found from the equation

$$A + B + C = 180°.$$

There is this advantage, however, in computing all three angles by the formulas, that we may then use the sum of the angles as a test of the accuracy of the results.

In case it is desired to compute all the angles, the formulas for the tangent may be put in a more convenient form.

The value of $\tan \frac{1}{2} A$ may be written

$$\sqrt{\frac{(s-a)(s-b)(s-c)}{s(s-a)^2}}$$

or

$$\frac{1}{s-a}\sqrt{\frac{(s-a)(s-b)(s-c)}{s}}.$$

Hence, if we put

$$\sqrt{\frac{(s-a)(s-b)(s-c)}{s}} = r, \qquad [31]$$

we have

$$\tan \frac{1}{2} A = \frac{r}{s-a}. \qquad [32]$$

Likewise, $\tan \frac{1}{2} B = \dfrac{r}{s-b}$, $\tan \frac{1}{2} C = \dfrac{r}{s-c}$.

EXAMPLE 1. Given $a = 3.41$, $b = 2.60$, $c = 1.58$; find the angles.

Using Formula [30] and the corresponding formula for $\tan \frac{1}{2} B$, we may arrange the work as follows:

$a = 3.41$	$\operatorname{colog} s = 9.42079$	$\operatorname{colog} s = 9.42079 - 10$
$b = 2.60$	$\operatorname{colog}(s-a) = 0.41454$	$\log(s-a) = 9.58546 - 10$
$c = 1.58$	$\log(s-b) = 0.07737$	$\operatorname{colog}(s-b) = 9.92263 - 10$
$2s = 7.59$	$\log(s-c) = 0.34537$	$\log(s-c) = 0.34537$
$s = 3.795$	$2\overline{)0.25807}$	$2\overline{)19.27425 - 20}$
$s - a = 0.385$	$\log \tan \frac{1}{2} A = 0.12903$	$\log \tan \frac{1}{2} B = 9.63713 - 10$
$s - b = 1.195$	$\frac{1}{2} A = 53° 23' 20''$	$\frac{1}{2} B = 23° 26' 37''$
$s - c = 2.215$	$A = 106° 46' 40''$	$B = 46° 53' 14''$

$\therefore A + B = 153° 39' 54''$, and $C = 26° 20' 6''$.

EXAMPLE 2.   Solve Example 1 by finding all three angles by the use of Formulas [31] and [32].

Here the work may be compactly arranged as follows, if we find log tan ¼ A, etc., by *subtracting* log (s − a), etc., from log r instead of adding the cologarithm :

a = 3.41	log (s − a) = 9.58546	log tan ¼ A = 10.12903
b = 2.60	log (s − b) = 0.07737	log tan ¼ B = 9.63713
c = 1.58	log (s − c) = 0.34537	log tan ¼ C = 9.36912
2s = 7.59	colog s = 9.42079	¼ A = 53° 23′ 20″
s = 3.795	log r² = 9.42899	¼ B = 23° 26′ 37″
s − a = 0.385	log r = 9.71450	¼ C = 13° 10′ 3″
s − b = 1.195		A = 106° 46′ 40″
s − c = 2.215		B = 46° 53′ 14″
2s = 7.590 (check).		C = 26° 20′ 6″

Check, A + B + C = 180° 0′ 0″

NOTE.   Even if no mistakes are made in the work, the sum of the three angles found as above may differ very slightly from 180° in consequence of the fact that logarithmic computation is at best only a method of close approximation.   When a difference of this kind exists, it should be divided among the angles according to the probable amount of error for each angle.

EXERCISE XX

Solve the following triangles, taking the three sides as the given parts :

	a	b	c	A	B	C
1	51	65	20	38° 52′ 48″	126° 52′ 12″	14° 15′
2	78	101	29	32° 10′ 55″	136° 23′ 50″	11° 25′ 15″
3	111	145	40	27° 20′ 32″	143° 7′ 48″	9° 31′ 40″
4	21	26	31	42° 6′ 13″	56° 6′ 36″	81° 47′ 11″
5	19	34	49	16° 25′ 36″	30° 24′	133° 10′ 24″
6	43	50	57	46° 49′ 35″	57° 59′ 44″	75° 10′ 41″
7	37	58	79	26° 0′ 29″	43° 25′ 20″	110° 34′ 11″
8	73	82	91	49° 34′ 58″	58° 46′ 58″	71° 38′ 4″
9	14.493	55.4363	66.9129	8° 20′ 1″	33° 40′ 5″	137° 59′ 54″
10	√5	√6	√7	51° 53′ 12″	59° 31′ 48″	68° 35′

11. Given $a = 6$, $b = 8$, $c = 10$; find the angles.

12. Given $a = 6$, $b = 6$, $c = 10$; find the angles.

13. Given $a = 6$, $b = 6$, $c = 6$; find the angles.

14. Given $a = 6$, $b = 9$, $c = 12$; find the angles.

15. Given $a = 2$, $b = \sqrt{6}$, $c = \sqrt{3} - 1$; find the angles.

16. Given $a = 2$, $b = \sqrt{6}$, $c = \sqrt{3} + 1$; find the angles.

17. The distances between three cities, A, B, and C, are as follows: $AB = 165$ miles, $AC = 72$ miles, and $BC = 185$ miles. B is due east from A. In what direction is C from A? What two answers are admissible?

18. Under what visual angle is an object 7 feet long seen by an observer whose eye is 5 feet from one end of the object and 8 feet from the other end?

19. When Formula [28] is used for finding the value of an angle, why does the ambiguity that occurs in Case II not exist?

20. If the sides of a triangle are 3, 4, and 6, find the sine of the largest angle.

21. Of three towns, A, B, and C, A is 200 miles from B and 184 miles from C, B is 150 miles due north from C. How far is A north of C?

22. The sides of a triangle are 78.9, 65.4, 97.3, respectively. Find the largest angle.

23. The sides of a triangle are 487.25, 512.33, 544.37, respectively. Find the smallest angle.

24. Find the angles of a triangle whose sides are $\dfrac{\sqrt{3} + 1}{2\sqrt{2}}$, $\dfrac{\sqrt{3} - 1}{2\sqrt{2}}$, $\dfrac{\sqrt{3}}{2}$, respectively.

25. The sides of a triangle are 14.6 inches, 16.7 inches, and 18.8 inches, respectively. Find the length of the perpendicular from the vertex of the largest angle upon the opposite side.

SECTION XLII

AREA OF A TRIANGLE

CASE I

When two sides and the included angle are given.

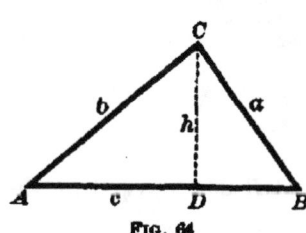

FIG. 64

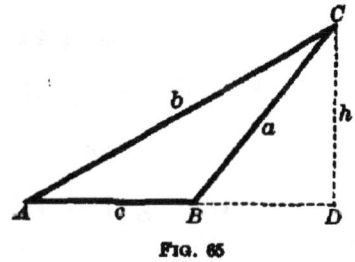

FIG. 65

In the triangle ABC (Fig. 64 or 65),

$$F = \tfrac{1}{2} c \times CD.$$

Now, $CD = a \sin B.$

Therefore, $\mathbf{F = \tfrac{1}{2} \, ac \, \sin B.}$ [33]

Also, $F = \tfrac{1}{2} ab \sin C,$ and $F = \tfrac{1}{2} bc \sin A.$

CASE II

When a side and the two adjacent angles are given.

$$\sin A : \sin C = a : c.   \text{(Sect. XXXIV, p. 65.)}$$

Therefore, $c = \dfrac{a \sin C}{\sin A}.$

Putting this value of c in Formula [33],

$$F = \frac{a^2 \sin B \sin C}{2 \sin A}.$$

But $\sin (B + C) = \sin (180° - A) = \sin A.$ (Sect. XXV, p. 48.)

Hence, $\mathbf{F = \dfrac{a^2 \sin B \sin C}{2 \sin (B + C)}.}$ [34]

Case III

When the three sides of a triangle are given.

By Formula [12], p. 58,

$$\sin B = 2 \sin \tfrac{1}{2} B \times \cos \tfrac{1}{2} B.$$

Now, by Formula [28], p. 81,

$$\sin \tfrac{1}{2} B = \sqrt{\frac{(s-a)(s-c)}{ac}},$$

and by Formula [29], p. 81,

$$\cos \tfrac{1}{2} B = \sqrt{\frac{s(s-b)}{ac}}.$$

By substituting these values of $\sin \tfrac{1}{2} B$ and $\cos \tfrac{1}{2} B$ in the above equation, we have

$$\sin B = \frac{2}{ac} \sqrt{s(s-a)(s-b)(s-c)}.$$

By putting this value of $\sin B$ in [33], we have

$$\mathbf{F} = \sqrt{s(s-a)(s-b)(s-c)}. \qquad [35]$$

Case IV

When the three sides and the radius of the circumscribed circle or the radius of the inscribed circle are given.

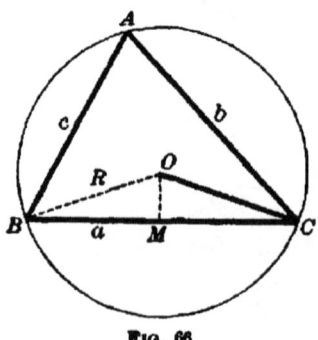

Fig. 66

If R denotes the radius of the circumscribed circle, we have, from Sect. XXXIV, p. 65,

$$\sin B = \frac{b}{2R}.$$

By putting this value of $\sin B$ in [33], we have

$$\mathbf{F} = \frac{abc}{4R}. \qquad [36]$$

If r denotes the radius of the inscribed circle, divide the triangle into three triangles by lines from the centre of this circle to the vertices; then the altitude of each of the three triangles is equal to r. Therefore,

$$\mathbf{F} = \tfrac{1}{2}\,\mathbf{r}\,(\mathbf{a} + \mathbf{b} + \mathbf{c}) = \mathbf{rs}. \qquad [37]$$

By putting in this formula the value of F given in [35],

$$r = \sqrt{\frac{(s-a)(s-b)(s-c)}{s}}\,;$$

whence r, in [31], Sect. XLI, p. 82, is equal to the radius of the inscribed circle.

EXERCISE XXI

Find the area:

1. Given $a = 4474.5$, $b = 2164.5$, $C = 116° 30' 20''$.
2. Given $b = 21.66$, $c = 36.94$, $A = 66° 4' 19''$.
3. Given $a = 510$, $c = 173$, $B = 162° 30' 28''$.
4. Given $a = 408$, $b = 41$, $c = 401$.
5. Given $a = 40$, $b = 13$, $c = 37$.
6. Given $a = 624$, $b = 205$, $c = 445$.
7. Given $b = 149$, $A = 70° 42' 30''$, $B = 39° 18' 28''$.
8. Given $a = 215.9$, $c = 307.7$, $A = 25° 9' 31''$.
9. Given $b = 8$, $c = 5$, $A = 60°$.
10. Given $a = 7$, $c = 3$, $A = 60°$.

11. Given $a = 60$, $B = 40° 35' 12''$, area $= 12$; find the radius of the inscribed circle.

12. Obtain a formula for the area of a parallelogram in terms of two adjacent sides and the included angle.

13. Obtain a formula for the area of an isosceles trapezoid in terms of the two parallel sides and an acute angle.

14. Two sides and included angle of a triangle are 2416, 1712, and 30°; and two sides and included angle of another triangle are 1948, 2848, and 150°. Find the sum of their areas.

15. The base of an isosceles triangle is 20, and its area is $100 + \sqrt{3}$; find its angles.

16. Show that the area of a quadrilateral is equal to one-half the product of its diagonals into the sine of their included angle.

EXERCISE XXII

1. From a ship sailing down the English Channel the Eddystone was observed to bear N. 33° 45' W., and after the ship had sailed 18 miles S. 67° 30' W. it bore N. 11° 15' E. Find its distance from each position of the ship.

2. Two objects, A and B, were observed from a ship to be at the same instant in a line bearing N. 15° E. The ship then sailed northwest 5 miles, when it was found that A bore due east and B bore northeast. Find the distance from A to B.

3. A castle and a monument stand on the same horizontal plane. The angles of depression of the top and the bottom of the monument viewed from the top of the castle are 40° and 80°; the height of the castle is 140 feet. Find the height of the monument.

4. If the sun's altitude is 60°, what angle must a stick make with the horizon in order that its shadow in a horizontal plane may be the longest possible?

5. If the sun's altitude is 30°, find the length of the longest shadow cast on a horizontal plane by a stick 10 feet in length.

6. In a circle with the radius 3 find the area of the part comprised between parallel chords whose lengths are 4 and 5. (Two solutions.)

CHAPTER V

MISCELLANEOUS EXAMPLES

PROBLEMS IN PLANE TRIGONOMETRY

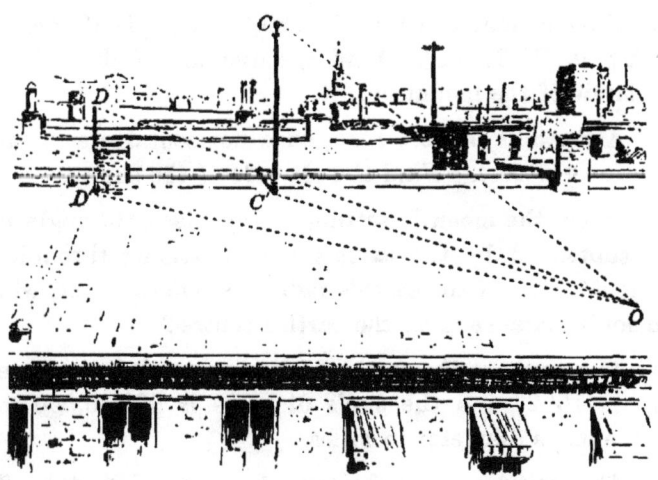

FIG. 67

If two objects are not in the same horizontal plane with each other or with the point of observation, we may suppose vertical lines to be passed through the two objects and to meet the horizontal plane of the point of observation in two points. The angular distance of these two points is the *bearing* of either of the objects from the other. Thus, the angle $C'OD'$ (Fig. 67) is the bearing of C from D.

NOTE. "Problems in Plane Trigonometry" are selected from those published by Mr. Charles W. Seaver, Cambridge, Mass. The full set can be obtained from him in pamphlet form.

RIGHT TRIANGLES

1. The angle of elevation of a tower is 48° 19′ 14″, and the distance of the base from the point of observation is 95 feet. Find the height of the tower and the distance of its top from the point of observation.

2. From a mountain 1000 feet high, the angle of depression of a ship is 77° 35′ 11″. Find the distance of the ship from the summit of the mountain.

3. A flagstaff 90 feet high, on a horizontal plane, casts a shadow of 117 feet. Find the altitude of the sun.

4. When the moon is setting at any place, the angle at the moon subtended by the earth's radius passing through that place is 57′ 3″. If the earth's radius is 3956.2 miles, what is the moon's distance from the earth's centre ?

5. The angle at the earth's centre subtended by the sun's radius is 16′ 2″ and the sun's distance is 92,400,000 miles. Find the sun's diameter in miles.

6. The latitude of Cambridge, Mass., is 42° 22′ 49″. What is the length of the radius of that parallel of latitude ?

7. At what latitude is the circumference of the parallel of latitude half of that of the equator ?

8. In a circle with a radius of 6.7 is inscribed a regular polygon of thirteen sides. Find the length of one of its sides.

9. A regular heptagon, one side of which is 5.73, is inscribed in a circle. Find the radius of the circle.

10. A tower 93.97 feet high is situated on the bank of a river. The angle of depression of an object on the opposite bank is 25° 12′ 54″. Find the breadth of the river.

11. From a tower 58 feet high the angles of depression of two objects situated in the same horizontal line with the base of the tower, and on the same side, are 30° 13′ 18″ and 45° 46′ 14″. Find the distance between these two objects.

12. Standing directly in front of one corner of a flat-roofed house, which is 150 feet in length, I observe that the horizontal angle which the length subtends has for its cosine $\sqrt{\frac{1}{5}}$, and that the vertical angle subtended by its height has for its sine $\dfrac{3}{\sqrt{34}}$. What is the height of the house?

13. A regular pyramid, with a square base, has a lateral edge 150 feet long, and a side of its base is 200 feet. Find the inclination of the face of the pyramid to the base.

14. From one edge of a ditch 36 feet wide, the angle of elevation of a wall on the opposite edge is 62° 39′ 10″. Find the length of a ladder that will just reach from the point of observation to the top of the wall.

15. The top of a flagstaff has been partly broken off and touches the ground at a distance of 15 feet from the foot of the staff. If the length of the broken part is 39 feet, find the length of the whole staff.

16. From a balloon, which is directly above one town, is observed the angle of depression of another town, 10° 14′ 9″. The towns being 8 miles apart, find the height of the balloon.

17. From the top of a mountain 3 miles high the angle of depression of the most distant object which is visible on the earth's surface is found to be 2° 13′ 50″. Find the diameter of the earth.

18. A ladder 40 feet long reaches a window 33 feet high, on one side of a street. Being turned over upon its foot, it reaches another window 21 feet high, on the opposite side of the street. Find the width of the street.

19. The height of a house subtends a right angle at a window on the other side of the street; and the angle of elevation of the top of the house, from the same point, is 60°. The street is 30 feet wide. How high is the house?

20. A lighthouse 54 feet high is situated on a rock. The angle of elevation of the top of the lighthouse, as observed from a ship, is 4° 52', and the angle of elevation of the top of the rock is 4° 2'. Find the height of the rock and its distance from the ship.

21. A man in a balloon observes the angle of depression of an object on the ground, bearing south, to be 35° 30'; the balloon drifts 2½ miles east at the same height, when the angle of depression of the same object is 23° 14'. Find the height of the balloon.

22. A man standing south of a tower, on the same horizontal plane, observes its angle of elevation to be 54° 16'; he goes east 100 yards, and then finds its angle of elevation is 50° 8'. Find the height of the tower.

23. The angle of elevation of a tower at a place A, south of it, is 30°; and at a place B, west of A, and at a distance of a from it, the angle of elevation is 18°. Show that the height of the tower is $\dfrac{a}{\sqrt{2 + 2\sqrt{5}}}$; the tangent of 18° being $\dfrac{\sqrt{5} - 1}{\sqrt{10 + 2\sqrt{5}}}$:

24. A pole is fixed on the top of a mound, and the angles of elevation of the top and the bottom of the pole are 60° and 30°, respectively. Prove that the length of the pole is twice the height of the mound.

25. At a distance a from the foot of a tower, the angle of elevation A of the top of the tower is the complement of the angle of elevation of a flagstaff on top of it. Show that the length of the staff is $2\,a \cot 2\,A$.

26. *A line of true level* is a line every point of which is equally distant from the centre of the earth. A line drawn tangent to a line of true level at any point is a line of *apparent* level. If at any point both these lines are drawn, and extended one mile, find the distance they are then apart.

27. In Problem 1, page 90, determine the effect upon the computed height of the tower of an error in either the angle of elevation or the measured distance.

OBLIQUE TRIANGLES

28. To determine the height of an inaccessible object situated on a horizontal plane, by observing its angles of elevation at two points in the same line with its base, and measuring the distance between these two points.

29. The angle of elevation of an inaccessible tower situated on a horizontal plane is $63°\ 26'$; at a point 500 feet farther from the base of the tower the angle of elevation of its top is $32°\ 14'$. Find the height of the tower.

30. A tower is situated on the bank of a river. From the opposite bank the angle of elevation of the tower is $60°\ 13'$, and from a point 40 feet more distant the angle of elevation is $50°\ 19'$. Find the breadth of the river.

31. A ship sailing north sees two lighthouses 8 miles apart, in a line due west; after an hour's sailing, one lighthouse bears S.W., and the other S.S.W. Find the ship's rate.

32. To determine the height of an accessible object situated on an inclined plane.

33. At the distance of 40 feet from the foot of a tower on an inclined plane, the tower subtends an angle of $41°\ 19'$; at a point 60 feet farther away, the angle subtended by the tower is $23°\ 45'$. Find the height of the tower.

34. A tower makes an angle of 113° 12' with the inclined plane on which it stands ; and at a distance of 89 feet from its base, measured down the plane, the angle subtended by the tower is 23° 27'. Find the height of the tower.

35. From the top of a house 42 feet high the angle of elevation of the top of a pole is 14° 13'; at the bottom of the house it is 23° 19'. Find the height of the pole.

36. The sides of a triangle are 17, 21, 28. Prove that the length of a line bisecting the greatest side and drawn from the opposite angle is 13.

37. A privateer, 10 miles S.W. of a harbor, sees a ship sail from it in a direction S. 80° E., at a rate of 9 miles an hour. In what direction, and at what rate, must the privateer sail in order to come up with the ship in 1½ hours?

38. A person goes 70 yards up a slope of 1 in 3½ from the edge of a river and observes the angle of depression of an object on the opposite bank to be 2¼°. Find the breadth of the river.

39. The length of a lake subtends, at a certain point, an angle of 46° 24', and the distances from this point to the two extremities of the lake are 346 and 290 feet. Find the length of the lake.

40. Two ships are a mile apart. The angular distance of the first ship from a fort on shore, as observed from the second ship, is 35° 14' 10"; the angular distance of the second ship from the fort, observed from the first ship, is 42° 11' 53". Find the distance in feet from each ship to the fort.

41. Along the bank of a river is drawn a base line of 500 feet. The angular distance of one end of this line from an object on the opposite side of the river, as observed from the other end of the line, is 53°; that of the second extremity from the same object, observed at the first, is 79° 12'.   Find the breadth of the river.

42. A vertical tower stands on a declivity inclined 15° to the horizon.   A man ascends the declivity 80 feet from the base of the tower, and finds the angle then subtended by the tower to be 30°.   Find the height of the tower.

43. The angle subtended by a tower on an inclined plane is, at a certain point, 42° 17'; 325 feet farther down it is 21° 47'. The inclination of the plane is 8° 53'.   Find the height of the tower.

44. A cape bears north by east, as seen from a ship.   The ship sails northwest 30 miles, and then the cape bears east. How far is it from the second point of observation?

45. Two observers, stationed on *opposite* sides of a cloud, observe its angles of elevation to be 44° 56' and 36° 4'.   Their distance from each other is 700 feet.   What is the height of the cloud ?

46. From a point *B* at the foot of a mountain, the angle of elevation of the top *A* is 60°.   After ascending the mountain one mile, at an inclination of 30° to the horizon, and reaching a point *C*, the angle *ACB* is found to be 135°.   Find the height of the mountain in feet.

47. From a ship two rocks are seen in the same right line with the ship, bearing N. 15° E.   After the ship has sailed northwest 5 miles, the first rock bears east, and the second northeast.   Find the distance between the rocks.

48. From a window on a level with the bottom of a steeple the angle of elevation of the steeple is 40°, and from a second window 18 feet higher the angle of elevation is 37° 30'.   Find the height of the steeple.

49. To determine the distance between two inaccessible objects by observing angles at the extremities of a line of known length.

50. Wishing to determine the distance between a church A and a tower B, on the opposite side of a river, I measure a line CD along the river (C being nearly opposite A), and observe the angles ACB, $58°\ 20'$; ACD, $95°\ 20'$; ADB, $53°\ 30'$; BDC, $98°\ 45'$. CD is 600 feet. What is the distance required?

51. Wishing to find the height of a summit A, I measure a horizontal base line CD, 440 yards. At C, the angle of elevation of A is $37°\ 18'$, and the horizontal angle between D and the summit is $76°\ 18'$; at D, the horizontal angle between C and the summit is $67°\ 14'$. Find the height.

52. A balloon is observed from two stations 3000 feet apart. At the first station the horizontal angle of the balloon and the other station is $75°\ 25'$, and the angle of elevation of the balloon is $18°$. The horizontal angle of the first station and the balloon, measured at the second station, is $64°\ 30'$. Find the height of the balloon.

53. Two forces, one of 410 pounds, and the other of 320 pounds, make an angle of $51°\ 37'$. Find the intensity and the direction of their resultant.

54. An unknown force, combined with one of 128 pounds, produces a resultant of 200 pounds, and this resultant makes an angle of $18°\ 24'$ with the known force. Find the intensity and direction of the unknown force.

55. At two stations, the height of a kite subtends the same angle A. The angle which the line joining one station and the kite subtends at the other station is B; and the distance between the two stations is a. Show that the height of the kite is $\frac{1}{2} a \sin A \sec B$.

56. Two towers on a horizontal plane are 120 feet apart. A person standing successively at their bases observes that the angle of elevation of one is double that of the other; but, when he is half-way between them, the angles of elevation are complementary. Prove that the heights of the towers are 90 and 40 feet.

57. To find the distance of an inaccessible point C from either of two points A and B, having no instruments to measure angles. Prolong CA to a, and CB to b, and join AB, Ab, and Ba. Measure AB, 500; aA, 100; aB, 560; bB, 100; and Ab, 550. Compute the distances AC and BC.

58. Two inaccessible points A and B are visible from D, but no other point can be found whence both are visible. Take some point C, whence A and D can be seen, and measure CD, 200 feet; ADC, 89°; ACD, 50° 30'. Then take some point E, whence D and B are visible, and measure DE, 200 feet; BDE, 54° 30'; BED, 88° 30'. At D measure ADB, 72° 30'. Compute the distance AB.

59. To compute the horizontal distance between two inaccessible points A and B, when no point can be found whence both can be seen. Take two points C and D, distant 200 yards, so that A can be seen from C, and B from D. From C measure CF, 200 yards to F, whence A can be seen; and from D measure DE, 200 yards to E, whence B can be seen. Measure AFC, 83°; ACD, 53° 30'; ACF, 54° 31'; BDE, 54° 30'; BDC, 156° 25'; DEB, 88° 30'.

60. A column in the north temperate zone is east-southeast of an observer, and at noon the extremity of its shadow is northeast of him. The shadow is 80 feet in length, and the elevation of the column, at the observer's station, is 45°. Find the height of the column.

61. From the top of a hill the angles of depression of two objects situated in the horizontal plane of the base of the hill are 45° and 30°; and the horizontal angle between the two objects is 30°. Show that the height of the hill is equal to the distance between the objects.

62. Wishing to know the breadth of a river from A to B, I take AC, 100 yards in the prolongation of BA, and then take CD, 200 yards at right angles to AC. The angle BDA is $37° 18' 30''$. Find AB.

63. The sum of the sides of a triangle is 100. The angle at A is double that at B, and the angle at B is double that at C. Determine the sides.

64. If $\sin^2 A + 5 \cos^2 A = 3$, find A.

65. If $\sin^2 A = m \cos A - n$, find $\cos A$.

66. Given $\sin A = m \sin B$, and $\tan A = n \tan B$; find $\sin A$ and $\cos B$.

67. If $\tan^2 A + 4 \sin^2 A = 6$, find A.

68. If $\sin A = \sin 2 A$, find A.

69. If $\tan 2 A = 3 \tan A$, find A.

70. Prove that $\tan 50° + \cot 50° = 2 \sec 10°$.

71. Given a regular polygon of n sides, and calling one of them a, find expressions for the radii of the inscribed and the circumscribed circles in terms of n and a.

If P, H, D are the sides of a regular inscribed pentagon, hexagon, and decagon, prove $P^2 = H^2 + D^2$.

AREAS

72. Obtain the formula for the area of a triangle, given two sides b, c, and the included angle A.

73. Obtain the formula for the area of a triangle, given two angles A and B, and included side c.

74. Obtain the formula for the area of a triangle, given the three sides.

75. If a is the side of an equilateral triangle, show that its area is $\dfrac{a^2 \sqrt{3}}{4}$.

76. Two consecutive sides of a rectangle are 52.25 chains and 38.24 chains. Find the area.

77. Two sides of a parallelogram are 59.8 chains and 37.05 chains, and the included angle is 72° 10′. Find the area.

78. Two sides of a parallelogram are 15.36 chains and 11.46 chains, and the included angle is 47° 30′. Find the area.

79. Two sides of a triangle are 12.38 chains and 6.78 chains, and the included angle is 46° 24′. Find the area.

80. Two sides of a triangle are 18.37 chains and 13.44 chains, and they form a right angle. Find the area.

81. Two angles of a triangle are 76° 54′ and 57° 33′ 12″, and the included side is 9 chains. Find the area.

82. Two sides of a triangle are 19.74 chains and 17.34 chains. The first bears N. 82° 30′ W.; the second S. 24° 15′ E. Find the area.

83. The three sides of a triangle are 49 chains, 50.25 chains, and 25.69 chains. Find the area.

84. The three sides of a triangle are 10.64 chains, 12.28 chains, and 9 chains. Find the area.

85. The sides of a triangular field, of which the area is 14 acres, are in the ratio of 3, 5, 7. Find the sides.

86. In the quadrilateral $ABCD$ we have AB, 17.22 chains; AD, 7.45 chains; CD, 14.10 chains; BC, 5.25 chains; and the diagonal AC, 15.04 chains. Find the area.

87. The diagonals of a quadrilateral are a and b, and they intersect at an angle D. Show that the area of the quadrilateral is $\frac{1}{2} ab \sin D$.

88. The diagonals of a quadrilateral are 34 and 56, intersecting at an angle of 67°. Find the area.

89. The diagonals of a quadrilateral are 75 and 49, intersecting at an angle of 42°. Find the area.

90. Show that the area of a regular polygon of n sides, of which one is a, is $\frac{na^2}{4} \cot \frac{180°}{n}$.

91. One side of a regular pentagon is 25. Find the area.

92. One side of a regular hexagon is 32. Find the area.

93. One side of a regular decagon is 46. Find the area.

94. Find the area of a circle whose circumference is 74 feet.

95. Find the area of a circle whose radius is 125 feet.

96. In a circle with a diameter of 125 feet find the area of a sector with an arc of 22°.

97. In a circle with a radius of 44 feet find the area of a sector with an arc of 25°.

98. In a circle with a diameter of 50 feet find the area of a segment with an arc of 280°.

99. Find the area of a segment (less than a semicircle) of which the chord is 20, and the distance of the chord from the middle point of the smaller arc is 2.

100. If r is the radius of a circle, the area of a regular circumscribed polygon of n sides is $nr^2 \tan \frac{180°}{n}$.

The area of a regular inscribed polygon is $\frac{n}{2} r^2 \sin \frac{360°}{n}$.

101. If a is a side of a regular polygon of n sides, the area of the inscribed circle is $\dfrac{\pi a^2}{4} \cot^2 \dfrac{180°}{n}$.

The area of the circumscribed circle is $\dfrac{\pi a^2}{4} \csc^2 \dfrac{180°}{n}$.

102. The area of a regular polygon inscribed in a circle is to that of the circumscribed regular polygon of the same number of sides as 3 to 4. Find the number of sides.

103. The area of a regular polygon inscribed in a circle is the geometric mean between the areas of an inscribed and a circumscribed regular polygon of half the number of sides.

104. The area of a circumscribed regular polygon is the harmonic mean between the areas of an inscribed regular polygon of the same number of sides and of a circumscribed regular polygon of half that number.

105. The perimeter of a circumscribed regular triangle is double that of the inscribed regular triangle.

106. The square described about a circle is four-thirds the inscribed regular dodecagon.

107. Two sides of a triangle are 3 and 12, and the included angle is 30°. Find the hypotenuse of an isosceles right triangle of equal area.

PLANE SAILING

Plane Sailing is that branch of Navigation in which the surface of the earth is considered a plane. The problems which arise are therefore solved by the methods of Plane Trigonometry.

The *difference of latitude* of two places is the arc of a meridian comprehended between the parallels of latitude passing through those places.

The *departure* between two meridians is the arc of a parallel of latitude comprehended between those meridians. It diminishes as the distance from the equator increases.

When a ship sails in such a manner as to cross successive meridians at the same angle, it is said to sail on a *rhumb-line*. This angle is called the *course*, and the *distance* between two places is measured on a rhumb-line.

If we consider the distance, departure, and difference of latitude of two places to be straight lines, lying in one plane, they form a right triangle, called *the triangle of plane sailing*. If ABC is a plane triangle, right-angled at B, and BC represents the difference of latitude of B and C, ACB will be the course from C to A, CA the distance, and D the departure, measured from B, between the meridian of A and that of B.

108. Taking the earth's equatorial diameter to be 7925.6 miles, find the length in feet of the arc of one minute of a great circle.*

109. A ship sails from latitude 43° 45' S., on a course N. by E. 2345 miles. Find the latitude reached, and the departure made.

110. A ship sails from latitude 1° 45' N., on a course S.E. by E., and reaches latitude 2° 31' S. Find the distance, and the departure.

111. A ship sails from latitude 13° 17' S., on a course N.E. by E. ¼ E., until the departure is 207 miles. Find the distance, and the latitude reached.

112. A ship sails on a course between S. and E. 244 miles, leaving latitude 2° 52' S., and reaching latitude 5° 8' S. Find the course, and the departure.

---

* The length of the arc of one minute of a great circle of the earth is called a *geographical mile* or a *knot*. In the following problems, this is the distance meant by the term " mile," unless otherwise stated.

113. A ship sails from latitude 32° 18′ N., on a course between N. and W., a distance of 344 miles, and a departure of 103 miles. Find the course, and the latitude reached.

114. A ship sails on a course between S. and E., making a difference of latitude 136 miles, and a departure 203 miles. Find the distance, and the course.

115. A ship sails due north 15 *statute* miles an hour, for one day. What is the distance, in a straight line, from the point left to the point reached? (Take earth's radius, 3962.8 statute miles.)

PARALLEL AND MIDDLE LATITUDE SAILING

The *difference of longitude* of two places is the angle at the pole made by the meridians of these two places; or, it is the arc of the equator comprehended between these two meridians.

FIG. 68

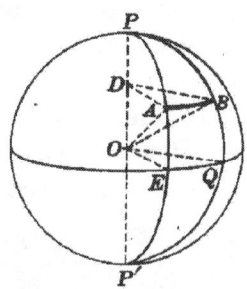

FIG. 69

In **Parallel Sailing** a vessel is supposed to sail due east or due west. The distance sailed is the departure made; and the difference of longitude is found as follows:

116. Given the departure between any two meridians at any latitude; find the difference of longitude of any point on one meridian from any point on the other.

SOLUTION.   In rt. $\triangle ODA$, $\angle AOD = 90° -$ lat.

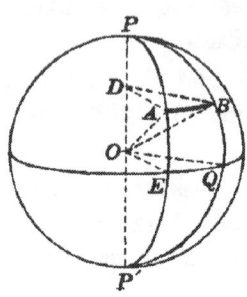

Hence, $\dfrac{DA}{OA} = \sin(90° - \text{lat.}) = \cos$ lat.

The $\triangle DAB$ and OEQ are similar.

Therefore, $\dfrac{DA}{OE} = \dfrac{AB}{EQ}$, or $\dfrac{DA}{OA} = \dfrac{AB}{EQ}$.

Hence, $\cos$ lat. $= \dfrac{AB}{EQ}$.

Therefore, $EQ = \dfrac{AB}{\cos \text{lat.}} = AB \times \sec$ lat.

FIG. 70   That is,   **Diff. long. = depart. $\times$ sec lat.**

117. A ship in latitude 42° 16′ N., longitude 72° 16′ W., sails due east a distance of 149 miles. What is the position of the point reached?

118. A ship in latitude 44° 49′ S., longitude 119° 42′ E., sails due west until it reaches longitude 117° 16′ E. Find the distance made.

In **Middle Latitude Sailing** the departure between two places is measured on that parallel of latitude which lies midway between the parallels of the two places. Except in very high latitudes or excessive runs, this assumption produces no great error. Hence, in middle latitude sailing,

Diff. long. = depart. $\times$ sec mid. lat.

FIG. 71

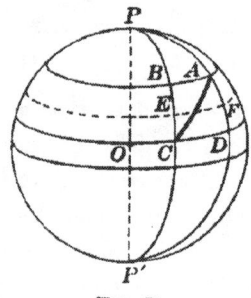

FIG. 72

119. A ship leaves latitude 31° 14′ N., longitude 42° 19′ W., and sails E.N.E. 325 miles. Find the position reached.

120. Find the bearing and distance of Cape Cod from Havana. (Cape Cod 42° 2′ N., 70° 3′ W.; Havana, 23° 9′ N., 82° 22′ W.)

121. Leaving latitude 49° 57′ N., longitude 15° 16′ W., a ship sails between S. and W. till the departure is 194 miles, and the latitude is 47° 18′ N. Find the course, distance, and longitude reached.

122. Leaving latitude 42° 30′ N., longitude 58° 51′ W., a ship sails S.E. by S. 300 miles. Find the position reached.

123. Leaving latitude 49° 57′ N., longitude 30° W., a ship sails S. 39° W., and reaches latitude 47° 44′ N. Find the distance, and longitude reached.

124. Leaving latitude 37° N., longitude 32° 16′ W., a ship sails between N. and W. 300 miles, and reaches latitude 41° N. Find the course, and longitude reached.

125. Leaving latitude 50° 10′ S., longitude 30° E., a ship sails E.S.E., making a departure of 160 miles. Find the distance, and position reached.

126. Leaving latitude 49° 30′ N., longitude 25° W., a ship sails between S. and E. 215 miles, making a departure of 167 miles. Find the course, and position reached.

127. Leaving latitude 43° S., longitude 21° W., a ship sails 273 miles, and reaches latitude 40° 17′ S. What are the *two* courses and longitudes which will satisfy the data?

128. Leaving latitude 17° N., longitude 119° E., a ship sails 219 miles, making a departure of 162 miles. What four sets of answers do we get?

129. A ship in latitude 30° sails due east 360 statute miles. What is the shortest distance from the point left to the point reached? Solve the same problem for latitude 45°. 60°.

TRAVERSE SAILING

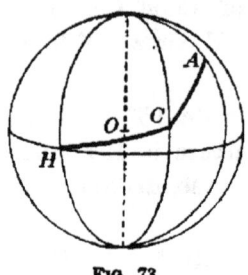

FIG. 73

Traverse Sailing is the application of the principles of Plane and Middle Latitude Sailing to cases when the ship sails from one point to another on two or more different courses. Each course is worked by itself, and these independent results are combined, as may be seen in the solution of the following examples.

130. Leaving latitude 37° 16' S., longitude 18° 42' W., a ship sails N.E. 104 miles, then N.N.W. 60 miles, then W. by S. 216 miles. Find the position reached, and its bearing and distance from the point left.

We have, for the first course, difference of latitude 73.5 N., departure 73.5 E.; for the second course, difference of latitude 55.4 N., departure 23 W.; for the third course, difference of latitude 42.1 S., departure 211.8 W.

On the whole, then, the ship has made 128.9 miles of north latitude, and 42.1 miles of south latitude. The place reached is therefore on a parallel of latitude 86.8 miles to the north of the parallel left, that is, in latitude 35° 49.2' S.

The departure is, in the same way, found to be 161.3 miles W.; and the middle latitude is 36° 32.6'. With these data, and the formula after Example 118, we find the difference of longitude to be 201', or 3° 21' W. Hence, the longitude reached is 22° 3' W.

With the difference of latitude 86.8 miles, and the departure 161.3 miles, we find the course to be N. 61° 43' W., and the distance 183.2 miles. The ship has reached the same point that it would have reached if it had sailed directly on a course N. 61° 43' W. for a distance of 183.2 miles.

131. A ship leaves Cape Cod (Example 120), and sails S.E. by S. 114 miles, N. by E. 94 miles, W.N.W. 42 miles. Solve as in Example 130.

132. A ship leaves Cape of Good Hope (latitude 34° 22′ S., longitude 18° 30′ E.), and sails N.W. 126 miles, N. by E. 84 miles, W.S.W. 217 miles. Solve as in Example 130.

EXERCISE XXIV

PROBLEMS IN GONIOMETRY

Prove that:

1. $\sin x + \cos x = \sqrt{2} \cos (x - \frac{1}{4} \pi)$.

2. $\sin x - \cos x = - \sqrt{2} \cos (x + \frac{1}{4} \pi)$.

3. $\sin x + \sqrt{3} \cos x = 2 \sin (x + \frac{1}{3} \pi)$.

4. $\sin (x + \frac{1}{3} \pi) + \sin (x - \frac{1}{3} \pi) = \sin x$.

5. $\cos (x + \frac{1}{6} \pi) + \cos (x - \frac{1}{6} \pi) = \sqrt{3} \cos x$.

6. $\tan x + \sec x = \tan (\frac{1}{2} x + \frac{1}{4} \pi)$.

7. $\tan x + \sec x = \dfrac{1}{\sec x - \tan x}$.

8. $\dfrac{1 - \tan x}{1 + \tan x} = \dfrac{\cot x - 1}{\cot x + 1}$.

9. $\dfrac{\sin x}{1 + \cos x} + \dfrac{1 + \cos x}{\sin x} = 2 \csc x$.

10. $\tan x + \cot x = 2 \csc 2 x$.

11. $\cot x - \tan x = 2 \cot 2 x$.

12. $1 + \tan x \tan 2 x = \sec 2 x$.

13. $\sec 2 x = \dfrac{\sec^2 x}{2 - \sec^2 x}$.

14. $2 \sec 2 x = \sec (x + 45°) \sec (x - 45°)$.

15. $\tan 2x + \sec 2x = \dfrac{\cos x + \sin x}{\cos x - \sin x}.$

16. $\sin 2x = \dfrac{2 \tan x}{1 + \tan^2 x}.$

17. $2 \sin x + \sin 2x = \dfrac{2 \sin^3 x}{1 - \cos x}.$

18. $\sin 3x = \dfrac{\sin^2 2x - \sin^2 x}{\sin x}.$

19. $\tan 3x = \dfrac{3 \tan x - \tan^3 x}{1 - 3 \tan^2 x}.$

20. $\dfrac{\tan 2x + \tan x}{\tan 2x - \tan x} = \dfrac{\sin 3x}{\sin x}.$

21. $\sin (x + y) + \cos (x - y) = 2 \sin (x + \tfrac{1}{4} \pi) \sin (y + \tfrac{1}{4} \pi).$

22. $\sin (x + y) - \cos (x - y) = -2 \sin (x - \tfrac{1}{4} \pi) \sin (y - \tfrac{1}{4} \pi).$

23. $\tan x + \tan y = \dfrac{\sin (x + y)}{\cos x \cos y}.$

24. $\tan (x + y) = \dfrac{\sin 2x + \sin 2y}{\cos 2x + \cos 2y}.$

25. $\dfrac{\sin x + \cos y}{\sin x - \cos y} = \dfrac{\tan \{\tfrac{1}{2} (x + y) + 45°\}}{\tan \{\tfrac{1}{2} (x - y) - 45°\}}.$

26. $\sin 2x + \sin 4x = 2 \sin 3x \cos x.$

27. $\sin 4x = 4 \sin x \cos x - 8 \sin^3 x \cos x$
$\qquad = 8 \cos^3 x \sin x - 4 \cos x \sin x.$

28. $\cos 4x = 1 - 8 \cos^2 x + 8 \cos^4 x = 1 - 8 \sin^2 x + 8 \sin^4 x.$

29. $\cos 2x + \cos 4x = 2 \cos 3x \cos x.$

30. $\sin 3x - \sin x = 2 \cos 2x \sin x.$

31. $\sin^3 x \sin 3x + \cos^3 x \cos 3x = \cos^3 2x.$

32. $\cos^4 x - \sin^4 x = \cos 2x.$

33. $\cos^4 x + \sin^4 x = 1 - \frac{1}{2} \sin^2 2 x.$

34. $\cos^6 x - \sin^6 x = (1 - \sin^2 x \cos^2 x) \cos 2 x.$

35. $\cos^6 x + \sin^6 x = 1 - 3 \sin^2 x \cos^2 x.$

36. $\dfrac{\sin 3 x + \sin 5 x}{\cos 3 x - \cos 5 x} = \cot x.$

37. $\dfrac{\sin 3 x + \sin 5 x}{\sin x + \sin 3 x} = 2 \cos 2 x.$

38. $\csc x - 2 \cot 2 x \cos x = 2 \sin x.$

39. $(\sin 2 x - \sin 2 y) \tan (x + y) = 2 (\sin^2 x - \sin^2 y).$

40. $(1 + \cot x + \tan x) (\sin x - \cos x) = \dfrac{\sec x}{\csc^2 x} - \dfrac{\csc x}{\sec^2 x}.$

41. $\sin x + \sin 3 x + \sin 5 x = \dfrac{\sin^2 3 x}{\sin x}.$

42. $\dfrac{3 \cos x + \cos 3 x}{3 \sin x - \sin 3 x} = \cot^3 x.$

43. $\sin 3 x = 4 \sin x \sin (60° + x) \sin (60° - x).$

44. $\sin 4 x = 2 \sin x \cos 3 x + \sin 2 x.$

45. $\sin x + \sin (x - \frac{2}{3} \pi) + \sin (\frac{4}{3} \pi - x) = 0.$

46. $\cos x \sin (y - z) + \cos y \sin (z - x) + \cos z \sin (x - y) = 0.$

47. $\cos (x + y) \sin y - \cos (x + z) \sin z$
$\qquad = \sin (x + y) \cos y - \sin (x + z) \cos z.$

48. $\cos (x + y + z) + \cos (x + y - z) + \cos (x - y + z)$
$\qquad + \cos (y + z - x) = 4 \cos x \cos y \cos z.$

49. $\sin (x + y) \cos (x - y) + \sin (y + z) \cos (y - z)$
$\qquad + \sin (z + x) \cos (z - x) = \sin 2 x + \sin 2 y + \sin 2 z.$

50. $\dfrac{\sin 75° + \sin 15°}{\sin 75° - \sin 15°} = \tan 60°.$

51. $\cos 20° + \cos 100° + \cos 140° = 0.$

52. $\cos 36° + \sin 36° = \sqrt{2} \cos 9°.$

53. $\tan 11° 15' + 2 \tan 22° 30' + 4 \tan 45° = \cot 11° 15'.$

If A, B, C are the angles of a plane triangle, prove that:

54. $\sin 2 A + \sin 2 B + \sin 2 C = 4 \sin A \sin B \sin C.$

55. $\cos 2 A + \cos 2 B + \cos 2 C = -1 - 4 \cos A \cos B \cos C.$

56. $\sin 3 A + \sin 3 B + \sin 3 C = -4 \cos \dfrac{3A}{2} \cos \dfrac{3B}{2} \cos \dfrac{3C}{2}.$

57. $\cos^2 A + \cos^2 B + \cos^2 C = 1 - 2 \cos A \cos B \cos C.$

If $A + B + C = 90°$, prove that:

58. $\tan A \tan B + \tan B \tan C + \tan C \tan A = 1.$

59. $\sin^2 A + \sin^2 B + \sin^2 C = 1 - 2 \sin A \sin B \sin C.$

60. $\sin 2 A + \sin 2 B + \sin 2 C = 4 \cos A \cos B \cos C.$

Prove that:

61. $\sin (\sin^{-1} x + \sin^{-1} y) = x \sqrt{1 - y^2} + y \sqrt{1 - x^2}.$

62. $\tan (\tan^{-1} x + \tan^{-1} y) = \dfrac{x + y}{1 - xy}.$

63. $2 \tan^{-1} x = \tan^{-1} \dfrac{2 x}{1 - x^2}.$

64. $2 \sin^{-1} x = \sin^{-1} (2 x \sqrt{1 - x^2}).$

65. $2 \cos^{-1} x = \cos^{-1} (2 x^2 - 1).$

66. $3 \tan^{-1} x = \tan^{-1} \dfrac{3 x - x^3}{1 - 3 x^2}.$

67. $\sin^{-1} \sqrt{\dfrac{x}{y}} = \tan^{-1} \sqrt{\dfrac{x}{y - x}}.$

68. $\sin^{-1} \sqrt{\dfrac{x - y}{x - z}} = \tan^{-1} \sqrt{\dfrac{x - y}{y - z}}.$

69. $\sin^{-1}x = \sec^{-1}\dfrac{1}{\sqrt{1-x^2}}.$

70. $2\sec^{-1}x = \tan^{-1}\dfrac{2\sqrt{x^2-1}}{2-x^2}.$

71. $\tan^{-1}\frac{1}{2} + \tan^{-1}\frac{1}{3} = 45°.$

72. $\tan^{-1}\frac{1}{3} + \tan^{-1}\frac{1}{5} = \tan^{-1}\frac{4}{7}.$

73. $\sin^{-1}\frac{3}{5} + \sin^{-1}\frac{12}{13} = \sin^{-1}\frac{63}{65}.$

74. $\sin^{-1}\dfrac{1}{\sqrt{82}} + \sin^{-1}\dfrac{4}{\sqrt{41}} = 45°.$

75. $\sec^{-1}\frac{5}{3} + \sec^{-1}\frac{13}{12} = 75°\ 45'.$

76. $\tan^{-1}(2+\sqrt{3}) - \tan^{-1}(2-\sqrt{3}) = \sec^{-1}2.$

77. $\tan^{-1}\frac{1}{3} + \tan^{-1}\frac{1}{5} + \tan^{-1}\frac{1}{7} + \tan^{-1}\frac{1}{8} = 45°.$

78. $\tan^{-1}\dfrac{1}{1-2x+4x^2} + \tan^{-1}\dfrac{1}{1+2x+4x^2} = \tan^{-1}\dfrac{1}{2x^2}.$

79. Given $\cos x = \frac{3}{5}$; find $\sin \frac{1}{2}x$ and $\cos \frac{1}{2}x.$

80. Given $\tan x = \frac{1}{2}$; find $\tan \frac{1}{2}x.$

81. Given $\sin x + \cos x = \sqrt{\frac{7}{5}}$; find $\cos 2x.$

82. Given $\tan 2x = \frac{24}{7}$; find $\sin x.$

83. Given $\cos 3x = \frac{23}{27}$; find $\tan x.$

84. Given $2\csc x - \cot x = \sqrt{3}$; find $\sin \frac{1}{2}x.$

85. Find $\sin 18°$ and $\cos 36°.$

Find the value of:

86. $a \sec x + b \csc x,$ when $\tan x = \sqrt[3]{\dfrac{b}{a}}.$

87. $\sin 3x,$ when $\sin 2x = \sqrt{1-m^2}.$

88. $\sin x,$ when $\tan^2 x + 3\cot^2 x = 4.$

89. $\dfrac{\csc^2 x - \sec^2 x}{\csc^2 x + \sec^2 x}$, when $\tan x = \sqrt{\tfrac{1}{7}}$.

90. $\cos x$, when $5 \tan x + \sec x = 5$.

91. $\sec x$, when $\tan x = \dfrac{a}{\sqrt{2\,a + 1}}$.

Simplify the following expressions :

92. $\dfrac{(\cos x + \cos y)^2 + (\sin x + \sin y)^2}{\cos^2 \tfrac{1}{2}(x - y)}$.

93. $\dfrac{\sin (x + 2\,y) - 2 \sin (x + y) + \sin x}{\cos (x + 2\,y) - 2 \cos (x + y) + \cos x}$.

94. $\dfrac{\sin (x - z) + 2 \sin x + \sin (x + z)}{\sin (y - z) + 2 \sin y + \sin (y + z)}$.

95. $\dfrac{\cos 6\,x - \cos 4\,x}{\sin 6\,x + \sin 4\,x}$.

96. $\tan^{-1}(2\,x + 1) + \tan^{-1}(2\,x - 1)$.

97. $\dfrac{1}{1 + \sin^2 x} + \dfrac{1}{1 + \cos^2 x} + \dfrac{1}{1 + \sec^2 x} + \dfrac{1}{1 + \csc^2 x}$.

98. $2 \sec^2 x - \sec^4 x - 2 \csc^2 x + \csc^4 x$.

SOLUTION OF SINGLE EQUATIONS

To **solve** a single equation that involves different functions of the same angle, or the same or different functions of related angles, first transform the equation, if necessary, into an equivalent equation that involves a single function of the same angle.

Employ the method of factoring, if possible, in the algebraic part of the solution.

Completely solve each equation, and check the results by substitution in the given equation.

Solve $\cos x = \sin 2x$.

By [12], p. 58, $\qquad \sin 2x = 2 \sin x \cos x.$

$$\therefore \cos x = 2 \sin x \cos x.$$

$$\therefore (1 - 2 \sin x) \cos x = 0.$$

$$\therefore \cos x = 0, \text{ or } 1 - 2 \sin x = 0.$$

$$\therefore x = 90° \text{ or } 270°, \text{ or } 30° \text{ or } 150°.$$

Each of these values satisfies the given equation.

Solve the following equations:

99. $\sin x = 2 \sin (\tfrac{1}{3} \pi + x)$.

100. $\sin 2x = 2 \cos x$.

101. $\cos 2x = 2 \sin x$.

102. $\sin x + \cos x = 1$.

103. $\sin x + \cos 2x = 4 \sin^2 x$.

104. $4 \cos 2x + 3 \cos x = 1$.

105. $\sin x + \sin 2x = \sin 3x$.

106. $\sin 2x = 3 \sin^2 x - \cos^2 x$.

107. $\cot \theta = \tfrac{1}{3} \tan \theta$.

108. $2 \sin \theta = \cos \theta$.

109. $2 \sin^2 x + 5 \sin x = 3$.

110. $\tan x \sec x = \sqrt{2}$.

111. $\sin x = \cos 2x$.

112. $\tan x \tan 2x = 2$.

113. $\sec x = 4 \csc x$.

114. $\cos \theta + \cos 2\theta = 0$.

115. $\cot \tfrac{1}{2} \theta + \csc \theta = 2$.

116. $\cot x \tan 2x = 3$.

117. $\sin x \sec 2x = 1$.

118. $\sin^2 x + \sin 2x = 1$.

119. $\cos x \sin 2x \csc x = 1$.

120. $\cot x \tan 2x = \sec 2x$.

121. $\sin 2x = \cos 4x$.

122. $\sin 2z \cot z - \sin^2 z = \tfrac{1}{2}$.

123. $\tan x + \tan 2x = \tan 3x$.

124. $\cot x - \tan x = \sin x + \cos x$.

125. $\tan^2 x = \sin 2x$.

126. $\tan x + \cot x = \tan 2x$.

127. $\dfrac{1 - \tan x}{1 + \tan x} = \cos 2x$.

128. $\sin x + \sin 2x = 1 - \cos 2x$.

129. $\sec 2x + 1 = 2 \cos x$.

130. $\tan 2x + \tan 3x = 0$.

131. $\tan(\frac{1}{4}\pi + x) + \tan(\frac{1}{4}\pi - x) = 4$.

132. $\sqrt{1 + \sin x} - \sqrt{1 - \sin x} = 2 \cos x$.

133. $\tan x \tan 3x = -\frac{2}{3}$.

134. $\sin(45° + x) + \cos(45° - x) = 1$.

135. $\tan x + \sec x = a$.

136. $\cos 2x = a(1 - \cos x)$.

137. $(1 - \tan x)\cos 2x = a(1 + \tan x)$.

138. $\sin^6 x + \cos^6 x = \frac{7}{12}\sin^2 2x$.

139. $\cos 3x + 8 \cos^3 x = 0$.

140. $\sec(x + 120°) + \sec(x - 120°) = 2 \cos x$.

141. $\csc x = \cot x + \sqrt{3}$.

142. $4 \cos 2x + 6 \sin x = 5$

143. $\cos x - \cos 2x = 1$.

144. $\sin 4x - \sin 2x = \sin x$.

145. $2 \sin^2 x + \sin^2 2x = 2$.

146. $\cos 5x + \cos 3x + \cos x = 0$.

147. $\sec x - \cot x = \csc x - \tan x$.

148. $\tan^2 x + \cot^2 x = \frac{10}{3}$.

149. $\sin 4x - \cos 3x = \sin 2x$.

150. $\sin x + \cos x = \sec x$.

151. $2 \cos x \cos 3x + 1 = 0$.

152. $\cos 3x - 2 \cos 2x + \cos x = 0$.

153. $\tan 2x \tan x = 1$.

154. $\sin (x + 12°) + \sin (x - 8°) = \sin 20°$.

155. $\tan (60° + x) \tan (60° - x) = -2$.

156. $\sin (x + 120°) + \sin (x + 60°) = \frac{3}{2}$.

157. $\sin (x + 30°) \sin (x - 30°) = \frac{1}{2}$.

158. $\sin^4 x + \cos^4 x = \frac{5}{8}$.

159. $\sin^4 x - \cos^4 x = \frac{7}{25}$.

160. $\tan (x + 30°) = 2 \cos x$.

161. $\sec x = 2 \tan x + \frac{1}{4}$.

162. $\sin 11x \sin 4x + \sin 5x \sin 2x = 0$.

163. $\cos x + \cos 3x + \cos 5x + \cos 7x = 0$.

164. $\sin (x + 12°) \cos (x - 12°) = \cos 33° \sin 57°$.

165. $\sin^{-1} x + \sin^{-1} \frac{1}{2} x = 120°$.

166. $\tan^{-1} x + \tan^{-1} 2x = \tan^{-1} 3 \sqrt{3}$.

167. $\sin^{-1} x + 2 \cos^{-1} x = \frac{2}{3} \pi$.

168. $\sin^{-1} x + 3 \cos^{-1} x = 210°$.

169. $\tan^{-1} x + 2 \cot^{-1} x = 135°$.

170. $\tan^{-1} (x + 1) + \tan^{-1} (x - 1) = \tan^{-1} 2x$.

171. $\tan^{-1} \dfrac{x + 2}{x + 1} + \tan^{-1} \dfrac{x - 2}{x - 1} = \frac{3}{4} \pi$.

172. $\tan^{-1} \dfrac{2x}{1 - x^2} = 60°$.

173. $\cos 2\theta \sec \theta + \sec \theta + 1 = 0$.

174. $\sin x \cos 2x \tan x \cot 2x \sec x \csc 2x = 1$.

175. $\sin \frac{1}{2} x (\cos 2x - 2)(1 - \tan^2 x) = 0$.

Hint. Equate to 0 each factor except the second. The second factor cannot equal 0.

176. $\sin 3x = \cos 2x - 1.$ **178.** $\sin 2\theta = \cos 3\theta.$

177. $\tan x + \tan 2x = 0.$ **179.** $(3 - 4\cos^2 x)\sin 2x = 0.$

180. $\sin x + \sin 2x + \sin 3x = 0.$

181. $\sin \theta + 2\sin 2\theta + 3\sin 3\theta = 0.$

182. $\sin^2 x \cos^2 x - \cos^2 x - \sin^2 x + 1 = 0.$

183. $\sin x + \sin 3x = \cos x - \cos 3x.$

184. $(1 - \sqrt{1 - \tan^2 x})\cos 2x \operatorname{vers} 3x = 0.$

185. $\tan(\theta + 45°) = 8\tan \theta.$

186. $\sin(x - 30°) = \tfrac{1}{2}\sqrt{3}\sin x.$

187. $\tan(\theta + 45°)\tan \theta = 2.$

188. $\sin^{-1}\tfrac{1}{2}x = 30°.$

SYSTEMS OF EQUATIONS

189. Solve for x and y the system

$$x \sin \alpha + y \sin \beta = a, \qquad (1)$$
$$x \cos \alpha + y \cos \beta = b. \qquad (2)$$

$(1) \times \cos \alpha,$ $\quad x \sin \alpha \cos \alpha + y \sin \beta \cos \alpha = a \cos \alpha. \qquad (3)$

$(2) \times \sin \alpha,$ $\quad x \sin \alpha \cos \alpha + y \cos \beta \sin \alpha = b \sin \alpha. \qquad (4)$

$(3) - (4),$ $\quad y(\sin \beta \cos \alpha - \cos \beta \sin \alpha) = a \cos \alpha - b \sin \alpha. \qquad (5)$

$$\therefore y = \frac{a \cos \alpha - b \sin \alpha}{\sin(\beta - \alpha)}.$$

Similarly,
$$x = \frac{b \sin \beta - a \cos \beta}{\sin(\beta - \alpha)}.$$

190. Solve for x and y the system

$$\sin x + \sin y = a, \qquad (1)$$
$$\cos x + \cos y = b. \qquad (2)$$

Transform (1) and (2), by Sect. XXXII,

by [20], p. 59, $\qquad 2 \sin \frac{1}{2}(x + y) \cos \frac{1}{2}(x - y) = a,$ $\qquad$ (3)

by [22], p. 59, $\qquad 2 \cos \frac{1}{2}(x + y) \cos \frac{1}{2}(x - y) = b.$ $\qquad$ (4)

(3) ÷ (4), $\qquad \tan \frac{1}{2}(x + y) = \frac{a}{b}.$ $\qquad$ (5)

$\qquad \therefore \sin \frac{1}{2}(x + y) = \dfrac{a}{\sqrt{a^2 + b^2}}.$ $\qquad$ (6)

Substitute value of $\sin \frac{1}{2}(x + y)$ in (3),

$\qquad \cos \frac{1}{2}(x - y) = \frac{1}{2} \sqrt{a^2 + b^2}.$ $\qquad$ (7)

From (5), $\qquad x + y = 2 \tan^{-1} \dfrac{a}{b}.$ $\qquad$ (8)

From (7), $\qquad x - y = 2 \cos^{-1} \frac{1}{2} \sqrt{a^2 + b^2}.$ $\qquad$ (9)

Whence $\qquad x = \tan^{-1} \dfrac{a}{b} + \cos^{-1} \frac{1}{2} \sqrt{a^2 + b^2},$

and $\qquad y = \tan^{-1} \dfrac{a}{b} - \cos^{-1} \frac{1}{2} \sqrt{a^2 + b^2}.$

191. Solve for r and θ the system

$$r \sin \theta = a, \qquad (1)$$
$$r \cos \theta = b. \qquad (2)$$

(1) ÷ (2), $\qquad \tan \theta = \dfrac{a}{b}.$ $\qquad$ (3)

From (3), $\qquad \theta = \tan^{-1} \dfrac{a}{b}.$ $\qquad$ (4)

Square (1) and (2) and add,

$$r^2 (\sin^2 \theta + \cos^2 \theta) = a^2 + b^2.$$
$$\therefore r = \sqrt{a^2 + b^2}.$$

192. Solve for r and θ the system

$$r \sin (\theta + \alpha) = a, \qquad (1)$$
$$r \cos (\theta + \beta) = b. \qquad (2)$$

Expand (1) and (2),

$$r \sin \theta \cos \alpha + r \cos \theta \sin \alpha = a. \qquad (3)$$
$$r \cos \theta \cos \beta - r \sin \theta \sin \beta = b. \qquad (4)$$

Now solve (3) and (4) for $r \sin \theta$ and $r \cos \theta$, as in Example 189. Then solve for r and θ, as in Example 191.

193. Solve for r, θ, and ϕ the system

$$r \cos \phi \sin \theta = a, \tag{1}$$
$$r \cos \phi \cos \theta = b, \tag{2}$$
$$r \sin \phi = c. \tag{3}$$

(1) ÷ (2), $\quad\quad \tan \theta = \dfrac{a}{b}. \quad \therefore \theta = \tan^{-1}\dfrac{a}{b}. \tag{4}$

Square (1) and (2) and add,

$$r^2 \cos^2 \phi = a^2 + b^2. \tag{5}$$

(3) ÷ (5), $\quad\quad \tan \phi = \dfrac{c}{\sqrt{a^2 + b^2}}. \quad \therefore \phi = \tan^{-1}\dfrac{c}{\sqrt{a^2 + b^2}}. \tag{6}$

Square (3) and add to (5),

$$r^2 = a^2 + b^2 + c^2.$$
$$\therefore r = \sqrt{a^2 + b^2 + c^2}. \tag{7}$$

Solve the following systems for r, θ, ϕ, x, and y:

194. $x \sin 21° + y \cos 44° = 179.70,$
$x \cos 21° + y \sin 44° = 232.30.$

195. $\sin x - \sin y = 0.7038,$
$\cos x - \cos y = -0.7245.$

196. $r \sin \theta = 92.344,$
$r \cos \theta = 205.309.$

197. $r \sin (\theta - 19° 18') = 59.4034,$
$r \cos (\theta - 30° 54') = 147.9347.$

198. $r \cos \phi \cos \theta = -46.7654,$
$r \sin \phi \cos \theta = 81,$
$r \sin \theta = -54.$

199. Eliminate θ from the system

$$x = r(\theta - \sin \theta),$$
$$y = r(1 - \cos \theta).$$

HINT. $1 - \cos \theta = \text{vers } \theta. \quad \therefore \theta = \text{vers}^{-1}\dfrac{y}{r}.$

CHAPTER VI

CONSTRUCTION OF TABLES

SECTION XLIII

LOGARITHMS

Properties of Logarithms. Any positive number except unity being selected as a *base*, the index or exponent which the base must have to produce a given number is the logarithm of that number to the given base.

Thus, if $a^n = N$, then $n = \log_a N$.

$n = \log_a N$ is read, n is equal to log N to the base a.

Let a be the base, M and N any positive numbers, m and n their logarithms to the base a; so that

$$a^m = M, \qquad a^n = N,$$
$$m = \log_a M, \qquad n = \log_a N.$$

Then, in any system of logarithms:

1. *The logarithm of* 1 *is* 0.

For, $a^0 = 1.$ $\therefore 0 = \log_a 1.$

2. *The logarithm of the base itself is* 1.

For, $a^1 = a.$ $\therefore 1 = \log_a a.$

3. *The logarithm of the reciprocal of a positive number is the negative of the logarithm of the number.*

For, if $a^n = N$, then $\dfrac{1}{N} = \dfrac{1}{a^n} = a^{-n}.$

$$\therefore \log_a \left(\frac{1}{N} \right) = -n = -\log_a N.$$

119

4. *The logarithm of the product of two or more positive numbers is found by adding together the logarithms of the several factors.*

For, $$M \times N = a^m \times a^n = a^{m+n}.$$

$$\therefore \log_a (M \times N) = m + n = \log_a M + \log_a N.$$

Similarly for the product of three or more factors.

5. *The logarithm of the quotient of two positive numbers is found by subtracting the logarithm of the divisor from the logarithm of the dividend.*

For, $$\frac{M}{N} = \frac{a^m}{a^n} = a^{m-n}.$$

$$\therefore \log_a \left(\frac{M}{N} \right) = m - n = \log_a M - \log_a N.$$

6. *The logarithm of a power of a positive number is found by multiplying the logarithm of the number by the exponent of the power.*

For, $$N^p = (a^n)^p = a^{np}.$$

$$\therefore \log_a (N^p) = np = p \log_a N.$$

7. *The logarithm of the real positive value of a root of a positive number is found by dividing the logarithm of the number by the index of the root.*

For, $$\sqrt[r]{N} = \sqrt[r]{a^n} = a^{\frac{n}{r}}.$$

$$\therefore \log_a \sqrt[r]{N} = \frac{n}{r} = \frac{\log_a N}{r}.$$

Change of System. Logarithms to any base a may be converted into logarithms to any other base b as follows:

Let N be any number, and let

$$n = \log_a N \text{ and } m = \log_b N.$$

Then, $$N = a^n \text{ and } N = b^m.$$

$$\therefore a^n = b^m.$$

Taking logarithms to any base whatever,

$$n \log a = m \log b,$$

or, $$\log a \times \log_a N = \log b \times \log_b N,$$

from which $\log_b N$ may be found when $\log a$, $\log b$, and $\log_a N$ are given; and conversely, $\log_a N$ may be found when $\log a$, $\log b$, and $\log_b N$ are given.

Two Important Systems. Although the number of different systems of logarithms is unlimited, there are but two systems which are in common use. These are:

1. The common system, also called the Briggs, denary, or decimal system, of which the base is 10.

2. The natural system of which the base is the fixed value which the sum of the series

$$1 + \frac{1}{1} + \frac{1}{1.2} + \frac{1}{1.2.3} + \frac{1}{1.2.3.4} + \cdots$$

approaches as the number of terms is indefinitely increased. This fixed value, correct to seven places of decimals, is 2.7182818, and is denoted by the letter e.

The common system is used in actual calculation; the natural system is used in the higher mathematics.

EXERCISE XXV

1. Given $\log_{10} 2 = 0.30103$, $\log_{10} 3 = 0.47712$, $\log_{10} 7 = 0.84510$; find $\log_{10} 6$, $\log_{10} 14$, $\log_{10} 21$, $\log_{10} 4$, $\log_{10} 12$, $\log_{10} 5$, $\log_{10} \frac{1}{2}$, $\log_{10} \frac{1}{4}$, $\log_{10} \frac{7}{8}$, $\log_{10} \frac{21}{16}$.

2. With the data of Example 1, find $\log_2 10$, $\log_2 5$, $\log_8 5$, $\log_7 \frac{1}{2}$, $\log_5 \frac{2}{3\cdot3}$.

3. Given $\log_{10} e = 0.43429$; find $\log_e 2$, $\log_e 3$, $\log_e 5$, $\log_e 7$, $\log_e 8$, $\log_e 9$, $\log_e \frac{2}{3}$, $\log_e \frac{4}{5}$, $\log_e \frac{14}{27}$, $\log_e \frac{7}{10}$.

4. Find x from the equations $5^x = 12$, $16^x = 10$, $27^x = 4$.

SECTION XLIV

EXPONENTIAL AND LOGARITHMIC SERIES

Exponential Series. By the binomial theorem,

$$\left(1+\frac{1}{n}\right)^{nx} = 1 + nx \times \frac{1}{n} + \frac{nx\,(nx-1)}{1\cdot 2} \times \frac{1}{n^2}$$

$$+ \frac{nx\,(nx-1)\,(nx-2)}{1\cdot 2\cdot 3} \times \frac{1}{n^3} + \cdots$$

$$= 1 + x + \frac{x\left(x-\dfrac{1}{n}\right)}{\underline{|2}} + \frac{x\left(x-\dfrac{1}{n}\right)\left(x-\dfrac{2}{n}\right)}{\underline{|3}} + \cdots \quad (1)$$

This equation is true for all real values of x, since the binomial theorem may be extended to the case of incommensurable exponents (Wentworth's *College Algebra*, § 299); it is, however, true only for values of n numerically greater than 1, since $\frac{1}{n}$ must be numerically less than 1 (*College Algebra*, § 418).

As (1) is true for all values of x, it is true when $x = 1$.

$$\therefore \left(1+\frac{1}{n}\right)^{n} = 1 + 1 + \frac{1-\dfrac{1}{n}}{\underline{|2}} + \frac{\left(1-\dfrac{1}{n}\right)\left(1-\dfrac{2}{n}\right)}{\underline{|3}} + \cdots \quad (2)$$

But $$\left[\left(1+\frac{1^n}{n}\right)\right]^{x} = \left(1+\frac{1}{n}\right)^{nx}.$$

Hence, from (1) and (2),

$$\left[1 + 1 + \frac{1-\dfrac{1}{n}}{\underline{|2}} + \frac{\left(1-\dfrac{1}{n}\right)\left(1-\dfrac{2}{n}\right)}{\underline{|3}} + \cdots\right]^{x}$$

$$= 1 + x + \frac{x\left(x-\dfrac{1}{n}\right)}{\underline{|2}} + \frac{x\left(x-\dfrac{1}{n}\right)\left(x-\dfrac{2}{n}\right)}{\underline{|3}} + \cdots$$

This last equation is true for all values of n numerically greater than 1. Taking the limits of the two members as n increases without limit, we obtain

$$\left(1 + 1 + \frac{1}{\lfloor 2} + \frac{1}{\lfloor 3} + \cdots\right)^x = 1 + x + \frac{x^2}{\lfloor 2} + \frac{x^3}{\lfloor 3} + \cdots, \quad (3)$$

and this is true for all values of x. It is easily seen that both series are convergent for all values of x.

The sum of the infinite series in parenthesis is the natural base e.

Hence, by (3), $e^x = 1 + x + \dfrac{x^2}{\lfloor 2} + \dfrac{x^3}{\lfloor 3} + \cdots$ $\qquad\qquad$ (4)

To calculate the value of e, we proceed as follows:

	1.000000
2	1.000000
3	0.500000
4	0.166667
5	0.041667
6	0.008333
7	0.001388
8	0.000198
9	0.000025
	0.000003

Adding, $\qquad\qquad\qquad e = 2.71828.$

To ten places, $\qquad\qquad e = 2.7182818284.$

Limit of $\left(1 + \dfrac{x}{n}\right)^n$. By the binomial theorem,

$$\left(1 + \frac{x}{n}\right)^n = 1 + n \times \frac{x}{n} + \frac{n(n-1)}{1 \cdot 2} \times \frac{x^2}{n^2}$$

$$+ \frac{n(n-1)(n-2)}{1 \cdot 2 \cdot 3} \times \frac{x^3}{n^3} + \cdots$$

$$= 1 + x + \frac{1 - \dfrac{1}{n}}{\lfloor 2} x^2 + \frac{\left(1 - \dfrac{1}{n}\right)\left(1 - \dfrac{2}{n}\right)}{\lfloor 3} x^3 + \cdots$$

This equation is true for all values of n greater than x (*College Algebra*, § 418). Take the limit as n increases without limit, x remaining finite; then

$$\underset{n \doteq \infty}{\text{limit}} \left(1 + \frac{x}{n}\right)^n = 1 + x + \frac{x^2}{\lfloor 2} + \frac{x^3}{\lfloor 3} + \cdots$$

$$= e^x = \underset{n \doteq \infty}{\text{limit}} \left(1 + \frac{1}{n}\right)^{nx}. \qquad (5)$$

Logarithmic Series.

Let
$$y = \log_e(1 + x);$$

then
$$1 + x = e^y = \underset{n \doteq \infty}{\text{limit}} \left(1 + \frac{y}{n}\right)^n.$$

If n is merely a large number, but not infinite,

$$\left(1 + \frac{y}{n}\right)^n = 1 + x + \epsilon,$$

where ϵ is a variable number which approaches the limit 0, when n increases without limit. Hence,

$$1 + \frac{y}{n} = \sqrt[n]{1 + x + \epsilon},$$

$$y = n \sqrt[n]{1 + x + \epsilon} - n.$$

If n increases without limit, and consequently ϵ approaches 0 as a limit, we have

$$y = \underset{n \doteq \infty}{\text{limit}} [n \sqrt[n]{1 + x} - n].$$

If x is less than 1, we may expand the right-hand member of this equation by the binomial theorem. The result is

$$y = \underset{n \doteq \infty}{\text{limit}} \left[n\left\{ 1 + \frac{1}{n}x + \frac{1}{n}\left(\frac{1}{n} - 1\right)\frac{x^2}{\lfloor 2} + \cdots \right\} - n \right]$$

$$= \underset{n \doteq \infty}{\text{limit}} \left[x + \left(\frac{1}{n} - 1\right)\frac{x^2}{\lfloor 2} + \left(\frac{1}{n} - 1\right)\left(\frac{1}{n} - 2\right)\frac{x^3}{\lfloor 3} + \cdots \right]$$

$$= x - \frac{x^2}{\lfloor 2} + \frac{\lfloor 2\, x^3}{\lfloor 3} - \frac{\lfloor 3\, x^4}{\lfloor 4} + \cdots$$

$$\therefore \log_e(1 + x) = x - \frac{x^2}{2} + \frac{x^3}{3} - \frac{x^4}{4} + \cdots$$

This series is known as the *logarithmic series.* It is convergent only if x lies between -1 and $+1$, or is equal to $+1$. Even within these limits it converges rather slowly, and for these reasons it is not well adapted to the computation of logarithms. A more convenient series is obtained as follows.

Calculation of Logarithms.   The equation

$$\log_e(1 + y) = y - \frac{y^2}{2} + \frac{y^3}{3} - \frac{y^4}{4} + \cdots \tag{1}$$

holds true for all values of y numerically less than 1; therefore, if it holds true for any particular value of y less than 1, it will hold true when we put $-y$ for y; this gives

$$\log_e(1 - y) = -y - \frac{y^2}{2} - \frac{y^3}{3} - \frac{y^4}{4} - \cdots \tag{2}$$

Subtracting (2) from (1), since

$$\log_e(1 + y) - \log_e(1 - y) = \log_e\left(\frac{1 + y}{1 - y}\right),$$

we find $\log_e\left(\dfrac{1 + y}{1 - y}\right) = 2\left(y + \dfrac{y^3}{3} + \dfrac{y^5}{5} + \cdots\right).$

Put $y = \dfrac{1}{2z + 1};$

then $\dfrac{1 + y}{1 - y} = \dfrac{z + 1}{z},$

and $\log_e\left(\dfrac{z + 1}{z}\right) = \log_e(z + 1) - \log_e z$

$$= 2\left(\frac{1}{2z + 1} + \frac{1}{3(2z + 1)^3} + \frac{1}{5(2z + 1)^5} + \cdots\right).$$

This series is convergent for all positive values of z.

Logarithms to any base a can be calculated by the series:

$$\log_a(z+1) - \log_a z$$
$$= \frac{2}{\log_e a}\left(\frac{1}{2z+1} + \frac{1}{3(2z+1)^3} + \frac{1}{5(2z+1)^5} + \cdots\right).$$

EXAMPLE. Calculate $\log_e 2$ to five places of decimals.

Let $z = 1$; then $z + 1 = 2$, $2z + 1 = 3$,

and $\log_e 2 = \dfrac{2}{3} + \dfrac{2}{3 \times 3^3} + \dfrac{2}{5 \times 3^5} + \dfrac{2}{7 \times 3^7} + \cdots$

The work may be arranged as follows:

```
3 | 2.000000
9 | 0.666667 ÷  1 = 0.666667
9 | 0.074074 ÷  3 = 0.024691
9 | 0.008230 ÷  5 = 0.001646
9 | 0.000914 ÷  7 = 0.000131
9 | 0.000102 ÷  9 = 0.000011
    0.000011 ÷ 11 = 0.000001
          log_e 2 = 0.693147
```

NOTE. In calculating logarithms the accuracy of the work may be tested every time we come to a composite number by adding the logarithms of the several factors. In fact, the logarithms of composite numbers are best found by addition, and then only the logarithms of prime numbers need be computed by the series.

EXERCISE XXVI

1. Calculate to five places of decimals $\log_e 3$.

2. Calculate to five places of decimals $\log_e 5$.

3. Calculate to five places of decimals $\log_e 7$.

4. Calculate to ten places of decimals $\log_e 10$.

5. Calculate to five places of decimals $\log_{10} 2$, $\log_{10} e$, $\log_{10} 11$.

SECTION XLV

TRIGONOMETRIC FUNCTIONS OF SMALL ANGLES

Let AOP (Fig. 74) be any angle less than 90° and x its circular measure. Describe a
circle of unit radius about
O as a centre and take
$\angle AOP' = -\angle AOP$. Draw
the tangents to the circle at
P and P', meeting OA in T.
Then, from Geometry,

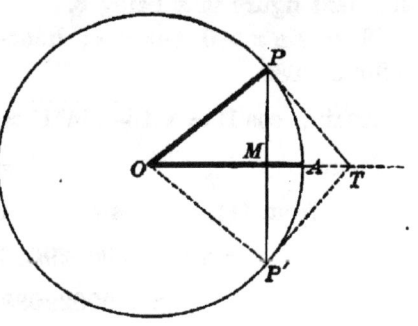

FIG. 74

chord $PP' <$ arc PP'

$$< PT + P'T,$$

or, by dividing by 2,

$$MP < \text{arc } AP < PT,$$

or $\qquad\qquad \sin x < x < \tan x.$

Hence, dividing by $\sin x$,

$$1 < \frac{x}{\sin x} < \sec x,$$

$$1 > \frac{\sin x}{x} > \cos x. \tag{1}$$

Then $\dfrac{\sin x}{x}$ lies between $\cos x$ and 1.

If now the angle x is constantly diminished, $\cos x$ approaches
the value 1.

Accordingly, the limit of $\dfrac{\sin x}{x}$, as x approaches 0, is 1.
In other words, if x is a very small angle, then $\dfrac{\sin x}{x}$ differs
from 1 by a small value ϵ; and this small value ϵ approaches
0 as x approaches 0.

EXAMPLE. To find the sine and cosine of 1'.

If x is the circular measure of 1',

$$x = \frac{2\pi}{360 \times 60} = \frac{3.14159+}{10800} = 0.00029088+,$$

the next figure in x being 8.

Now $\sin x > 0$ but $< x$; hence, sin 1' lies between 0 and 0.000290889.

Again, $\cos 1' = \sqrt{1 - \sin^2 1'} > \sqrt{1 - (0.0003)^2} > 0.9999999$.

Hence, $\cos 1' = 0.9999999+$.

But, from (1), $\sin x > x \cos x$.

$$\therefore \sin 1' > 0.000290888 \times 0.9999999$$

$$> 0.000290888\,(1 - 0.0000001)$$

$$> 0.000290888 - 0.0000000000290888$$

$$> 0.000290887.$$

Hence, sin 1' lies between 0.000290887 and 0.000290889; that is, to eight places of decimals

$$\sin 1' = 0.00029088+,$$

the next figure being 7 or 8.

EXERCISE XXVII

Given $\pi = 3.141592653589$:

1. Compute sin 1', cos 1', and tan 1' to eleven places of decimals.

2. Compute sin 2' by the same method, and also by the formula $\sin 2x = 2 \sin x \cos x$. Carry the operations to nine places of decimals. Do the two results agree?

3. Compute sin 1° to four places of decimals.

4. From the formula $\cos x = 1 - 2 \sin^2 \frac{x}{2}$, show that

$$\cos x > 1 - \frac{x^2}{2}.$$

5. Show by aid of a table of natural sines that $\sin x$ and x agree to four places of decimals for all angles less than $4° \, 40'$.

6. If the values of $\log x$ and $\log \sin x$ agree to five decimal places, find from a table the greatest value x can have.

SECTION XLVI

SIMPSON'S METHOD OF CONSTRUCTING A TRIGONOMETRIC TABLE

By Sect. XXXII, p. 59,

$$\sin (A + B) + \sin (A - B) = 2 \sin A \cos B.$$

If we put $\qquad A = x + 2y, \;\; B = y,$

we have $\quad \sin (x + 3y) + \sin (x + y) = 2 \sin (x + 2y) \cos y,$

or $\qquad \sin (x + 3y) = 2 \sin (x + 2y) \cos y - \sin (x + y).$

Similarly, $\cos (x + 3y) = 2 \cos (x + 2y) \cos y - \cos (x + y). \;\; (1)$

If $y = 1'$, the last two equations become

$$\sin (x + 3') = 2 \sin (x + 2') \cos 1' - \sin (x + 1'),$$
$$\cos (x + 3') = 2 \cos (x + 2') \cos 1' - \cos (x + 1').$$

Hence, taking x successively equal to $-1', 0', 1', 2', \cdots$, we obtain

$$\sin 2' = 2 \sin 1' \cos 1',$$
$$\sin 3' = 2 \sin 2' \cos 1' - \sin 1',$$
$$\sin 4' = 2 \sin 3' \cos 1' - \sin 2',$$
$$\cdot \quad \cdot \quad \cdot \quad \cdot \quad \cdot \quad \cdot$$
$$\cos 2' = 2 \cos^2 1' - 1,$$
$$\cos 3' = 2 \cos 2' \cos 1' - \cos 1',$$
$$\cos 4' = 2 \cos 3' \cos 1' - \cos 2',$$

Since sin 1' and cos 1' are known, these equations enable us to compute step by step the sine and cosine of any angle. The tangent may then be found in each case as the quotient of the sine divided by the cosine.

This process need be carried only as far as 30°. For

$$\sin(30° + x) + \sin(30° - x) = \quad 2 \sin 30° \cos x = \quad \cos x,$$
$$\cos(30° + x) - \cos(30° - x) = - 2 \sin 30° \sin x = - \sin x.$$
$$\therefore \sin(30° + x) = \quad \cos x - \sin(30° - x),$$
$$\cos(30° + x) = - \sin x + \cos(30° - x).$$

Moreover, the sines and cosines need be calculated only to 45°, since

$$\sin(45° + x) = \cos(45° - x),$$
$$\cos(45° + x) = \sin(45° - x).$$

In using this method, the multiplication by cos 1', which occurs at each step, can be simplified by noting that

$$\cos 1' = 0.9999999 = 1 - 0.0000001.$$

Note. Simpson's method is superseded in actual practice by much more rapid and convenient processes in which we employ the expansions of the trigonometric functions in infinite series.

EXERCISE XXVIII

1. Compute the sine and cosine of 6' to seven decimal places.

In Formula (1) let $y = 1°$. Assuming
$$\sin 1° = 0.017454+, \quad \cos 1° = 0.999848+ :$$

2. Compute the sine and cosine of two degrees.

3. Compute the sine and cosine of three degrees.

4. Compute the sine and cosine of four degrees.

5. Compute the sine and cosine of five degrees.

SECTION XLVII

DE MOIVRE'S THEOREM

Expressions of the form

$$\cos x + i \sin x,$$

when $i = \sqrt{-1}$, play an important part in modern analysis.
Given two such expressions,

$$\cos x + i \sin x, \quad \cos y + i \sin y$$

their product is

$$(\cos x + i \sin x)(\cos y + i \sin y)$$
$$= \cos x \cos y - \sin x \sin y + i(\cos x \sin y + \sin x \cos y)$$
$$= \cos(x + y) + i \sin(x + y).$$

Hence, the product of two expressions of the form

$$\cos x + i \sin x, \quad \cos y + i \sin y$$

is an expression of the same form in which x or y is replaced
by $x + y$. In other words, the angle which enters into such a
product is the sum of the angles of the factors.

If x and y are equal, we have at once, from the preceding,

$$(\cos x + i \sin x)^2 = \cos 2x + i \sin 2x;$$

and again,

$$(\cos x + i \sin x)^3 = (\cos x + i \sin x)^2(\cos x + i \sin x)$$
$$= (\cos 2x + i \sin 2x)(\cos x + i \sin x)$$
$$= \cos 3x + i \sin 3x.$$

Similarly,

$$(\cos x + i \sin x)^4 = \cos 4x + i \sin 4x,$$

and in general, if n is a positive integer,

$$(\cos x + i \sin x)^n = \cos nx + i \sin nx. \tag{1}$$

Hence,

To raise the expression cos x *+ i sin* x *to the nth power when* n *is a positive integer, we have only to multiply the angle* x *by* n.

Again, if n is a positive integer as before,

$$\left(\cos \frac{x}{n} + i \sin \frac{x}{n} \right)^n = \cos x + i \sin x.$$

$$\therefore (\cos x + i \sin x)^{\frac{1}{n}} = \cos \frac{x}{n} + i \sin \frac{x}{n}.$$

Since, however, x may be increased by any integral multiple of 2π without changing $\cos x + i \sin x$, it follows that all the n expressions,

$$\cos \frac{x}{n} + i \sin \frac{x}{n}, \quad \cos \frac{x + 2\pi}{n} + i \sin \frac{x + 2\pi}{n},$$

$$\cos \frac{x + 4\pi}{n} + i \sin \frac{x + 4\pi}{n}, \cdots,$$

$$\cos \frac{x + (n-1)2\pi}{n} + i \sin \frac{x + (n-1)2\pi}{n},$$

are nth roots of $\cos x + i \sin x$. There are no other roots, since

$$\cos \frac{x + n 2\pi}{n} + i \sin \frac{x + n 2\pi}{n}$$

$$= \cos \left(\frac{x}{n} + 2\pi \right) + i \sin \left(\frac{x}{n} + 2\pi \right) = \cos \frac{x}{n} + i \sin \frac{x}{n},$$

and $\cos \dfrac{x + (n+1)2\pi}{n} + i \sin \dfrac{x + (n+1)2\pi}{n}$

$$= \cos \left(\frac{x + 2\pi}{n} + 2\pi \right) + i \sin \left(\frac{x + 2\pi}{n} + 2\pi \right)$$

$$= \cos \frac{x + 2\pi}{n} + i \sin \frac{x + 2\pi}{n},$$

and so on.

Hence, if n is a positive integer,

$$(\cos x + i \sin x)^{\frac{1}{n}}$$

$$= \cos \frac{x + 2\,k\pi}{n} + i \sin \frac{x + 2\,k\pi}{n} \quad (k = 0, 1, 2, \cdots n-1). \quad (2)$$

From (1) and (2) it follows at once that if m and n are positive integers,

$$(\cos x + i \sin x)^{\frac{m}{n}} = \{(\cos x + i \sin x)^{\frac{1}{n}}\}^{m}$$

$$= \cos \frac{m}{n}(x + 2\,k\pi) + i \sin \frac{m}{n}(x + 2\,k\pi)$$

$$(k = 0, 1, 2, \cdots n-1). \quad (3)$$

Finally, if $-\dfrac{m}{n}$ is a negative fraction,

$$(\cos x + i \sin x)^{-\frac{m}{n}} = \frac{1}{(\cos x + i \sin x)^{\frac{m}{n}}}.$$

But
$$\frac{1}{\cos x + i \sin x} = \frac{\cos x - i \sin x}{(\cos x + i \sin x)(\cos x - i \sin x)}$$

$$= \frac{\cos x - i \sin x}{\cos^2 x + \sin^2 x}$$

$$= \cos x - i \sin x$$

$$= \cos(-x) + i \sin(-x).$$

Hence, $(\cos x + i \sin x)^{-\frac{m}{n}} = \{\cos(-x) + i \sin(-x)\}^{\frac{m}{n}}$

$$= \cos \frac{m}{n}(-x + 2\,k\pi) + i \sin \frac{m}{n}(-x + 2\,k\pi),$$

$$(k = 0, 1, 2, \cdots n-1)$$

$$= \cos\left\{-\frac{m}{n}(x + 2\,k\pi)\right\} + i \sin\left\{-\frac{m}{n}(x + 2\,k\pi)\right\},$$

$$(k = 0, 1, 2, \cdots n-1). \quad (4)$$

Consequently, if n is a positive or negative integer or fraction,

$$(\cos x + i \sin x)^n = \cos[n(x + 2k\pi)] + i \sin[n(x + 2k\pi)],$$
$$(k = 0, 1, 2, \cdots n - 1). \quad (5)$$

EXAMPLE. Find the three cube roots of -1.

We have $-1 = \cos 180° + i \sin 180°$.

$$\therefore (-1)^{\frac{1}{3}} = \cos \frac{180° + 2k\pi}{3} + i \sin \frac{180° + 2k\pi}{3} (k = 0, 1, 2).$$

For the three cube roots of -1 we find, therefore,

$$\cos 60° + i \sin 60°, \quad \cos 180° + i \sin 180°, \quad \cos 300° + i \sin 300°,$$

or $$\frac{1 + i\sqrt{3}}{2}, \quad -1, \quad \frac{1 - i\sqrt{3}}{2}.$$

By aid of De Moivre's Theorem, we may express $\sin n\theta$ and $\cos n\theta$, when n is an integer, in terms of $\sin \theta$ and $\cos \theta$.

Thus, $$\cos n\theta + i \sin n\theta = (\cos \theta + i \sin \theta)^n$$

$$= \cos^n \theta + in \cos^{n-1}\theta \sin \theta + i^2 \frac{n(n-1)}{\lfloor 2} \cos^{n-2}\theta \sin^2\theta$$

$$+ i^3 \frac{n(n-1)(n-2)}{\lfloor 3} \cos^{n-3}\theta \sin^3\theta + \cdots$$

Or, since $i^2 = -1$, $i^3 = -i$, $i^4 = +1$, $\cdots$,

$$\cos n\theta + i \sin n\theta = \cos^n \theta + in \cos^{n-1}\theta \sin \theta$$

$$- \frac{n(n-1)}{\lfloor 2} \cos^{n-2}\theta \sin^2\theta - i \frac{n(n-1)(n-2)}{\lfloor 3} \cos^{n-3}\theta \sin^3\theta + \cdots$$

Equating now the real parts and the imaginary parts separately, we obtain

$$\cos n\theta = \cos^n \theta - \frac{n(n-1)}{\lfloor 2} \cos^{n-2}\theta \sin^2\theta \ .$$

$$+ \frac{n(n-1)(n-2)(n-3)}{\lfloor 4} \cos^{n-4}\theta \sin^4\theta - \cdots,$$

$$\sin n\theta = n \cos^{n-1}\theta \sin \theta - \frac{n(n-1)(n-2)}{\lfloor 3} \cos^{n-8}\theta \sin^8\theta$$

$$+ \frac{n(n-1)(n-2)(n-3)(n-4)}{\lfloor 5} \cos^{n-5}\theta \sin^5\theta - \cdots$$

EXERCISE XXIX

1. Find the six 6th roots of -1; of $+1$.
2. Find the three cube roots of i.
3. Find the four 4th roots of $-i$.
4. Express $\sin 4\theta$ and $\cos 4\theta$ in terms of $\sin \theta$ and $\cos \theta$.

SECTION XLVIII

EXPANSION OF SIN X, COS X, AND TAN X IN INFINITE SERIES

Let one radian be denoted simply by 1, and let

$$\cos 1 + i \sin 1 = k.$$

Then $\qquad \cos x + i \sin x = (\cos 1 + i \sin 1)^x = k^x$,

and, putting $-x$ for x,

$$\cos(-x) + i \sin(-x) = \cos x - i \sin x = k^{-x}.$$

That is, $\qquad \cos x + i \sin x = k^x$,

and $\qquad \cos x - i \sin x = k^{-x}$.

By taking the sum and difference of these two equations, and dividing the sum by 2 and the difference by $2i$, we have

$$\cos x = \frac{1}{2}(k^x + k^{-x}), \qquad \sin x = \frac{1}{2i}(k^x - k^{-x}).$$

But $\qquad k^x = (e^{\log k})^x = e^{x \log k}, \qquad k^{-x} = e^{-x \log k},$

and $\qquad e^{x \log k} = 1 + x \log k + \frac{x^2 (\log k)^2}{\lfloor 2} + \frac{x^3 (\log k)^3}{\lfloor 3} + \cdots,$

$$e^{-x \log k} = 1 - x \log k + \frac{x^2 (\log k)^2}{\lfloor 2} - \frac{x^3 (\log k)^3}{\lfloor 3} + \cdots$$

$$\therefore \cos x = \frac{1}{2}(k^x + k^{-x}) = 1 + \frac{x^2 (\log k)^2}{\lfloor 2} + \frac{x^4 (\log k)^4}{\lfloor 4} + \cdots,$$

$$\sin x = \frac{1}{i}\left\{ x \log k + \frac{x^3 (\log k)^3}{\lfloor 3} + \frac{x^5 (\log k)^5}{\lfloor 5} + \cdots \right\}.$$

It only remains to find the value of k, and this can be obtained by dividing the last equation through by x and letting x approach 0 indefinitely.

Then we have

$$\underset{x \doteq 0}{\text{limit}}\left(\frac{\sin x}{x} \right) = \frac{1}{i} \log k.$$

But

$$\underset{x \doteq 0}{\text{limit}}\left(\frac{\sin x}{x} \right) = 1.$$

$$\therefore \log k = i.$$

$$\therefore k = e^i.$$

Therefore, we have

$$\cos x = \frac{1}{2}(e^{xi} + e^{-xi}) = 1 - \frac{x^2}{\lfloor 2} + \frac{x^4}{\lfloor 4} - \frac{x^6}{\lfloor 6} + \cdots,$$

$$\sin x = \frac{1}{2i}(e^{xi} - e^{-xi}) = x - \frac{x^3}{\lfloor 3} + \frac{x^5}{\lfloor 5} - \frac{x^7}{\lfloor 7} \cdots$$

From the last two series we obtain, by division,

$$\tan x = \frac{\sin x}{\cos x} = x + \frac{x^3}{3} + \frac{2 x^5}{15} + \frac{17 x^7}{315} \cdots$$

By the aid of these series the trigonometric functions of any angle are readily calculated.

In the computation it must be remembered that x is the *circular measure* of the given angle.

Verify by the series just obtained that:

1. $\sin^2 x + \cos^2 x = 1$.

2. $\sin(-x) = -\sin x$ and $\cos(-x) = \cos x$.

3. $\sin 2x = 2 \sin x \cos x$.

4. $\cos 2x = 1 - 2\sin^2 x$.

5. Find the series for $\sec x$ as far as the term containing the 6th power of x.

6. Find the series for $x \cot x$, noting that

$$x \cot x = \frac{x}{\sin x} \cos x.$$

7. Calculate $\sin 10°$ and $\cos 10°$ to five places of decimals.

8. Calculate $\tan 15°$ to five places of decimals.

9. From the exponential value of $\cos x$ show that

$$\cos 3x = 4 \cos^3 x - 3 \cos x.$$

10. From the exponential value of $\sin x$ show that

$$\sin 3x = 3 \sin x - 4 \sin^3 x.$$

SPHERICAL TRIGONOMETRY

CHAPTER VII

THE RIGHT SPHERICAL TRIANGLE

SECTION XLIX

INTRODUCTION

The object of *Spherical Trigonometry* is to explain the method of solving spherical triangles. To *solve* a spherical triangle is to compute any three of its parts when the other three parts are given.

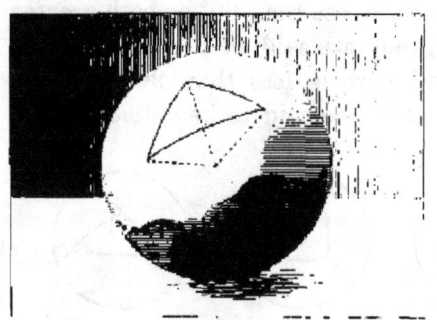

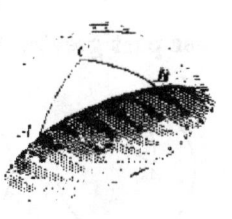

FIG. 75   FIG. 76

The sides of a spherical triangle are arcs of great circles. Thus, AB (Fig. 76) is an arc of a great circle. The sides of a spherical triangle are measured in degrees, minutes, and seconds, and therefore by the plane angles formed by radii of the sphere drawn to the vertices of the triangle. Therefore, the measures of the sides are independent of the length of the radius, which may be assumed to have any convenient numerical value; as, for example, unity.

The angles of a spherical triangle are measured by the dihedral angles made by the planes of the sides. Each angle is ·also measured by the number of degrees in the arc of a great circle, described from the vertex of the angle as a pole, and included between the sides of the angle.

The sides may have any value from 0° to 360°; but in this work only sides that are less than 180° will be considered. The angles may have any value from 0° to 180°.

A **right spherical triangle** may have one, two, or three right angles.

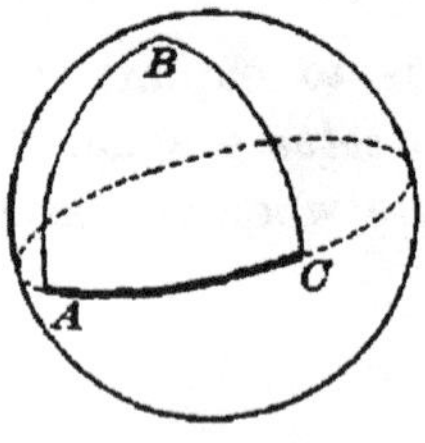

When a spherical triangle has one or more of its sides equal to a quadrant it is called a **quadrantal triangle** (Fig. 77).

If any two parts of a spherical triangle are either both less than 90° or both greater than 90°, they are said to be **alike in kind**; but if one part is less than 90°, and the other part greater than 90°, they are said to be **unlike in kind.**

Fig. 77

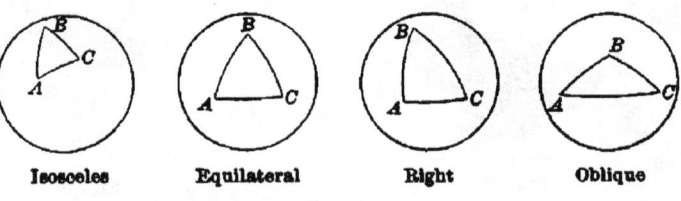

| Isosceles | Equilateral | Right | Oblique |

Fig. 78

Spherical triangles are named isosceles, equilateral, equiangular, right, and oblique, under the same conditions as plane triangles are named isosceles, equilateral, equiangular, right, and oblique.

THE RIGHT SPHERICAL TRIANGLE

The following propositions are proved in Geometry
Wentworth's *Geometry*, §§ 815, 790, 795, 793.)

1. *If two angles of a spherical triangle are unequal, the
sides opposite are unequal, and the greater side is
the greater angle; and conversely.*

2. *The sum of the sides of a spherical triangle are more
than 360°.*

Fig. 79

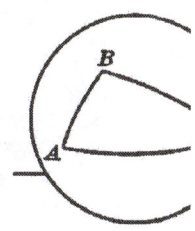

Fig. 80

3. *The sum of the angles of a spherical triangle is greater
than 180° and less than 540°.*

4. *If, from the vertices of a spherical triangle a
arcs of great circles are drawn, another triangle is for
related to the first that each angle of either triangle
supplement of the side opposite it in the other triangl*

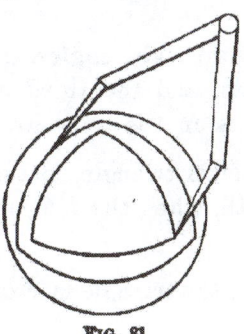

Fig. 81

Fig. 82

Two spherical triangles, drawn as explained in Theorem 4, p. 141, are called **polar triangles,** or *supplemental triangles.*

Let A, B, C (Fig. 83) denote the angles of one triangle; a, b, c the sides opposite these angles, respectively; and let A', B', C' and a', b', c' denote the corresponding angles and the corresponding sides of the polar triangle. Then Theorem 4 gives the six following equations:

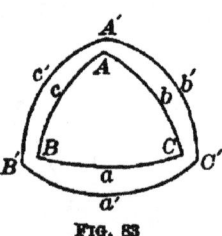

Fic. 83

$$A + a' = 180°,$$
$$B + b' = 180°,$$
$$C + c' = 180°,$$
$$A' + a = 180°,$$
$$B' + b = 180°,$$
$$C' + c = 180°.$$

EXERCISE XXXI

1. The angles of a triangle are 70°, 80°, and 100°. Find the sides of the polar triangle.

2. The sides of a triangle are 40°, 90°, and 125°. Find the angles of the polar triangle.

3. Show that, if a triangle has three right angles, the sides of the triangle are quadrants.

4. Show that, if a triangle has two right angles, the sides opposite these angles are quadrants, and the third angle is measured by the number of degrees in the opposite side.

5. How can the sides of a spherical triangle, measured in degrees, be found in units of length, when the length of the radius of the sphere is known?

6. Find the lengths of the sides of the triangle in Example 2 if the radius of the sphere is 4 feet.

SECTION L

FORMULAS RELATING TO RIGHT SPHERICAL TRIANGLES

As is evident from Examples 3 and 4, Exercise XXXI, the only kind of right spherical triangle that requires further investigation is that which contains *only one* right angle.

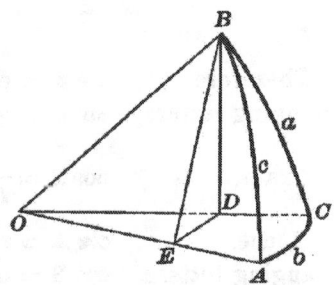

FIG. 84   FIG. 85

Let *ABC* (Fig. 85) be a right spherical triangle having only one right angle; and let *A*, *B*, *C* denote the angles of the triangle; *a*, *b*, *c*, respectively, the opposite sides.

Let *C* be the right angle; and for the present suppose that each of the other parts is less than 90°, and that the radius of the sphere is 1.

Let planes be passed through the sides, intersecting in the radii *OA*, *OB*, and *OC*.

Also, let a plane ⊥ to *OA* be passed through *B*, cutting *OA* at *E* and *OC* at *D*. Draw *BE*, *BD*, and *DE*.

BE and *DE* are each ⊥ to *OA* (Wentworth's *Geometry*, § 501); therefore, ∠ *BED* = *A*. The plane *BDE* is ⊥ to the plane *AOC* (Wentworth's *Geometry*, § 554); hence, *BD*, which is the intersection of the planes *BDE* and *BOC*, is ⊥ to the plane *AOC* (Wentworth's *Geometry*, § 556), and therefore ⊥ to *OC* and *DE*.

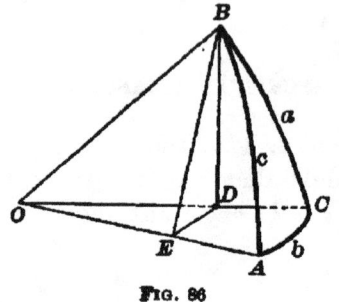

Fig. 86

Now,

$$\cos c = OE = OD \times \cos b,$$

and $OD = \cos a.$

Therefore,

$$\cos c = \cos a \cos b. \qquad [38]$$

Again, $\sin a = BD = BE \times \sin A,$

and $BE = \sin c.$

Therefore, $\sin a = \sin c \sin A$ $\Big\}$

changing letters, $\sin b = \sin c \sin B$ $\qquad [39]$

Again, $\cos A = \dfrac{DE}{BE} = \dfrac{OE \tan b}{OE \tan c}.$

Hence, $\cos A = \tan b \cot c$ $\Big\}$

changing letters, $\cos B = \tan a \cot c$ $\qquad [40]$

Again, $\cos A = \dfrac{DE}{BE} = \dfrac{OD \sin b}{\sin c} = \cos a \dfrac{\sin b}{\sin c}.$

By substituting for $\dfrac{\sin b}{\sin c}$ its value from [39], we obtain

$$\cos A = \cos a \sin B \Big\}$$

changing letters, $\cos B = \cos b \sin A$ $\qquad [41]$

Again, $\sin b = \dfrac{DE}{OD} = \dfrac{BD \cot A}{OD} = \tan a \cot A.$

Hence, $\sin b = \tan a \cot A$ $\Big\}$

changing letters, $\sin a = \tan b \cot B$ $\qquad [42]$

If in [38] we substitute for $\cos a$ and $\cos b$ their values from [41], we obtain

$$\cos c = \cot A \cot B. \qquad [43]$$

NOTE. In order to deduce the second formulas in [39]–[42] geometrically, the auxiliary plane must be passed through $A \perp$ to OB.

These ten formulas are sufficient for the solution of any right spherical triangle. In deducing these formulas, all the parts of the triangle, except the right angle, were assumed to be less than 90°. But the formulas hold when this hypothesis is not true.

Let one of the legs a be greater than 90°, and construct a figure for this case (Fig. 88) in the same manner as Fig. 85.

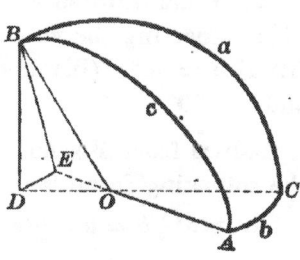

FIG. 87   FIG. 88

The auxiliary plane BDE will now cut both CO and AO produced beyond the centre O; and we have

$$\cos c = -\ OE = -\ OD \cos DOE$$
$$= -\ (-\cos a)\cos b = \cos a \cos b.$$

Likewise, the other formulas, [39]–[43], hold true in this case.

Again, suppose that both the legs a and b are greater than 90°. In this case the plane BDE will cut CO produced beyond O, and AO between A and O; and we have

$$\cos c = OE = OD \cos DOE$$
$$= (-\cos a)(-\cos b) = \cos a \cos b,$$

a result agreeing with [38].

Likewise the other formulas, [39]–[43], hold true in this case.

Like results may be obtained in all cases.

In other words, Formulas [38]–[43] are universally true.

1. Show, by aid of Formula [38], p. 144, that the hypotenuse of a right spherical triangle is *less than* or *greater than* 90°, according as the two legs are *alike* or *unlike* in kind.

2. Show, by aid of Formula [41], that in a right spherical triangle each leg and the opposite angle are always alike in kind.

3. What inferences may be drawn from Formulas [38]–[43] respecting the values of the other parts: (i) if $c = 90°$; (ii) if $a = 90°$; (iii) if $c = 90°$ and $a = 90°$; (iv) if $a = 90°$ and $b = 90°$?

Deduce from Formulas [38]–[43] and Formulas [18]–[23] the following formulas:

4. $\tan^2 \tfrac{1}{2} b = \tan \tfrac{1}{2} (c - a) \tan \tfrac{1}{2} (c + a).$

Hint. Substitute in Formula [18] the value of $\cos b$ from [38].

5. $\tan^2 (45° - \tfrac{1}{2} A) = \tan \tfrac{1}{2} (c - a) \cot \tfrac{1}{2} (c + a).$

6. $\tan^2 \tfrac{1}{2} B = \sin (c - a) \csc (c + a).$

7. $\tan^2 \tfrac{1}{2} c = - \cos (A + B) \sec (A - B).$

8. $\tan^2 \tfrac{1}{2} a = \tan \left[\tfrac{1}{2} (A + B) - 45° \right] \tan \left[\tfrac{1}{2} (A - B) + 45° \right].$

9. $\tan^2 (45° - \tfrac{1}{2} c) = \tan \tfrac{1}{2} (A - a) \cot \tfrac{1}{2} (A + a).$

10. $\tan^2 (45° - \tfrac{1}{2} b) = \sin (A - a) \csc (A + a).$

11. $\tan^2 (45° - \tfrac{1}{2} B) = \tan \tfrac{1}{2} (A - a) \tan \tfrac{1}{2} (A + a).$

SECTION LI

NAPIER'S RULES

The ten formulas deduced in Sect. L express the relations of five parts of a right triangle, the three sides and the two oblique angles. All these relations may be shown to follow from two very useful rules, devised by Baron Napier, the inventor of logarithms.

For this purpose the right angle (not entering the formulas) is not taken into account, and instead of the hypotenuse and the two oblique angles their respective *complements* are employed; so that the five parts considered by Napier's Rules are: *a, b, Co. A, Co. c, Co. B.*

Any one of these parts may be called a *middle* part; and then the two parts immediately adjacent are called *adjacent* parts, and the other two are called *opposite* parts. Napier's Rules are

RULE I. *The sine of any middle part is equal to the product of the tangents of the adjacent parts.*

RULE II. *The sine of any middle part is equal to the product of the cosines of the opposite parts.*

These rules are easily remembered by the expressions **tan. ad.** and **cos. op.**

The correctness of these rules may be easily shown by taking in turn each of the five parts as middle part, and comparing the resulting equations with the equations contained in Formulas [38]–[43], p. 144.

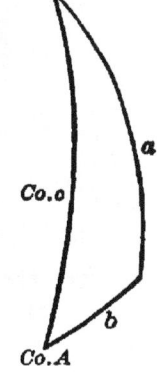

FIG. 89

For example, let *Co. c* be taken as middle part; then *Co. A* and *Co. B* are the adjacent parts, and *a* and *b* the opposite parts, as is very plainly seen in Fig. 89. Then, by Napier's Rules,

$$\sin (Co.\ c) = \tan (Co.\ A) \tan (Co.\ B),$$

or $$\cos c = \cot A \cot B;$$

$$\sin (Co.\ c) = \cos a \cos b,$$

or $$\cos c = \cos a \cos b.$$

These results agree respectively with Formulas [43] and [38], p. 144.

1. Show that Napier's Rules lead to the equations contained in Formulas [39], [40], [41], and [42].

2. What will Napier's Rules become if we take as the five parts of the triangle the hypotenuse, the two oblique angles, and the *complements* of the two legs?

SECTION LII

SOLUTION OF THE RIGHT SPHERICAL TRIANGLE

By means of Formulas [38]–[43], p. 144, we can solve a right triangle in all possible cases. In every case two parts besides the right angle must be given.

FIG. 90

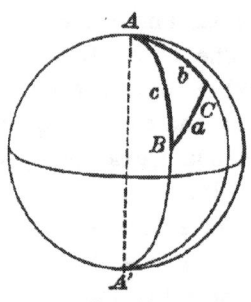

FIG. 91

CASE I

Given the two legs a *and* b.

From Formulas [38] and [42], p. 144, we obtain

$$\cos c = \cos a \cos b,$$
$$\tan A = \tan a \csc b,$$
$$\tan B = \tan b \csc a.$$

For a check use cos c = cot A cot B, [43], p. 144.

EXAMPLE.   Given $a = 27° 28' 36''$, $b = 51° 12' 8''$; solve
the triangle.

log cos a = 9.94802	log tan a = 9.71605
log cos b = 9.79697	log csc b = 0.10826
log cos c = 9.74499	log tan A = 9.82431
$c = 56° 13' 41''$	$A = 33° 42' 51''$

Check.

log tan b = 10.09476	log cot A = 10.17569
log csc a = 0.33594	log cot B = 9.56930
log tan B = 10.43070	log cos c = 9.74499
$B = 69° 38' 54''$	

CASE II

Given the hypotenuse c *and the leg* a.

From Formulas [38], [39], and [40], p. 144, we obtain

$$\cos b = \cos c \sec a,$$
$$\sin A = \sin a \csc c,$$
$$\cos B = \tan a \cot c.$$

For a check use cos B = cos b sin A, [41], p. 144.

Although two angles in general correspond to sin A, one
acute, the other obtuse, yet in this case it is easy to determine
whether A is acute or obtuse since A and a must be alike in
kind.   (See Example 2, Exercise XXXII, p. 146.)

CASE III

Given the leg a *and the opposite angle* A.

From Formulas [39], [42], and [41], we obtain

$$\sin c = \sin a \csc A,$$
$$\sin b = \tan a \cot A,$$
$$\sin B = \sec a \cos A.$$

Or, from [38] and [40], p. 144, we obtain

$$\cos b = \cos c \sec a,$$
$$\cos B = \tan a \cot c.$$

For a check use $\sin b = \sin c \sin B$, [39], p. 144.

When c has been computed, b and B are determined by these values of their cosines; but, since c must be found from its sine, c may have, in general, two values which are supplements of each other. This case, therefore, really admits of two solutions.

In fact, if the sides b and c are extended until they meet in A' (Fig. 91), the two right triangles ABC and $A'BC$ have the side a in common, and the angle $A = A'$. Also, $A'C = 180° - b$, $A'B = 180° - c$, and $\angle A'BC = 180° - B$. Hence, if ABC is one solution, $A'BC$ is the other.

Case IV

Given the leg a *and the adjacent angle* B.

From Formulas [40], [42], and [41], p. 144, we obtain

$$\tan c = \tan a \sec B,$$
$$\tan b = \sin a \tan B,$$
$$\cos A = \cos a \sin B.$$

For a check use $\cos A = \tan b \cot c$, [40], p. 144.

Case V

Given the hypotenuse c *and the angle* A.

From Formulas [39], [40], and [43], p. 144, we obtain

$$\sin a = \sin c \sin A,$$
$$\tan b = \tan c \cos A,$$
$$\cot B = \cos c \tan A.$$

Here a is determined by $\sin a$, since a and A must be alike in kind. (See Example 2, Exercise XXXII, p. 146.)

For a check use $\sin a = \tan b \cot B$, [42], p. 144.

Case VI

Given the two angles A and B.

From Formulas [43] and [41], p. 144, we obtain

$$\cos c = \cot A \cot B,$$
$$\cos a = \cos A \csc B,$$
$$\cos b = \cos B \csc A.$$

For a check use $\cos c = \cos a \cos b$, [38], p. 144.

NOTE 1.   In Case I (a and b given), if c is very near $0°$ or $180°$, it may be found with greater accuracy by first computing B, and then computing c, as in Case IV.

NOTE 2.   In Case II (c and a given), if b is very near $0°$ or $180°$, it may be computed more accurately by means of the derived formula

$$\tan^2 \tfrac{1}{2} b = \tan \tfrac{1}{2}(c-a) \tan \tfrac{1}{2}(c+a). \qquad \text{(Ex. 4, Sect. L)}$$

If A is so near $90°$ that it cannot be found accurately in the tables, it may be computed from the derived formula

$$\tan^2(45° - \tfrac{1}{2} A) = \tan \tfrac{1}{2}(c-a) \cot \tfrac{1}{2}(c+a). \qquad \text{(Ex. 5, Sect. L)}$$

If B cannot be found accurately, we may use the formula

$$\tan^2 \tfrac{1}{2} B = \sin(c-a) \csc(c+a). \qquad \text{(Ex. 6, Sect. L)}$$

NOTE 3.   In Case III (a and A given), when the formulas do not give accurate results, we may employ the derived formulas

$$\tan^2(45° - \tfrac{1}{2} c) = \tan \tfrac{1}{2}(A-a) \cot \tfrac{1}{2}(A+a), \qquad \text{(Ex. 9, Sect. L)}$$
$$\tan^2(45° - \tfrac{1}{2} b) = \sin(A-a) \csc(A+a), \qquad \text{(Ex. 10, Sect. L)}$$
$$\tan^2(45° - \tfrac{1}{2} B) = \tan \tfrac{1}{2}(A-a) \tan \tfrac{1}{2}(A+a) \qquad \text{(Ex. 11, Sect. L)}$$

NOTE 4.   In Case IV (a and B given), if A is near $0°$ or $180°$, it may be more accurately found by first computing b and then finding A.

NOTE 5.   In Case V (c and A given), if a is near $90°$, it may be found by first computing b, and then computing a by Formula [42], p. 144.

NOTE 6.   In Case VI (A and B given), for unfavorable values of the sides greater accuracy may be obtained by means of the derived formulas

$$\tan^2 \tfrac{1}{2} c = -\cos(A+B)\sec(A-B), \qquad \text{(Ex. 7, Sect. L)}$$
$$\tan^2 \tfrac{1}{2} a = \tan[\tfrac{1}{2}(A+B) - 45°] \tan[45° + \tfrac{1}{2}(A-B)], \qquad \text{(Ex. 8, Sect. L)}$$
$$\tan^2 \tfrac{1}{2} b = \tan[\tfrac{1}{2}(A+B) - 45°] \tan[45° - \tfrac{1}{2}(A-B)].$$

Note 7. In Cases I, IV, and V, the solution is always possible. In the other Cases, in order that the solution may be possible, it is necessary and sufficient that in Case II $\sin a < \sin c$; in Case III, that a and A be alike in kind, and $\sin A > \sin a$; in Case VI, that $A + B + C > 180°$, and the difference between A and $B < 90°$.

Note 8. It is easy to trace analogies between the formulas for solving right spherical triangles and those for solving right plane triangles. The former become identical with the latter if we suppose the radius of the sphere to be infinite in length. Then the cosines of the sides become each equal to 1, and the ratios of the *sines* of the sides and of the *tangents* of the sides must be taken as equal to the ratios of the sides themselves.

Note 9. In solving spherical triangles, the algebraic sign of the functions must receive careful attention. Write the sign of each function just above the function. Then the sign of the function in the first member of the equation is + or − according as the law of signs makes the second member of the equation positive or negative. (See Example 1, p. 175.)

If the function is a cosine, tangent, or cotangent, the + sign shows the angle < 90°; the − sign shows the angle > 90°, and the *supplement* of the angle obtained from the table must be taken.

If the function is a sine, since the sine of an angle and of its supplement are the same, the acute angle obtained from the table and its supplement must be considered as solutions, unless there are other conditions that remove the ambiguity. For conditions that remove the ambiguity, see Examples 1 and 2 in Exercise XXXII, p. 146.

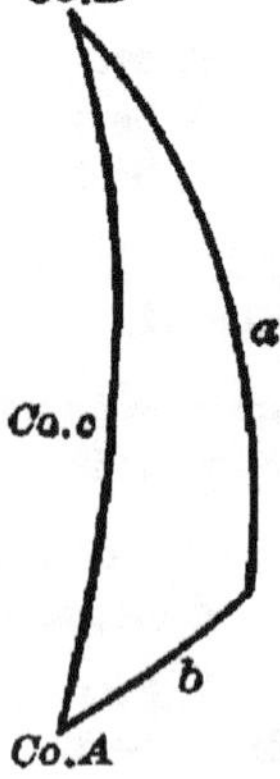

It is always easy to find the required formula by means of Napier's Rules. In applying these rules we must choose for the middle part that one of the three parts which will make the two given parts either adjacent parts or opposite parts.

Example: *Given* a *and* B; *solve the triangle.*

To find b, take a as the middle part; then b and *Co. B* are the adjacent parts; and, by Rule I,

$$\sin a = \tan b \cot B.$$

Whence, $$\tan b = \sin a \tan B.$$

Fig. 92

To find c, take $Co.\,B$ as the middle part; then a and $Co.\,c$ are the adjacent parts; and, by Rule I,

$$\cos B = \tan a \cot c.$$

Whence, $\tan c = \tan a \sec B.$

To find A, take $Co.\,A$ as the middle part; then a and $Co.\,B$ are the opposite parts; and, by Rule II,

$$\cos A = \cos a \sin B.$$

EXERCISE XXXIV

Solve the following right triangles, taking for the given parts in each case those printed in columns I and II:

	I	II	III	IV	V
	a	b	c	A	B
1	36° 27′	43° 32′ 31″	54° 20′	46° 59′ 43″	57° 59′ 19″
2	86° 40′	32° 40′	87° 11′ 40″	88° 11′ 58″	32° 42′ 39″
3	50°	36° 54′ 49″	59° 4′ 26″	63° 15′ 13″	44° 26′ 22″
4	120° 10′	150° 59′ 44″	63° 55′ 43″	105° 44′ 21″	147° 19′ 47″
	c	a	b	A	B
5	55° 9′ 32″	22° 15′ 7″	51° 53′	27° 28′ 38″	73° 27′ 11″
6	23° 49′ 51″	14° 16′ 35″	19° 17′	37° 36′ 49″	54° 49′ 28″
7	44° 33′ 17″	32° 9′ 17″	32° 41′	49° 20′ 16″	50° 19′ 16″
8	97° 13′ 4″	132° 14′ 12″	79° 13′ 38″	131° 43′ 50″	81° 58′ 53″
	a	A	c	b	B
9	77° 21′ 50″	83° 56′ 40″	78° 53′ 20″⎫ 101° 6′ 40″⎭	28° 14′ 31″⎫ 151° 45′ 29″⎭	28° 49′ 57″⎫ 151° 10′ 3″⎭
10	77° 21′ 50″	40° 40′ 40″	impossible		

Note. The values in the last three columns of Example 9 cannot be combined promiscuously with those given in columns I and II.

If $a < 90°$, with the value of $b > 90°$ must be taken $B > 90°$ and $c > 90°$; while with the value of $b < 90°$ must be taken, for the same reason, $B < 90°$ and $c < 90°$. (See Examples 1 and 2, Exercise XXXII, p. 146.)

	I	II	III	IV	V
	a	*B*	*c*	*b*	*A*
11	92° 47′ 32″	50° 2′ 1″	91° 47′ 40″	49° 59′ 58″	92° 8′ 23″
12	2° 0′ 55″	12° 40′	2° 3′ 56″	0° 27′ 10″	77° 20′ 28″
13	20° 20′ 20″	38° 10′ 10″	25° 14′ 38″	15° 16′ 50″	54° 35′ 17″
14	54° 30′	35° 30′	59° 51′ 21″	30° 8′ 39″	70° 17′ 35″
	c	*A*	*a*	*b*	*B*
15	69° 25′ 11″	54° 54′ 42″	50°	56° 50′ 49″	63° 25′ 4″
16	112° 48′	56° 11′ 56″	50°	127° 4′ 30″	120° 3′ 50″
17	46° 40′ 12″	37° 46′ 9″	26° 27′ 24″	39° 57′ 42″	62° 0′ 4″
18	118° 40′ 1″	128° 0′ 4″	136° 15′ 32″	48° 23′ 38″	58° 27′ 4″
	A	*B*	*a*	*b*	*c*
19	63° 15′ 12″	135° 33′ 39″	50° 0′ 4″	143° 5′ 12″	120° 55′ 34″
20	116° 43′ 12″	116° 31′ 25″	120° 10′ 3″	119° 59′ 46″	75° 26′ 58″
21	46° 59′ 42″	57° 59′ 17″	36° 27′	48° 32′ 30″	54° 20′
22	90°	88° 24′ 35″	90°	88° 24′ 35″	90°

23. Define a quadrantal triangle, and show how its solution may be reduced to that of the right triangle.

24. Solve the quadrantal triangle the sides of which are
$$a = 174° 12′ 49″, \ b = 94° 8′ 20″, \ c = 90°.$$

25. Solve the quadrantal triangle in which
$$c = 90°, \ A = 110° 47′ 50″, \ B = 135° 35′ 34.5″.$$

26. Given in a spherical triangle *A, C,* and *c* each equal to 90°; solve the triangle.

27. Given $A = 60°$, $C = 90°$, and $c = 90°$; solve the triangle.

28. In a right spherical triangle, given $A = 42° 24′ 9″$, $B = 9° 4′ 11″$; solve the triangle.

29. In a right spherical triangle, given $a = 119° 11′$, $B = 126° 54′$; solve the triangle.

30. In a right spherical triangle, given $c = 50°$, $b = 44°18'39''$; solve the triangle.

31. In a right spherical triangle, given $A = 156°\ 20'\ 30''$, $a = 65°\ 15'\ 45''$; solve the triangle.

32. In a right spherical triangle, given $A = 74°\ 12'\ 31''$, $c = 64°\ 28'\ 47''$; solve the triangle.

33. In a right spherical triangle, given $a = 112°\ 42'\ 38''$, $B = 44°\ 28'\ 44''$; solve the triangle.

34. In a right spherical triangle, given $b = 48°\ 12'\ 48''$, $A = 108°\ 14'\ 14''$; solve the triangle.

35. In a right spherical triangle, given $A = 122°\ 58!\ 47''$, $B = 104°\ 17'\ 55''$; solve the triangle.

36. If the legs a and b of a right spherical triangle are equal, show that $\cos a = \cot A = \sqrt{\cos c}$.

37. In a right spherical triangle show that
$$\cos^2 A \sin^2 c = \sin(c + a)\sin(c - a).$$

38. In a right spherical triangle show that
$$\tan a \cos c = \sin b \cot B.$$

39. In a right spherical triangle show that
$$\sin^2 A = \cos^2 B + \sin^2 a \sin^2 B.$$

40. In a right spherical triangle show that
$$\sin(b + c) = 2\cos^2 \tfrac{1}{2} A \cos b \sin c.$$

41. In a right spherical triangle show that
$$\sin(c - b) = 2\sin^2 \tfrac{1}{2} A \cos b \sin c.$$

42. If in a right spherical triangle, p denotes the arc of the great circle passing through the vertex of the right angle and perpendicular to the hypotenuse, m and n the segments of the hypotenuse made by this arc adjacent to the legs a and b, show that (i) $\tan^2 a = \tan c \tan m$, (ii) $\sin^2 p = \tan m \tan n$.

SECTION LIII

SOLUTION OF THE ISOSCELES SPHERICAL TRIANGLE

An arc of a great circle, passed through the vertex of an isosceles spherical triangle and the middle point of the base (Fig. 94), divides the triangle into two equivalent right spherical triangles. Hence, the solution of an isosceles spherical triangle may be reduced to that of a right spherical triangle.

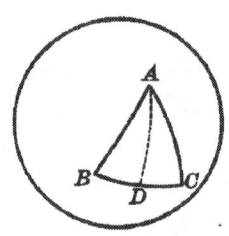

FIG. 93   FIG. 94

Likewise the solution of a regular spherical polygon may be reduced to that of a right spherical triangle. Arcs of great

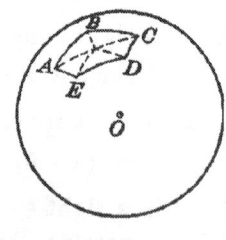

FIG. 95   FIG. 96

circles, passed through the centre of the polygon and the vertices (Fig. 96), divide the polygon into equal isosceles triangles;

and each one of these equal isosceles triangles may be divided into two equivalent
right triangles.

A regular spherical polygon (Fig. 97) is
the polygon formed by the intersections of
the spherical surface with the faces of a
regular pyramid whose vertex is at the
centre of the sphere.

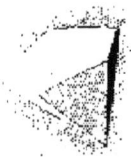

FIG. 97

EXERCISE XXXV

1. In an isosceles spherical triangle, given the base b and
the side a; find A the angle at the base, B the angle at the
vertex, and h the altitude.

2. In an equilateral spherical triangle, given the side a;
find the angle A.

3. Given the side a of a regular spherical polygon of n sides;
find the angle A of the polygon, the distance R from the centre
of the polygon to one of the vertices, and the distance r from
the centre to the middle point of one of the sides.

Tetrahedron   Hexahedron   Octahedron   Dodecahedron   Icosahedron

FIG. 98

4. Compute the dihedral angles made by the faces of the
five regular polyhedrons (Fig. 98).

5. A spherical square is a regular spherical quadrilateral.
Find the angle A of the square, having given the side a.

CHAPTER VIII

THE OBLIQUE SPHERICAL TRIANGLE

SECTION LIV

FUNDAMENTAL FORMULAS

Let ABC (Fig. 99) be an oblique spherical triangle, a, b, c its three sides, A, B, C the angles opposite a, b, c, respectively.

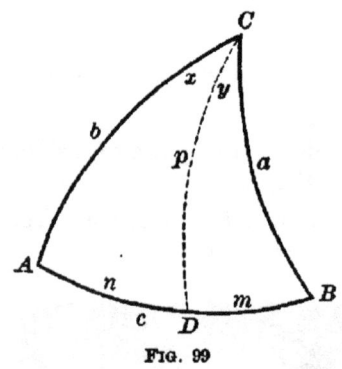

FIG. 99

Through C draw CD, an arc of a great circle, perpendicular to the side AB, meeting AB at D. For brevity let

$$CD = p, \quad AD = n, \quad BD = m,$$
$$\angle ACD = x, \quad \angle BCD = y.$$

1. By [39], p. 144, in the right triangles BDC and ADC,

$$\sin p = \sin a \sin B,$$
$$\text{and} \quad \sin p = \sin b \sin A.$$

Therefore,

$$\left. \begin{array}{l} \sin a \sin B = \sin b \sin A \\ \sin a \sin C = \sin c \sin A \\ \sin b \sin C = \sin c \sin B \end{array} \right\} . \qquad [44]$$

similarly, and

These equations may also be written in the form of proportions:

$$\sin a : \sin b : \sin c = \sin A : \sin B : \sin C.$$

That is, *the sines of the sides of a spherical triangle are proportional to the sines of the opposite angles.*

168

In Fig. 99 the arc CD cuts the side AB within the triangle. If CD falls without the triangle, for instance to the right of CB, $\sin(180° - B)$ would then be employed instead of $\sin B$. But $\sin(180° - B) = \sin B$ (Sect. XXV, p. 48). Hence, the Formulas [44] hold true in all cases.

2. In the right triangle BDC, by [38], p. 144,

$$\cos a = \cos p \cos m$$
$$= \cos p \cos(c - n)$$

by [9], p. 56, $= \cos p \cos c \cos n + \cos p \sin c \sin n.$

Now, in the right triangle ADC, by [38], p. 144,

$$\cos p \cos n = \cos b.$$

Whence, $\cos p = \cos b \sec n,$

and $\cos p \sin n = \cos b \tan n.$

By [40], p. 144, $\tan n = \tan b \cos A.$

$$\therefore \cos p \sin n = \cos b \tan b \cos A$$
$$= \sin b \cos A.$$

Substituting these values of $\cos p \cos n$ and $\cos p \sin n$ in the value of $\cos a$, we obtain

$$\left.\begin{array}{l} \cos a = \cos b \cos c + \sin b \sin c \cos A \\ \text{and similarly,}\quad \cos b = \cos a \cos c + \sin a \sin c \cos B \\ \cos c = \cos a \cos b + \sin a \sin b \cos C \end{array}\right\} . \quad [45]$$

3. In the right triangle ADC, by [41], p. 144,

$$\cos A = \cos p \sin x$$
$$= \cos p \sin(C - y)$$

by [8], p. 56, $= \cos p \sin C \cos y - \cos p \cos C \sin y.$

Now, by [41], p. 144,

$$\cos p \sin y = \cos B.$$

Therefore, $\cos p = \cos B \csc y,$

and $\cos p \cos y = \cos B \cot y$

by [43], p. 144, $= \cos B \tan B \cos a$
$$= \sin B \cos a.$$

Substituting these values of $\cos p \sin y$ and $\cos p \cos y$ in the value of $\cos A$, we obtain

$$\left.\begin{array}{l} \cos A = - \cos B \cos C + \sin B \sin C \cos a \\ \text{and similarly,} \quad \cos B = - \cos A \cos C + \sin A \sin C \cos b \\ \cos C = - \cos A \cos B + \sin A \sin B \cos c \end{array}\right\} . \quad [46]$$

Formulas [45] and [46] are also universally true; for the same equations are obtained when the arc CD cuts the side AB without the triangle.

EXERCISE XXXVI

1. What do Formulas [44] become if $A = 90°$? if $B = 90°$? if $C = 90°$? if $a = 90°$? if $A = B = 90°$? if $a = b = 90°$?

2. What do Formulas [45] become if $A = 90°$? if $B = 90°$? if $C = 90°$? if $A = B = C = 90$?

3. What does the first of Formulas [45] become if $A = 0°$? if $A = 90°$? if $A = 180°$?

4. From Formulas [45] deduce Formulas [46], by means of the relations between polar triangles (Theorem 4, p. 141).

SECTION LV

FORMULAS FOR THE HALF ANGLES AND SIDES

From the first equation of [45], p. 159,

$$\cos A = \frac{\cos a - \cos b \cos c}{\sin b \sin c}.$$

Therefore, $\quad 1 - \cos A = \dfrac{\sin b \sin c + \cos b \cos c - \cos a}{\sin b \sin c}$

by [9], p. 56, $\qquad = \dfrac{\cos (b - c) - \cos a}{\sin b \sin c}$

by [23], p. 59, $\qquad = \dfrac{-2 \sin \frac{1}{2}(a + b - c) \sin \frac{1}{2}(b - c - a)}{\sin b \sin c}.$

Also, $1 + \cos A = \dfrac{\sin b \sin c - \cos b \cos c + \cos a}{\sin b \sin c}$

by [5], p. 54, $= \dfrac{\cos a - \cos (b + c)}{\sin b \sin c}$

by [23], p. 59, $= \dfrac{-2 \sin \frac{1}{2}(a + b + c) \sin \frac{1}{2}(a - b - c)}{\sin b \sin c}.$

Since by [16], p. 58,

$$1 - \cos A = 2 \sin^2 \tfrac{1}{2} A,$$

$$\sin^2 \tfrac{1}{2} A = \sin \tfrac{1}{2}(a + b - c) \sin \tfrac{1}{2}(a - b + c) \csc b \csc c;$$

and since by [17], p. 58,

$$1 + \cos A = 2 \cos^2 \tfrac{1}{2} A,$$

$$\cos^2 \tfrac{1}{2} A = \sin \tfrac{1}{2}(a + b + c) \sin \tfrac{1}{2}(b + c - a) \csc b \csc c.$$

Now let $\tfrac{1}{2}(a + b + c) = s.$

Then, $\tfrac{1}{2}(b + c - a) = s - a,$

$\tfrac{1}{2}(a - b + c) = s - b,$

and $\tfrac{1}{2}(a + b - c) = s - c.$

Then, by substitution and extraction of the square root,

$$\left.\begin{array}{l}
\sin \tfrac{1}{2} A = \sqrt{\sin (s - b) \sin (s - c) \csc b \csc c} \\[4pt]
\cos \tfrac{1}{2} A = \sqrt{\sin s \sin (s - a) \csc b \csc c} \\[4pt]
\tan \tfrac{1}{2} A = \sqrt{\csc s \csc (s - a) \sin (s - b) \sin (s - c)}
\end{array}\right\} . \quad [47]$$

In like manner, it may be shown that

$$\left.\begin{array}{l}
\sin \tfrac{1}{2} B = \sqrt{\sin (s - a) \sin (s - c) \csc a \csc c} \\[4pt]
\cos \tfrac{1}{2} B = \sqrt{\sin s \sin (s - b) \csc a \csc c} \\[4pt]
\tan \tfrac{1}{2} B = \sqrt{\csc s \csc (s - b) \sin (s - a) \sin (s - c)} \\[10pt]
\sin \tfrac{1}{2} C = \sqrt{\sin (s - a) \sin (s - b) \csc a \csc b} \\[4pt]
\cos \tfrac{1}{2} C = \sqrt{\sin s \sin (s - c) \csc a \csc b} \\[4pt]
\tan \tfrac{1}{2} C = \sqrt{\csc s \csc (s - c) \sin (s - a) \sin (s - b)}
\end{array}\right\} . \quad [47]$$

Again, from the first equation of [46], p. 160,

$$\cos a = \frac{\cos B \cos C + \cos A}{\sin B \sin C}.$$

Therefore,

$$1 - \cos a = \frac{\sin B \sin C - \cos B \cos C - \cos A}{\sin B \sin C}$$

by [5], p. 54, $$= \frac{- \cos (B + C) - \cos A}{\sin B \sin C}$$

by [22], p. 59, $$= \frac{- 2 \cos \tfrac{1}{2}(B + C + A) \cos \tfrac{1}{2}(B + C - A)}{\sin B \sin C};$$

and $$1 + \cos a = \frac{\sin B \sin C + \cos B \cos C + \cos A}{\sin B \sin C}$$

by [9], p. 56, $$= \frac{\cos (B - C) + \cos A}{\sin B \sin C}$$

by [22], p. 59, $$= \frac{2 \cos \tfrac{1}{2}(B - C + A) \cos \tfrac{1}{2}(B - C - A)}{\sin B \sin C}.$$

Since by [16], p. 58,

$$1 - \cos a = 2 \sin^2 \tfrac{1}{2} a,$$

$$\sin^2 \tfrac{1}{2} a = - \cos \tfrac{1}{2}(B + C + A) \cos \tfrac{1}{2}(B + C - A) \csc B \csc C;$$

and since by [17], p. 58,

$$1 + \cos a = 2 \cos^2 \tfrac{1}{2} a,$$

$$\cos^2 \tfrac{1}{2} a = \cos \tfrac{1}{2}(B - C + A) \cos \tfrac{1}{2}(B - C - A) \csc B \csc C.$$

Now let $$\tfrac{1}{2}(A + B + C) = S.$$

Then, $$\tfrac{1}{2}(B + C - A) = S - A,$$

$$\tfrac{1}{2}(A - B + C) = S - B,$$

and $$\tfrac{1}{2}(A + B - C) = S - C.$$

Then, by substitution and extraction of the square root,

$$\left. \begin{aligned} &\sin \tfrac{1}{2} a = \sqrt{- \cos S \cos (S - A) \csc B \csc C} \\ &\cos \tfrac{1}{2} a = \sqrt{\cos (S - B) \cos (S - C) \csc B \csc C} \\ &\tan \tfrac{1}{2} a = \sqrt{- \cos S \cos (S - A) \sec (S - B) \sec (S - C)} \end{aligned} \right\} . \quad [48]$$

And, in like manner,

$$\left. \begin{aligned}
\sin \tfrac{1}{2} b &= \sqrt{-\cos S \cos (S - B) \csc A \csc C} \\
\cos \tfrac{1}{2} b &= \sqrt{\cos (S - A) \cos (S - C) \csc A \csc C} \\
\tan \tfrac{1}{2} b &= \sqrt{-\cos S \cos (S - B) \sec (S - A) \sec (S - C)} \\
\sin \tfrac{1}{2} c &= \sqrt{-\cos S \cos (S - C) \csc A \csc B} \\
\cos \tfrac{1}{2} c &= \sqrt{\cos (S - A) \cos (S - B) \csc A \csc B} \\
\tan \tfrac{1}{2} c &= \sqrt{-\cos S \cos (S - C) \sec (S - A) \sec (S - B)}
\end{aligned} \right\} . \quad [48]$$

SECTION LVI

GAUSS'S EQUATIONS AND NAPIER'S ANALOGIES

By [5], p. 54,

$$\cos \tfrac{1}{2} (A + B) = \cos \tfrac{1}{2} A \cos \tfrac{1}{2} B - \sin \tfrac{1}{2} A \sin \tfrac{1}{2} B;$$

or, by substituting for $\cos \tfrac{1}{2} A$, $\cos \tfrac{1}{2} B$, $\sin \tfrac{1}{2} A$, $\sin \tfrac{1}{2} B$, their values given in Formulas [47], p. 161, and reducing,

$$\cos \tfrac{1}{2} (A + B) = \sqrt{\frac{\sin s \sin (s - a)}{\sin b \sin c}} \times \sqrt{\frac{\sin s \sin (s - b)}{\sin a \sin c}}$$

$$- \sqrt{\frac{\sin (s - b) \sin (s - c)}{\sin b \sin c}} \times \sqrt{\frac{\sin (s - a) \sin (s - c)}{\sin a \sin c}}$$

$$= \frac{\sin s}{\sin c} \sqrt{\frac{\sin (s - a) \sin (s - b)}{\sin a \sin b}} - \frac{\sin (s - c)}{\sin c} \sqrt{\frac{\sin (s - a) \sin (s - b)}{\sin a \sin b}}$$

$$= \frac{\sin s - \sin (s - c)}{\sin c} \times \sqrt{\frac{\sin (s - a) \sin (s - b)}{\sin a \sin b}}.$$

By [21], p. 59,

$$\sin s - \sin (s - c) = 2 \cos \tfrac{1}{2} (s + s - c) \sin \tfrac{1}{2} (s - s + c)$$
$$= 2 \cos (s - \tfrac{1}{2} c) \sin \tfrac{1}{2} c.$$

By [12], p. 58, $\sin c = 2 \sin \tfrac{1}{2} c \cos \tfrac{1}{2} c.$

Again, by [47], p. 161,

$$\sqrt{\frac{\sin(s-a)\sin(s-b)}{\sin a \sin b}} = \sin \tfrac{1}{2} C.$$

Substituting in the value of $\cos \tfrac{1}{2}(A+B)$, we have

$$\cos \tfrac{1}{2}(A+B) = \frac{2 \sin \tfrac{1}{2} c \cos(s - \tfrac{1}{2} c)}{2 \sin \tfrac{1}{2} c \cos \tfrac{1}{2} c} \sin \tfrac{1}{2} C$$

$$= \frac{\cos(s - \tfrac{1}{2} c)}{\cos \tfrac{1}{2} c} \sin \tfrac{1}{2} C.$$

$\therefore \cos \tfrac{1}{2}(A+B)\cos \tfrac{1}{2} c = \cos(s - \tfrac{1}{2} c)\sin \tfrac{1}{2} C.$

Since $\qquad s - \tfrac{1}{2} c = \tfrac{1}{2}(a+b),$

$$\cos \tfrac{1}{2}(A+B)\cos \tfrac{1}{2} c = \cos \tfrac{1}{2}(a+b)\sin \tfrac{1}{2} C.$$

By proceeding in like manner with the values of

$$\sin \tfrac{1}{2}(A+B), \quad \cos \tfrac{1}{2}(A-B), \quad \text{and } \sin \tfrac{1}{2}(A-B),$$

three analogous equations are obtained.

The four equations,

$$\left.\begin{array}{l}
\cos \tfrac{1}{2}(A+B)\cos \tfrac{1}{2} c = \cos \tfrac{1}{2}(a+b)\sin \tfrac{1}{2} C \\
\sin \tfrac{1}{2}(A+B)\cos \tfrac{1}{2} c = \cos \tfrac{1}{2}(a-b)\cos \tfrac{1}{2} C \\
\cos \tfrac{1}{2}(A-B)\sin \tfrac{1}{2} c = \sin \tfrac{1}{2}(a+b)\sin \tfrac{1}{2} C \\
\sin \tfrac{1}{2}(A-B)\sin \tfrac{1}{2} c = \sin \tfrac{1}{2}(a-b)\cos \tfrac{1}{2} C
\end{array}\right\}, \qquad [49]$$

are called Gauss's Equations.

By dividing the second of Gauss's Equations by the first, the fourth by the third, the third by the first, and the fourth by the second, we obtain

$$\left.\begin{array}{l}
\tan \tfrac{1}{2}(A+B) = \dfrac{\cos \tfrac{1}{2}(a-b)}{\cos \tfrac{1}{2}(a+b)} \cot \tfrac{1}{2} C \\[2ex]
\tan \tfrac{1}{2}(A-B) = \dfrac{\sin \tfrac{1}{2}(a-b)}{\sin \tfrac{1}{2}(a+b)} \cot \tfrac{1}{2} C \\[2ex]
\tan \tfrac{1}{2}(a+b) = \dfrac{\cos \tfrac{1}{2}(A-B)}{\cos \tfrac{1}{2}(A+B)} \tan \tfrac{1}{2} c \\[2ex]
\tan \tfrac{1}{2}(a-b) = \dfrac{\sin \tfrac{1}{2}(A-B)}{\sin \tfrac{1}{2}(A+B)} \tan \tfrac{1}{2} c
\end{array}\right\}. \qquad [50]$$

There will be other forms in each case, according as other elements of the triangle are used.

These equations are called Napier's Analogies.

In the first equation the factors $\cos \frac{1}{2}(a - b)$ and $\cot \frac{1}{2} C$ are always positive; therefore, $\tan \frac{1}{2}(A + B)$ and $\cos \frac{1}{2}(a + b)$ must always have like signs.

Hence, if $a + b < 180°$, $\cos \frac{1}{2}(a + b) > 0$, and $\tan \frac{1}{2}(A + B) > 0$. Hence, $A + B < 180°$.

If $a + b > 180°$, then $A + B > 180°$.

If $a + b = 180°$, $\cos \frac{1}{2}(a + b) = 0$, and $\tan \frac{1}{2}(A + B) = \infty$. Hence, $\frac{1}{2}(A + B) = 90°$, and $A + B = 180°$.

Conversely, it may be shown from the third equation, that $a + b$ is less than, greater than, or equal to 180°, according as $A + B$ is less than, greater than, or equal to 180°.

SECTION LVII

Case I

Given two sides a *and* b *and the included angle* C.

The angles A and B may be found by the first two of Napier's Analogies:

$$\tan \tfrac{1}{2}(A + B) = \frac{\cos \frac{1}{2}(a - b)}{\cos \frac{1}{2}(a + b)} \cot \tfrac{1}{2} C;$$

$$\tan \tfrac{1}{2}(A - B) = \frac{\sin \frac{1}{2}(a - b)}{\sin \frac{1}{2}(a + b)} \cot \tfrac{1}{2} C.$$

After A and B have been found, the side c may be found by [44], p. 158, or by [50], p. 164; but it is better to use for this purpose Gauss's Equations, because they involve the functions of the same angles that occur in working Napier's Analogies. Any one of the equations may be used; for example,

$$\cos \tfrac{1}{2} c = \frac{\cos \tfrac{1}{2}(a + b)}{\cos \tfrac{1}{2}(A + B)} \sin \tfrac{1}{2} C.$$

EXAMPLE. Given $a = 73° 58' 54''$, $b = 38° 45'$, $C = 46° 33' 41''$; solve the triangle.

$a = 73° 58' 54''$,   Hence, $\tfrac{1}{2}(a - b) = 17° 36' 57''$,

$b = 38° 45' \ 0''$, $\tfrac{1}{2}(a + b) = 56° 21' 57''$,

$C = 46° 33' 41''$. $\tfrac{1}{2} C = 23° 16' 50.5''$.

$\log \cos \tfrac{1}{2}(a - b) = 9.97914$	$\log \sin \tfrac{1}{2}(a - b) = 9.48092$
$\log \sec \tfrac{1}{2}(a + b) = 0.25658$	$\log \csc \tfrac{1}{2}(a + b) = 0.07956$
$\log \cot \tfrac{1}{2} C = 0.36626$	$\log \cot \tfrac{1}{2} C = 0.36626$
$\log \tan \tfrac{1}{2}(A + B) = \overline{0.60198}$	$\log \tan \tfrac{1}{2}(A - B) = \overline{9.92674}$
$\log \sec \tfrac{1}{2}(A + B) = 0.61515$	$\tfrac{1}{2}(A + B) = \ \ 75° 57' 40.8''$
$\log \cos \tfrac{1}{2}(a + b) = 9.74342$	$\tfrac{1}{2}(A - B) = \ \ 40° 11' 25.4''$
$\log \sin \tfrac{1}{2} C = 9.59685$	$A = 116° \ \ 9' \ \ 6''$
$\log \cos \tfrac{1}{2} c = \overline{9.95542}$	$B = \ \ 35° 46' 15''$
$\tfrac{1}{2} c = 25° 31' 12''$	$c = \ \ 51° \ \ 2' 24''$

To test the accuracy of the work we may use the Rule of Sine Proportion given in Sect. LIV, p. 158.

If c only is desired, it may be found from [45], p. 159, without previously computing A and B. But the Formulas [45] are not adapted to logarithmic work. Instead of changing them to forms suitable for logarithms, we may use the following method, which leads to the same results and has the advantage that, in applying it, nothing has to be remembered except Napier's Rules:

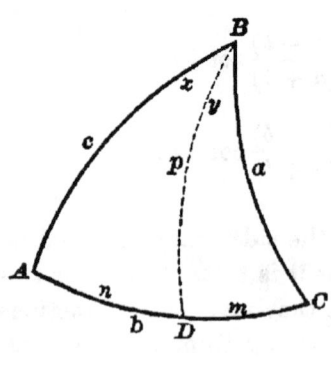

Through B (Fig. 100) draw an arc of a great circle perpendicular to AC, cutting AC at D. Let $BD = p$, $CD = m$, $AD = n$.

By Rule I,
$$\cos C = \tan m \cot a.$$

FIG. 100

Whence, $\tan m = \tan a \cos C$.

By Rule II,

$\cos a = \cos m \cos p$; whence, $\cos p = \cos a \sec m$.

$\cos c = \cos n \cos p$; whence, $\cos p = \cos c \sec n$.

Therefore, $\cos c \sec n = \cos a \sec m$.

Since $n = b - m$,

$$\cos c = \cos a \sec m \cos (b - m).$$

Now c may be computed from the two equations

$$\tan m = \tan a \cos C;$$

$$\cos c = \cos a \sec m \cos (b - m).$$

NOTE. If BD falls without the triangle, for instance to the right of BC, then $n = b + m$.

$$\therefore \cos c = \cos a \sec m \cos (b + m).$$

EXAMPLE. Given $a = 97° 30'$, $b = 55° 12'$, $C = 39° 58'$; find c.

$\log \tan a = 0.88057 \,(n)$	$\log \cos a = 9.11570 \,(n)$
$\log \cos C = 9.88447$	$\log \sec m = 0.77135 \,(n)$
$\log \tan m = 0.76504 \,(n)$	$\log \cos (b - m) = 9.85289$
$m = \quad 99° 44' 49''$	$\log \cos c = 9.73994$
$b - m = -44° 32' 49''$	$c = 56° 40' 9''$

EXERCISE XXXVII

1. What are the formulas for computing a when b, c, and A are given; and for computing b when a, c, and B are given?

2. Given $a = 88° 12' 20''$, $b = 124° 7' 17''$, $C = 50° 2' 1''$;
   find $A = 63° 15' 11''$, $B = 132° 17' 58''$, $c = 59° 4' 17''$.

3. Given $a = 120° 55' 35''$, $b = 88° 12' 20''$, $C = 47° 42' 1''$;
   find $A = 129° 58' 2''$, $B = 63° 15' 8''$, $c = 55° 52' 40''$.

4. Given $b = 63° 15' 12''$, $c = 47° 42' 1''$, $A = 59° 4' 25''$;
   find $B = 88° 12' 24''$, $C = 55° 52' 42''$, $a = 50° 1' 40''$.

5. Given $b = 69° 25' 11''$, $c = 109° 46' 19''$, $A = 54° 54' 42''$;
   find $B = 56° 11' 57''$, $C = 123° 21' 12''$, $a = 67° 11' 47''$.

SECTION LVIII

CASE II

Given two angles A *and* B, *and the included side* c.

The sides a and b may be found by the third and fourth of Napier's Analogies:

$$\tan \tfrac{1}{2}(a + b) = \frac{\cos \tfrac{1}{2}(A - B)}{\cos \tfrac{1}{2}(A + B)} \tan \tfrac{1}{2} c;$$

$$\tan \tfrac{1}{2}(a - b) = \frac{\sin \tfrac{1}{2}(A - B)}{\sin \tfrac{1}{2}(A + B)} \tan \tfrac{1}{2} c.$$

The angle C may be found by [44], p. 158, by Napier's second Analogy, [50], p. 164, or by one of Gauss's Equations, [49], p. 164. Thus, Gauss's second equation gives

$$\cos \tfrac{1}{2} C = \frac{\sin \tfrac{1}{2}(A + B)}{\cos \tfrac{1}{2}(a - b)} \cos \tfrac{1}{2} c.$$

EXAMPLE. Given $A = 107°47'7''$, $B = 38°58'27''$, $c = 51°41'14''$; solve the triangle.

$$A = 107° 47' \ 7'',$$
$$B = \ \ 38° 58' 27'',$$
$$c = \ \ 51° 41' 14''.$$

Hence, $\tfrac{1}{2}(A - B) = 34° 24' 20''$,
$\tfrac{1}{2}(A + B) = 73° 22' 47''$,
$\tfrac{1}{2} c = 25° 50' 37''$.

$\log \cos \tfrac{1}{2}(A - B) = 9.91648$	$\log \sin \tfrac{1}{2}(A - B) = 9.75208$
$\log \sec \tfrac{1}{2}(A + B) = 0.54359$	$\log \csc \tfrac{1}{2}(A + B) = 0.01854$
$\log \tan \tfrac{1}{2} c = \underline{9.68517}$	$\log \tan \tfrac{1}{2} c = 9.68517$
$\log \tan \tfrac{1}{2}(a + b) = \underline{0.14524}$	$\log \tan \tfrac{1}{2}(a - b) = \underline{9.45579}$
$\log \sin \tfrac{1}{2}(A + B) = 9.98146$	$\tfrac{1}{2}(a + b) = 54° 24' 24.4''$
$\log \sec \tfrac{1}{2}(a - b) = 0.01703$	$\tfrac{1}{2}(a - b) = 15° 56' 25.5''$
$\log \cos \tfrac{1}{2} c = \underline{9.95423}$	$\begin{cases} a = 70° 20' 50'' \\ b = 38° 27' 59'' \\ C = 52° 30' 20'' \end{cases}$
$\log \cos \tfrac{1}{2} C = 9.95272$	
$\tfrac{1}{2} C = 26° 15' 10''$	

If the angle C alone is wanted, we proceed as in Case I, p. 166, when the side c alone is desired.

Let (Fig. 101) $\angle ABD = x$, $\angle CBD = y$, $BD = p$; then,

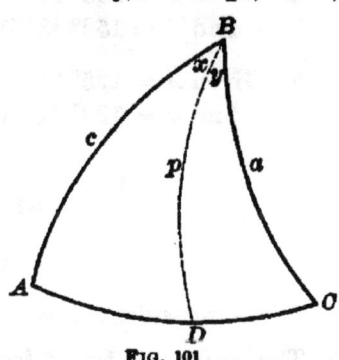

Rule I, $\cos c = \cot x \cot A$.

Whence, $\cot x = \tan A \cos c$.

Rule II, $\cos A = \cos p \sin x$.

Whence, $\cos p = \cos A \csc x$.

Rule II, $\cos C = \cos p \sin y$.

Whence, $\cos p = \cos C \csc y$.

$\therefore \cos C = \cos A \csc x \sin y$

$\qquad = \cos A \csc x \sin(B - x)$

Fig. 101

Now C may be computed from the equations

$$\cot x = \tan A \cos c;$$
$$\cos C = \cos A \csc x \sin(B - x).$$

NOTE.   When BD falls to the right of BC, the last equation becomes
$$\cos C = \cos A \csc x \sin(x - B).$$

EXAMPLE.   Given $A = 35° 46' 15''$, $B = 115° 9' 7''$, $c = 51° 2'$; find C.

$\log \tan A = 9.85760$	$\log \cos A = 9.90992$
$\log \cos c = 9.79856$	$\log \csc x = 0.04056$
$\log \cot x = 9.65616$	$\log \sin(B - x) = 9.88121$
$x = 65° 37' 35''$	$\log \cos C = 9.83069$
$\therefore B - x = 49° 31' 32''$	$C = 47° 22' 42''$

EXERCISE XXXVIII

1. What are the formulas for computing A when B, C, and a are given; and for computing B when A, C, and b are given?

2. Given $A = 26° 58' 46''$, $B = 39° 45' 10''$, $c = 154° 46' 48''$; find $a = 37° 14' 10''$, $b = 121° 28' 10''$, $C = 161° 22' 11''$.

3. Given $A = 128° 41' 49''$, $B = 107° 33' 20''$, $c = 124° 12' 31''$; find $a = 125° 41' 43''$, $b = 82° 47' 34''$, $C = 127° 22'$.

4. Given $B = 153° 17' 6''$, $C = 78° 43' 36''$, $a = 86° 15' 15''$; find $b = 152° 43' 51''$, $c = 88° 12' 21''$, $A = 78° 15' 48''$.

5. Given $A = 125° 41' 44''$, $C = 82° 47' 35''$, $b = 52° 37' 57''$; find $a = 128° 41' 46''$, $c = 107° 33' 20''$, $B = 55° 47' 40''$.

SECTION LIX

CASE III

Given two sides a *and* b, *and the angle* A *opposite* a.

The angle B is found from [44], p. 158; whence we have

$$\sin B = \sin A \sin b \csc a.$$

When B has been found, C and c may be found from the fourth and the second of Napier's Analogies:

$$\tan \tfrac{1}{2} c = \frac{\sin \tfrac{1}{2}(A + B)}{\sin \tfrac{1}{2}(A - B)} \tan \tfrac{1}{2}(a - b);$$

$$\cot \tfrac{1}{2} C = \frac{\sin \tfrac{1}{2}(a + b)}{\sin \tfrac{1}{2}(a - b)} \tan \tfrac{1}{2}(A - B).$$

The third and first of Napier's Analogies may also be used.

NOTE 1. Since B is determined from its sine, the problem in general has two solutions; and, moreover, in case $\sin B > 1$, the problem is impossible. By geometric construction it may be shown, as in the corresponding case in Plane Trigonometry, pp. 71-73, under what conditions the problem really has two solutions, one solution, or no solution. But in practical applications a general knowledge of the shape of the triangle is known beforehand; so that it is easy to see, without special investigation, which solution (if any) corresponds to the circumstances of the question.

It can be shown that there are *two solutions* when A and a are alike in kind and $\sin b > \sin a > \sin A \sin b$; *no solution* when A and a are unlike in kind (including the case in which either A or a is 90°) and $\sin b$

is greater than sin a or equal to sin a, or when sin a is less than sin A sin b; and *one solution* in every other case.

NOTE 2. The side c or the angle C may be computed, without first finding B, by means of the formulas

$$\tan m = \cos A \tan b, \text{ and } \cos (c - m) = \cos a \sec b \cos m;$$
$$\cot z = \tan A \cos b, \text{ and } \cos (C - z) = \cot a \tan b \cos z.$$

These formulas may be obtained by resolving the triangle into right triangles, and then applying Napier's Rules; m is equal to that part of the side c included between the vertex A and the foot of the perpendicular from C, and z is equal to the corresponding portion of the angle C.

NOTE 3. After the two values of B have been obtained, the number of solutions may be determined by Theorem 1, Sect. XLIX, p. 141.

If log sin B is positive, there is no solution.

EXAMPLE. Given $a = 57° 36'$, $b = 31° 14'$, $A = 104° 25' 30''$.

In this case, $A > 90°$,	log sin $A = 9.98609$
and $a + b < 180°$.	log sin $b = 9.71477$
Therefore, $A + B < 180°$,	log csc $a = \underline{0.07349}$
and $B < 90°$.	log sin $B = 9.77435$
Hence, there is only one solution.	$B = 36° 29' 46''$

$a + b = 88° 50'$	$\frac{1}{2}(a + b) = 44° 25'$
$a - b = 26° 22'$	$\frac{1}{2}(a - b) = 13° 11'$
$A + B = 140° 55' 16''$	$\frac{1}{2}(A + B) = 70° 27' 38''$
$A - B = 67° 55' 44''$	$\frac{1}{2}(A - B) = 33° 57' 52''$
log sin $\frac{1}{2}(A + B) = 9.97424$	log sin $\frac{1}{2}(a + b) = 9.84502$
log csc $\frac{1}{2}(A - B) = 0.25284$	log csc $\frac{1}{2}(a - b) = 0.64194$
log tan $\frac{1}{2}(a - b) = \underline{9.36966}$	log tan $\frac{1}{2}(A - B) = 9.82840$
log tan $\frac{1}{2}c = 9.59674$	log cot $\frac{1}{2}C = 0.31536$
$\frac{1}{2}c = 21° 33' 37''$	$\frac{1}{2}C = 25° 48' 58''$
$c = 43° 7' 14''$	$C = 51° 37' 56''$

EXERCISE XXXIX

1. Given $a = 73° 49' 38''$, $b = 120° 53' 35''$, $A = 88° 52' 42''$;
   find $B = 116° 42' 30''$, $c = 120° 57' 27''$, $C = 116° 47'$.

2. Given $a = 150° 57' 5''$, $b = 134° 15' 54''$, $A = 144° 22' 42''$;
   find $B_1 = 120° 47' 45''$, $c_1 = 55° 42' 8''$, $C_1 = 97° 42' 55''$;
   $B_2 = 59° 12' 15''$, $c_2 = 23° 57' 17''$, $C_2 = 29° 8' 39''$.

3. Given $a = 79° \ 0' \ 54''$, $b = 82° \ 17' \ 4''$, $A = 82° \ 9' \ 26''$;
find $B = 90°$, $c = 45° \ 12' \ 19''$, $C = 45° \ 44' \ 5''$.

4. Given $a = 30° \ 52' \ 37''$, $b = 31° \ 9' \ 16''$, $A = 87° \ 34' \ 12''$;
show that the triangle is impossible.

$\bullet$

SECTION LX

Case IV

Given two angles A *and* B, *and the side* a *opposite* A.

The side b is found from [44], p. 158; whence,

$$\sin b = \sin a \sin B \csc A.$$

The values of c and C may then be found by means of the fourth and second of Napier's Analogies:

$$\tan \tfrac{1}{2} c = \frac{\sin \tfrac{1}{2}(A + B)}{\sin \tfrac{1}{2}(A - B)} \tan \tfrac{1}{2}(a - b);$$

$$\cot \tfrac{1}{2} C = \frac{\sin \tfrac{1}{2}(a + b)}{\sin \tfrac{1}{2}(a - b)} \tan \tfrac{1}{2}(A - B).$$

Note 1. In this case the conditions for one solution, two solutions, or no solution can be deduced directly by the theory of polar triangles from the corresponding conditions of Case III, p. 170.   There are *two solutions* when A and a are alike in kind and $\sin B > \sin A > \sin a \sin B$; *no solution* when A and a are unlike in kind (including the case in which either A or a is 90°) and $\sin B$ is greater than $\sin A$ or equal to $\sin A$, or when $\sin A < \sin a \sin B$; and *one solution* in every other case.

Note 2.   By proceeding as indicated in Case III, Note 2, p. 171, formulas for computing c or C , independent of the side b , may be found; viz.,

$$\tan m = \tan a \cos B, \text{ and } \sin (c - m) = \cot A \tan B \sin m;$$

$$\cot x = \cos a \tan B, \text{ and } \sin (C - x) = \cos A \sec B \sin x.$$

In these formulas $m = BD$, $x = \angle BCD$, D being the foot of the perpendicular from the vertex C .

Note 3.   As in Case III, p. 171, only those values of b can be retained which are greater than or less than a , according as B is greater than or less than A .   If log sin b is positive, the triangle is impossible.

1. Given $A = 110°\ 10'$, $B = 133°\ 18'$, $a = 147°\ 5'\ 32''$;
   find $b = 155°\ 5'\ 18''$, $c = 33°\ 1'\ 37''$, $C = 70°\ 20'\ 40''$.

2. Given $A = 113°\ 39'\ 21''$, $B = 123°40'18''$, $a = 65°\ 39'\ 46''$;
   find $b = 124°\ 7'\ 20''$, $c = 159°50'15''$, $C = 159°\ 43'\ 34''$.

3. Given $A = 100°\ 2'11''$, $B = 98°\ 30'\ 28''$, $a = 95°\ 20'\ 39''$;
   find $b = 90°$, $c = 147°41'50''$, $C = 148°\ 5'\ 40''$.

4. Given $A = 24°\ 33'\ 9''$, $B = 38°\ 0'\ 12''$, $a = 65°\ 20'\ 13''$;
show that the triangle is impossible.

SECTION LXI

CASE V

Given the three sides a, b, *and* c.

The angles are computed by means of Formulas [47], p. 161, and the corresponding formulas for the angles B and C.

The formulas for the tangent are, in general, to be preferred.

If we multiply the equation

$$\tan \tfrac{1}{2} A = \sqrt{\csc s\ \csc (s - a) \sin (s - b) \sin (s - c)}$$

by the equation $1 = \dfrac{\sin (s - a)}{\sin (s - a)},$

and put $\tan r$ for $\sqrt{\csc s \sin (s - a) \sin (s - b) \sin (s - c)}$, at the same time making analogous changes in the equations for $\tan \tfrac{1}{2} B$ and $\tan \tfrac{1}{2} C$, we obtain

$$\tan \tfrac{1}{2} A = \tan r \csc (s - a),$$
$$\tan \tfrac{1}{2} B = \tan r \csc (s - b),$$
$$\tan \tfrac{1}{2} C = \tan r \csc (s - c),$$

which are the most convenient formulas to employ when all three angles have to be computed.

EXAMPLE 1.   Given $a = 50°\ 54'\ 32''$, $b = 37°\ 47'\ 18''$, $c = 74°\ 51'\ 50''$; find A.

$a =\ 50°\ 54'\ 32''$	$\log \csc s =\ 0.00448$
$b =\ 37°\ 47'\ 18''$	$\log \csc (s - a) =\ 0.28979$
$c =\ \underline{74°\ 51'\ 50''}$	$\log \sin (s - b) =\ 9.84171$
$2s = 163°\ 33'\ 40''$	$\log \sin (s - c) =\ \underline{9.08072}$
$s =\ 81°\ 46'\ 50''$	$2\ )\overline{19.21670}$
$s - a =\ 30°\ 52'\ 18''$	$\log \tan \tfrac{1}{2}A =\ 9.60835$
$s - b =\ 43°\ 59'\ 32''$	$\tfrac{1}{2}A = 22°\ 5'\ 20''$
$s - c =\ 6°\ 55'$	$A = 44°\ 10'\ 40''$

EXAMPLE 2.   Given $a = 124°\ 12'\ 31''$, $b = 54°\ 18'\ 16''$, $c = 97°\ 12'\ 25''$; find $A = 127°\ 22'\ 7''$, $B = 51°\ 18'\ 11''$, $C = 72°\ 26'\ 40''$.

$a = 124°\ 12'\ 31''$	$s - a = 13°\ 39'\ 5''$
$b =\ 54°\ 18'\ 16''$	$s - b = 83°\ 33'\ 20''$
$c =\ \underline{97°\ 12'\ 25''}$	$s - c = 40°\ 39'\ 11''$
$2s = 275°\ 43'\ 12''$	$\log \tan \tfrac{1}{2}A = 0.30577$
$s = 137°\ 51'\ 36''$	$\log \tan \tfrac{1}{2}B = 9.68145$
$\log \sin (s - a) = 9.37293$	$\log \tan \tfrac{1}{2}C = \underline{9.86480}$
$\log \sin (s - b) = 9.99725$	$\tfrac{1}{2}A =\ 63°\ 41'\ 3.8''$
$\log \sin (s - c) = 9.81390$	$\tfrac{1}{2}B = 25°\ 39'\ 5.6''$
$\log \csc s =\ \underline{0.17331}$	$\tfrac{1}{2}C = 36°\ 13'\ 20''$
$\log \tan^2 r = 9.35739$	$A = 127°\ 22'\ 7''$
$\log \tan r =\ 9.67870$	$B = 51°\ 18'\ 11''$
	$C = 72°\ 26'\ 40''$

EXERCISE XLI

1. Given $a = 120°\ 55'\ 35''$, $b = 59°\ 4'\ 25''$, $c = 106°\ 10'\ 22''$;
    find $A = 116°\ 44'\ 50''$, $B = 63°\ 15'\ 10''$, $C = 91°\ 7'\ 22''$.

2. Given $a = 50°\ 12'\ 4''$, $b = 116°\ 44'\ 48''$, $c = 129°\ 11'\ 42''$;
    find $A = 59°\ 4'\ 28''$, $B = 94°\ 23'\ 12''$, $C = 120°\ 4'\ 52''$.

3. Given $a = 131°\ 35'\ 4''$, $b = 108°\ 30'\ 14''$, $c = 84°\ 46'\ 34''$;
    find $A = 132°\ 14'\ 21''$, $B = 110°\ 10'\ 40''$, $C = 99°\ 42'\ 24''$.

4. Given $a = 20°\ 16'\ 38''$, $b = 56°\ 19'\ 40''$, $c = 66°\ 20'\ 44''$;
    find $A = 20°\ 9'\ 55''$, $B = 55°\ 52'\ 35''$, $C = 114°\ 20'\ 21''$.

SECTION LXII

CASE VI

Given the three angles A, B, *and* C.

The sides are computed by means of Formulas [48], p. 162. The formulas for the tangents are, in general, to be preferred. If we multiply the equation

$$\tan \tfrac{1}{2} a = \sqrt{- \cos S \cos (S - A) \sec (S - B) \sec (S - C)}$$

by the equation $1 = \dfrac{\sec (S - A)}{\sec (S - A)}$,

and put $\tan R$ for $\sqrt{- \cos S \sec (S - A) \sec (S - B) \sec (S - C)}$,

at the same time making analogous changes in the equations for $\tan \tfrac{1}{2} b$ and $\tan \tfrac{1}{2} c$, we obtain

$$\tan \tfrac{1}{2} a = \tan R \cos (S - A),$$
$$\tan \tfrac{1}{2} b = \tan R \cos (S - B),$$
$$\tan \tfrac{1}{2} c = \tan R \cos (S - C),$$

which are the most convenient formulas to use in case all three sides have to be computed.

EXAMPLE 1.   Given $A = 220°$, $B = 130°$, $C = 150°$; find a.

After we find the values of S, $S - A$, $S - B$, $S - C$, we write the formula for $\tan \tfrac{1}{2} a$ with the algebraic sign written above each function as follows:

$$\tan \tfrac{1}{2} a = \sqrt{\overset{-}{- \cos} S \overset{+}{\cos} (S - A) \overset{-}{\sec} (S - B) \overset{-}{\sec} (S - C)}.$$

$A = 220°$	$\log \cos S = 9.53405 \ (n)$
$B = 130°$	$\log \cos (S - A) = 9.93753$
$C = 150°$	$\log \sec (S - B) = 0.30103 \ (n)$
$2 S = 500°$	$\log \sec (S - C) = 0.76033 \ (n)$
$S = 250°$	$2 \,)\, 0.53294$
$S - A = 30°$	$\log \tan \tfrac{1}{2} a = 0.26647$
$S - B = 120°$	$\tfrac{1}{2} a = \ 61° \ 34' \ 6''$
$S - C = 100°$	$a = 123° \ 8' \ 12''$

NOTE. Here the effect, as regards algebraic sign, of three negative factors is canceled by the negative sign before the whole product.

EXAMPLE 2.   Given $A = 20° 9' 56''$, $B = 55° 52' 32''$, $C = 114°20'14''$; find $a = 20°16'38''$, $b = 56°19'41''$, $c = 66°20'43''$.

After we find the values of S, $S - A$, $S - B$, $S - C$, we write the formula for tan R with the algebraic sign written above each function as follows:

$$\tan R = \sqrt{\overset{-}{-\cos S}\ \overset{+}{\sec(S - A)}\ \overset{+}{\sec(S - B)}\ \overset{+}{\sec(S - C)}}.$$

$A = 20° 9' 56''$	$S = 95° 11' 21''$
$B = 55° 52' 32''$	$S - A = 75° 1' 25''$
$C = 114° 20' 14''$	$S - B = 39° 18' 49''$
$2 S = 190° 22' 42''$	$S - C = -19° 8' 53''$
$\log \cos S = 8.95638\ (n)$	$\log \tan \frac{1}{2} a = 9.25242$
$\log \sec(S - A) = 0.58768$	$\log \tan \frac{1}{2} b = 9.72867$
$\log \sec(S - B) = 0.11143$	$\log \tan \frac{1}{2} c = 9.81538$
$\log \sec(S - C) = 0.02472$	$\frac{1}{2} a = 10° 8' 18.9''$
$\log \tan^2 R = 9.68021$	$\frac{1}{2} b = 28° 9' 50.3''$
$\log \tan R = 9.84010$	$\frac{1}{2} c = 33° 10' 21.4''$
	$a = 20° 16' 38''$
	$b = 56° 19' 41''$
	$c = 66° 20' 43''$

EXERCISE XLII

1. Given $A = 130°$, $B = 110°$, $C = 80°$;
   find $a = 139°21'22''$, $b = 126° 57' 52''$, $c = 56° 51' 48''$.

2. Given $A = 59° 55' 10''$, $B = 85° 36' 50''$, $C = 59° 55' 10''$;
   find $a = 51° 17' 31''$, $b = 64° 2' 47''$, $c = 51° 17' 31''$.

3. Given $A = 102° 14' 12''$, $B = 54° 32' 24''$, $C = 89° 5' 46''$;
   find $a = 104° 25' 9''$, $b = 53° 49' 25''$, $c = 97°44'19''$.

4. Given $A = 4° 23' 35''$, $B = 8° 28' 20''$, $C = 172° 17' 56''$;
   find $a = 31° 9' 13''$, $b = 84° 18' 28''$, $c = 115° 10'$.

SECTION LXIII

AREA OF A SPHERICAL TRIANGLE

I. *When the three angles* A, B, C *are given.*

Let $R =$ the radius of the sphere,

$E =$ the spherical excess $= A + B + C - 180°$,

$F =$ the area of the triangle.

Three planes passed through the centre of a sphere, each perpendicular to the other two planes, divide the surface of the sphere into eight tri-rectangular triangles (*Geometry*, § 802).

It is convenient to divide each of these eight triangles into 90 equal parts, and to call each of these equal parts a spherical degree. The surface of every sphere, therefore, contains 720 spherical degrees.

Now in spherical degrees, the $\triangle ABC = E$ (*Geometry*, § 834), and the surface of the sphere is equal to 720 spherical degrees.

Hence, $\triangle ABC$: surface of the sphere $= E : 720$.

Since the surface of a sphere $= 4 \pi R^2$ (*Geometry*, § 824),

$$\triangle ABC : 4 \pi R^2 = E : 720.$$

Whence, $$\mathbf{F} = \frac{\pi \mathbf{R^2 E}}{180}.$$ [51]

II. *When the three sides* a, b, c *are given.*

A formula for computing the area is deduced as follows:
From the first of Formulas [49], p. 164,

$$\frac{\cos \tfrac{1}{2}(A + B)}{\sin \tfrac{1}{2}C} = \frac{\cos \tfrac{1}{2}(a + b)}{\cos \tfrac{1}{2}c}.$$

Now, $$\sin \tfrac{1}{2} C = \cos (90° - \tfrac{1}{2} C).$$

Therefore, $$\frac{\cos \tfrac{1}{2}(A + B)}{\cos (90° - \tfrac{1}{2} C)} = \frac{\cos \tfrac{1}{2}(a + b)}{\cos \tfrac{1}{2} c}.$$

Then, by division and composition,

$$\frac{\cos\frac{1}{2}(A+B)-\cos(90°-\frac{1}{2}C)}{\cos\frac{1}{2}(A+B)+\cos(90°-\frac{1}{2}C)} = \frac{\cos\frac{1}{2}(a+b)-\cos\frac{1}{2}c}{\cos\frac{1}{2}(a+b)+\cos\frac{1}{2}c}. \qquad \text{(a)}$$

Dividing [23], p. 59, by [22], p. 59,

$$\frac{\cos A-\cos B}{\cos A+\cos B} = -\tan\tfrac{1}{2}(A+B)\tan\tfrac{1}{2}(A-B). \qquad \text{(b)}$$

Substituting in (b) for A and B, $\frac{1}{2}(A+B)$ and $90°-\frac{1}{2}C$, respectively, we have

$$\frac{\cos\frac{1}{2}(A+B)-\cos(90°-\frac{1}{2}C)}{\cos\frac{1}{2}(A+B)+\cos(90°-\frac{1}{2}C)}$$

$$= -\tan\tfrac{1}{2}(\tfrac{1}{2}A+\tfrac{1}{2}B+90°-\tfrac{1}{2}C)\tan\tfrac{1}{2}(\tfrac{1}{2}A+\tfrac{1}{2}B-90°+\tfrac{1}{2}C)$$

$$= -\tan\tfrac{1}{4}(A+B-C+180°)\tan\tfrac{1}{4}(A+B+C-180°).$$

Now, $\qquad\qquad E = A+B+C-180°.$

$$\therefore \tan\tfrac{1}{4}(A+B-C+180°) = \tan\tfrac{1}{4}(360°-2C+A+B+C-180°)$$

$$= \tan\tfrac{1}{4}(360°-2C+E)$$

$$= \tan[90°-\tfrac{1}{4}(2C-E)]$$

$$= \cot\tfrac{1}{4}(2C-E).$$

Now, substituting E for $A+B+C-180°$ and $\cot\frac{1}{4}(2C-E)$ for $\tan\frac{1}{4}(A+B-C+180°)$, we have

$$\frac{\cos\frac{1}{2}(A+B)-\cos(90°-\frac{1}{2}C)}{\cos\frac{1}{2}(A+B)+\cos(90°-\frac{1}{2}C)} = -\cot\tfrac{1}{4}(2C-E)\tan\tfrac{1}{4}E. \qquad \text{(c)}$$

Again substituting, in equation (b), for A and B the values $\frac{1}{2}(a+b)$ and $\frac{1}{2}c$, and also substitute s for $\frac{1}{2}(a+b+c)$ and $s-c$ for $\frac{1}{2}(a+b-c)$, we have

$$\frac{\cos\frac{1}{2}(a+b)-\cos\frac{1}{2}c}{\cos\frac{1}{2}(a+b)+\cos\frac{1}{2}c} = -\tan\tfrac{1}{2}s\tan\tfrac{1}{2}(s-c). \qquad \text{(d)}$$

Comparing (a), (c), and (d), we obtain

$$\cot\tfrac{1}{4}(2C-E)\tan\tfrac{1}{4}E = \tan\tfrac{1}{2}s\tan\tfrac{1}{2}(s-c). \qquad \text{(e)}$$

By beginning with the second equation of [49], p. 164, and treating it in the same way, we obtain as the result

$$\tan \tfrac{1}{4}(2C - E)\tan \tfrac{1}{4}E = \tan \tfrac{1}{2}(s - a)\tan \tfrac{1}{2}(s - b). \quad \text{(f)}$$

By taking the product of (e) and (f) we obtain the elegant formula, known as l'Huilier's Formula,

$$\tan^2 \tfrac{1}{4}E = \tan \tfrac{1}{2}s \tan \tfrac{1}{2}(s - a)\tan \tfrac{1}{2}(s - b)\tan \tfrac{1}{2}(s - c). \quad [52]$$

By means of [52] E may be computed from the three sides, and then the area of the triangle may be found by [51], p. 177.

III. *In all other cases*, the area may be found by first solving the triangle so far as to obtain the angles or the sides, whichever may be more convenient, and then applying [51] or [52].

EXAMPLE 1.   Given $A = 102° 14' 12''$, $B = 54° 32' 24''$, $C = 89° 5' 46''$; find the area of the triangle.

$A = 102° 14' 12''$	$\log R^2 = \log R^2$
$B = 54° 32' 24''$	$\log E = 5.37501$
$C = 89° 5' 46''$	
$\overline{245° 52' 22''}$	$*\log \dfrac{\pi}{648000} = 4.68557 - 10$
$E = 65° 52' 22''$	$\log F = \overline{0.06058 + \log R^2}$
$= 237142''$	$F = 1.1497\, R^2$
$180° = 648000''$	

Hence, if we know the radius of the sphere, we can express the area of a spherical triangle in the ordinary units of area.

EXAMPLE 2.   Given $a = 133° 26' 19''$, $b = 64° 50' 53''$, $c = 144° 13' 45''$; find $E = 200° 46' 46''$.

$a = 133° 26' 19''$	$\log \tan \tfrac{1}{2}s = 1.11669$
$b = 64° 50' 53''$	$\log \tan \tfrac{1}{2}(s - a) = 9.53474$
$c = 144° 13' 45''$	$\log \tan \tfrac{1}{2}(s - b) = 0.12612$
$2s = \overline{342° 30' 57''}$	$\log \tan \tfrac{1}{2}(s - c) = 9.38083$
$s = 171° 15' 28.5''$	$\log \tan^2 \tfrac{1}{4}E = \overline{0.15838}$
$s - a = 37° 49' 9.5''$	$\log \tan \tfrac{1}{4}E = 0.07919$
$s - b = 106° 24' 35.5''$	$\tfrac{1}{4}E = 50° 11' 41.5''$
$s - c = 27° 1' 43.5''$	$E = 200° 46' 46''$

* See Wentworth & Hill's Logarithmic and Trigonometric Tables, p. 20.

1. Given $A = 84° 20' 19''$, $B = 27° 22' 40''$, $C = 75° 33'$; find $E = 26159''$, $F = 0.12682 R^2$.

2. Given $a = 69° 15' 6''$, $b = 120° 42' 47''$, $c = 159° 18' 33''$; find $E = 216° 40' 18''$.

3. Given $a = 33° 1' 45''$, $b = 155° 5' 18''$, $C = 110° 10'$; find $E = 133° 48' 53''$.

4. Given $c = 114° 27' 57''$, $A = 78° 42' 33''$, $B = 127° 13' 7''$; find the area.

5. Given $a = 76° 14' 47''$, $b = 82° 40' 15''$, $A = 60° 22' 44''$; find the area.

6. Given $A = 80° 12' 35''$, $B = 77° 38' 22''$, $a = 76° 42' 28''$; find the area.

7. Given $b = 44° 27' 40''$, $c = 15° 22' 44''$, $A = 167° 42' 27''$; find the area.

8. Given $b = 67° 15' 42''$, $A = 84° 55' 8''$, $C = 96° 18' 49''$; find the area.

9. Given $b = 72° 19' 38''$, $c = 54° 58' 52''$, $B = 77° 15' 14''$; find the area.

10. Given $B = 127° 16' 4''$, $C = 42° 34' 19''$, $b = 54° 47' 55''$; find the area.

11. Given $a = 128° 42' 56''$, $b = 107° 13' 48''$, $c = 88° 37' 51''$; find the area.

12. Given $A = 127° 22' 28''$, $B = 131° 45' 27''$, $C = 100° 52' 16''$; find the area.

13. Given $a = 116° 19' 45''$, $A = 160° 42' 24''$, $C = 171° 27' 15''$; find the area.

14. Find the area of a triangle on the earth's surface (regarded as spherical) if each side of the triangle is equal to 1°. (Radius of earth $= 3958$ miles.)

CHAPTER IX

APPLICATIONS OF SPHERICAL TRIGONOMETRY

SECTION LXIV

PROBLEM

To reduce to the horizon an angle measured in space.

Let O (Fig. 102) be the position of the eye; $DOC = h$, the angle measured in space; OD' and OC' the projections of the sides of the angle upon the horizontal plane; $DOD' = m$ and

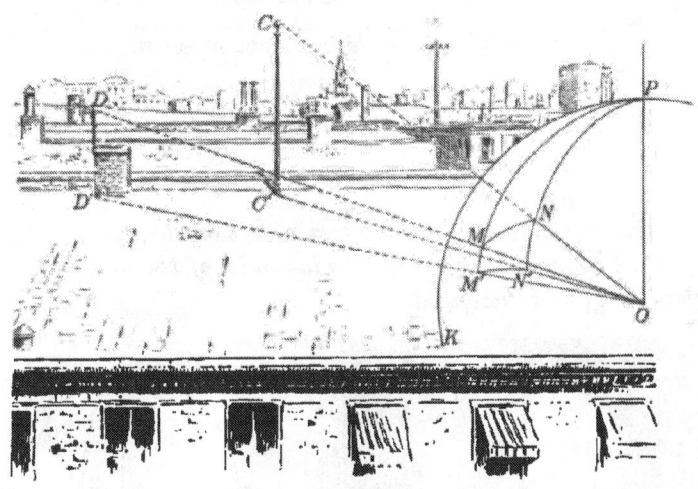

Fig. 102

$COC' = n$, the angles of inclination of OD and OC, respectively, to the horizon. Required the angle $D'OC' = x$ made by the projections on the horizon.

181

The planes of the angles DOD' and COC' intersect in the line OP, perpendicular to the horizontal plane (Wentworth's *Geometry*, § 556).

From O as a centre describe a sphere, and let its surface cut the edges of the trihedral angle $O-DCP$ in M, N, and P.

In the spherical triangle MNP, $MN = h$, $MP = 90° - m$, $NP = 90° - n$ are known; and $P = x$ is required.

By [47], p. 161,

$$\cos \tfrac{1}{2} x = \sqrt{\frac{\sin (90° + \tfrac{1}{2} h - \tfrac{1}{2} m - \tfrac{1}{2} n) \sin (90° - \tfrac{1}{2} h - \tfrac{1}{2} m - \tfrac{1}{2} n)}{\sin (90° - m) \sin (90° - n)}}.$$

Putting $\tfrac{1}{2}(h + m + n) = s$, we obtain

$$\cos \tfrac{1}{2} x = \sqrt{\frac{\sin [90° + (s - h)] \sin (90° - s)}{\cos m \cos n}}$$

$$= \sqrt{\cos (s - h) \cos s \sec m \sec n}.$$

SECTION LXV

PROBLEM

To find the distance between two places on the earth's surface (regarded as spherical), given the latitudes of the places and the difference of their longitudes.

FIG. 103

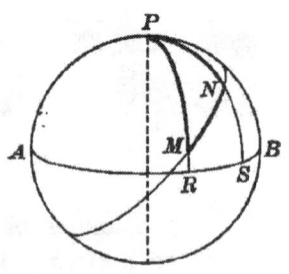

FIG. 104

Let M and N (Fig. 104) be the places. Then the distance MN is an arc of the great circle passing through the places. Let P be the pole, ARB the equator. The arcs MR and NS are the latitudes of the places, and the arc RS, or the angle MPN, is the difference of their longitudes. Let $MR = b$, $NS = a$, $RS = l$; then in the spherical triangle MNP two sides, $MP = 90° - b$, $NP = 90° - a$, and the included angle $MPN = l$ are given, and we have from Sect. LVII, p. 167,

$$\tan m = \cot a \cos l,$$
$$\cos MN = \sin a \sec m \sin (b + m).$$

From these equations first find m, then the arc MN, and then reduce MN to geographical miles, of which there are 60 in each degree.

SECTION LXVI

THE CELESTIAL SPHERE

The **Celestial Sphere** is an imaginary sphere of indefinite radius, upon the concave surface of which all the heavenly bodies appear to be situated.

The **Celestial Equator**, or **Equinoctial**, $AVBU$ (Fig. 105), is the great circle in which the plane of the earth's equator intersects the surface of the celestial sphere.

The **Poles**, P and P' (Fig. 105), of the celestial

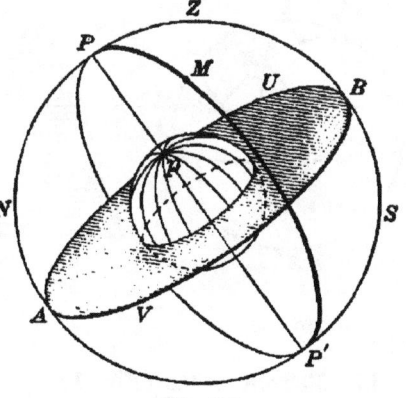

Fig. 105

equator are the points in which the earth's axis produced cuts the surface of the celestial sphere.

The **Celestial Meridian,** PBP' (Fig. 106), of an observer is the great circle in which the plane of his terrestrial meridian produced meets the surface of the celestial sphere.

Hour Circles, as PMP' (Fig. 106), or **Circles of Declination,** are great circles passing through the poles, perpendicular to the celestial equator.

The **Horizon,** $NWSE$ (Fig. 106), of an observer is the great circle in which the plane tangent to the earth's surface, at the place where he is, meets the surface of the celestial sphere.

The **Zenith,** Z (Fig. 106), of an observer is that pole of his horizon which is exactly above his head.

Vertical Circles, as $NPZS$ (Fig. 106), are great circles passing through the zenith of an observer, perpendicular to his horizon.

The vertical circle passing through the east and west points of the horizon is called the **Prime Vertical;** that passing through the north and south points coincides with the celestial meridian.

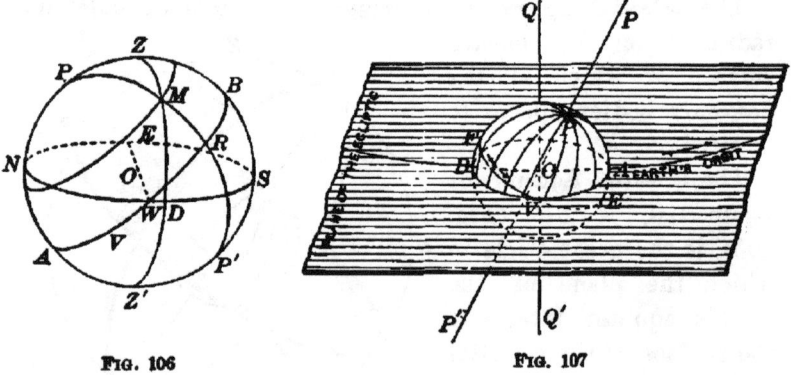

FIG. 106   FIG. 107

The **Ecliptic,** AVB (Fig. 107), is a great circle of the celestial sphere, apparently traversed by the sun in one year from west to east, in consequence of the motion of the earth around the sun.

The **Equinoxes** are the points where the ecliptic cuts the celestial equator. They are distinguished as the *Vernal* equinox and the *Autumnal* equinox; the sun in his annual journey passes through the former on March 21, and through the latter on September 21. The Vernal equinox is shown here at *V* (Fig. 108).

Circles of Latitude, as *QMT* (Fig. 108), are great circles passing through the poles of the ecliptic, perpendicular to the plane of the ecliptic.

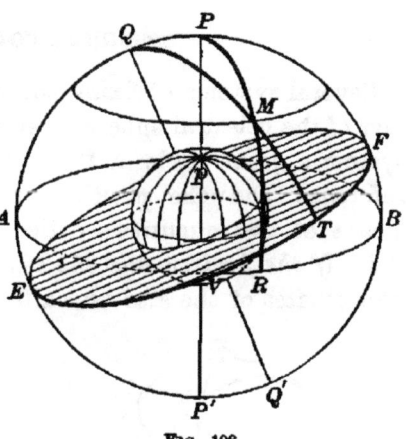

Fig. 108

The angle which the ecliptic makes with the celestial equator is called the **obliquity** of the ecliptic; it is equal to 23° 27', nearly, and is often denoted by the letter *e*.

The earth's diurnal motion causes all the heavenly bodies to appear to rotate from east to west at the uniform rate of 15° per hour. If in Fig. 106 we conceive the observer placed at the centre *O*, and his zenith, horizon, and celestial meridian fixed in position, and all the heavenly bodies rotating from east to west around *PP'* as an axis at the rate of 15° per hour, we form a correct idea of the apparent diurnal motions of these bodies. When the sun or a star in its diurnal motion crosses the meridian, it is said to make a *transit* across the meridian; when it passes across the part *NWS* of the horizon, it is said to *set;* and when it passes across the part *NES*, it is said to *rise* (the effect of refraction being here neglected). Each star, as *M*, describes daily a small circle of the sphere parallel to the celestial equator, and called the **diurnal circle** of the star. The nearer the star is to the pole, the smaller is the diurnal circle; and if there were stars at the poles P and P', they would have no diurnal motion. To an observer north of the equator the north pole P is *elevated* above the horizon (Fig. 108); to an observer south of the equator the south pole P' is the elevated pole.

SECTION LXVII

SPHERICAL CO-ORDINATES

Several systems of fixing the position of a star on the surface of the celestial sphere at any instant are in use. In each system a great circle and its pole are taken as standards of reference, and the position of the star is determined by means of two quantities called its *spherical co-ordinates*.

I. *If the horizon and the zenith* are chosen (Fig. 109), the co-ordinates of the star are called its altitude and its azimuth.

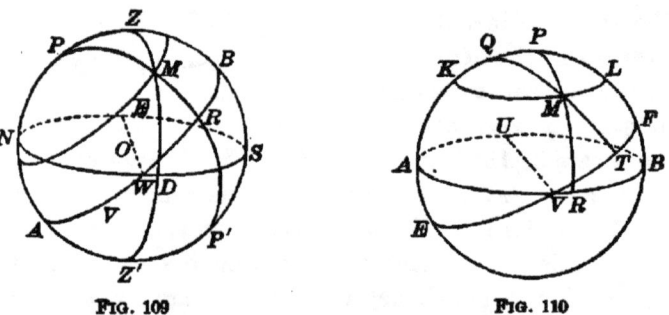

FIG. 109   FIG. 110

The **Altitude** of a star is its angular distance, DM, measured on a vertical circle, above the horizon. The complement, MZ, of the altitude is called the **Zenith Distance**.

The **Azimuth** of a star is the angle PZM at the zenith formed by the meridian of the observer and the vertical circle passing through the star, and is measured, therefore, by an arc of the horizon. It is usually reckoned from the north point of the horizon in north latitudes, and from the south point in south latitudes; and east or west according as the star is east or west of the meridian.

. II. *If the celestial equator and its pole* are chosen (Fig. 110), then the position of the star may be fixed by means of its declination and its hour angle.

The **Declination** of a star is its angular distance, *RM*, from the celestial equator, measured on an hour circle. The angular distance, *PM*, of the star, measured on the hour circle, *from the elevated pole*, is called its **Polar Distance**.

The declination of a star, like the latitude of a place on the earth's surface, may be either north or south; but, in practical problems, while latitude is always to be considered positive, declination, if of a different name from the latitude, must be regarded as *negative*.

If the declination is negative, the polar distance is equal numerically to 90° + the declination.

The **Hour Angle** of a star is the angle *MPQ* at the pole formed by the meridian, *APB*, of the observer and the hour circle, *PMR*, passing through the star. On account of the diurnal rotation the hour angle is constantly changing at the rate of 15° per hour. Hour angles are reckoned from the celestial meridian, positive towards the west, and negative towards the east.

III. *The celestial equator and its pole* being still retained, we may employ as the co-ordinates of the star its declination and its right ascension.

The **Right Ascension** of a star is the arc *VR* of the celestial equator included between the Vernal equinox and the point where the hour circle of the star cuts the celestial equator. Right ascension is reckoned from the Vernal equinox *eastward* from 0° to 360°.

IV. *The ecliptic, EVF, and its pole Q* may be taken as the standards of reference. The co-ordinates of the star are then called its *latitude* and its *longitude*.

The **Latitude** of a star is its angular distance, MT (Fig. 112), from the ecliptic measured on a circle of latitude.

The **Longitude** of a star is the arc VT (Fig. 112) of the ecliptic included between the Vernal equinox and the point where the circle of latitude through the star cuts the ecliptic. The longitude of a star is always measured *eastward* from the Vernal equinox.

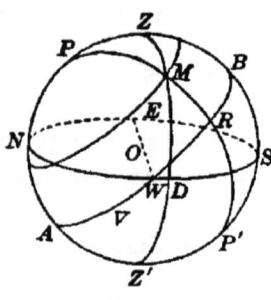

FIG. 111

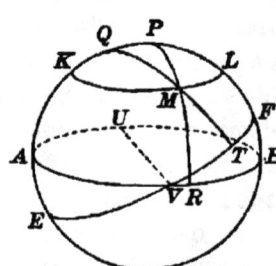

FIG. 112

For the star M (Figs. 111 and 112), let

$l =$ the latitude of the observer,

$h = \quad DM =$ the altitude of the star (Fig. 111),

$z = \quad ZM =$ the zenith distance of the star,

$a = \angle PZM =$ the azimuth of the star,

$t = \angle ZPM =$ the hour angle of the star,

$d = \quad RM =$ the declination of the star,

$p = \quad PM =$ the polar distance of the star,

$r = \quad VR =$ the right ascension of the star (Fig. 112),

$u = \quad MT =$ the latitude of the star,

$v = \quad VT =$ the longitude of the star,

$\quad NZS =$ the celestial meridian (Fig. 111),

$\quad ARB =$ the celestial equator (Fig. 112),

$\quad EVF =$ the ecliptic.

In many problems a simple way of representing the magnitudes involved is to project the sphere on the plane of the horizon, as shown in Figs. 114 and 115.

FIG. 113

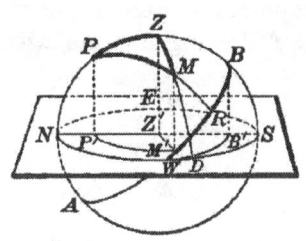

FIG. 114

NESW is the horizon, *Z* the zenith, *NZS* the celestial meridian, *WZE* the prime vertical, *WBE* an arc of the celestial equator, *P* the pole, *M* a star, *DM* its altitude, *ZM* its zenith distance, $\angle PZM$ its azimuth, *MR* its declination, *PM* its polar distance, $\angle ZPM$ its hour angle.

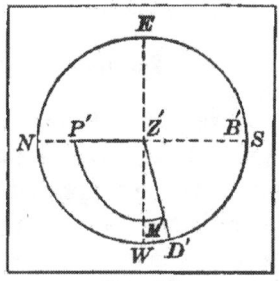

FIG. 115

SECTION LXVIII

THE ASTRONOMICAL TRIANGLE

The triangle *ZPM* (Fig. 111) is often called the *astronomical triangle*, on account of its importance in problems in Nautical Astronomy.

The side *PZ* is equal to the complement of the latitude of the observer. For (Fig. 111) since *O* is the centre of the sphere, the angle *ZOB* between the zenith of the observer and the celestial equator is obviously equal to his latitude, and the angle POZ is the complement of ZOB. Since the arc NP is the complement of PZ, *the altitude of the elevated pole is equal to the latitude of the place of observation.*

The triangle ZPM then (however much it may vary in shape for different positions of the star M) always contains the following five magnitudes:

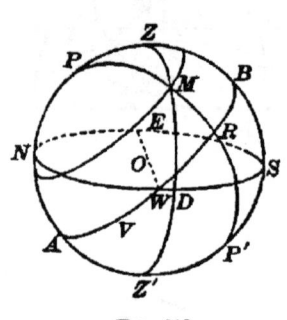

FIG. 116

$PZ = $ the co-latitude of observer
$= 90° - l,$

$ZM = $ the zenith distance of star
$= z,$

$\angle PZM = $ the azimuth of star
$= a,$

$PM = $ the polar distance of star
$= p,$

$\angle ZPM = $ the hour angle of star
$= t.$

A very simple relation exists between the hour angle of the sun and the local (apparent) time of day. The hourly rate at which the sun appears to move from east to west is 15°, and it is apparent noon at any place when the sun is on the meridian of that place. Hence, it is evident that if the hour angle is 0°, the time of day is noon; if the hour angle is 15°, the time of day is 1 o'clock P.M.; if the hour angle is 75°, the time of day is 5 o'clock P.M.; if the hour angle is − 15°, the time of day is 11 o'clock A.M.; if the hour angle is − 75°, the time of day is 7 o'clock A.M.; and so on.

In general, if t denotes the absolute value of the hour angle, when the sun is west of the meridian,

$$\text{the time of day is } \frac{t}{15} \text{ P.M.;}$$

when the sun is east of the meridian,

$$\text{the time of day is } 12 - \frac{t}{15} \text{ A.M.}$$

SECTION LXIX

PROBLEM

Given the latitude of the observer and the altitude and the azimuth of a star, to find its declination and its hour angle.

In the triangle ZPM (Fig. 117),
given

$PZ = 90° - l =$ the co-latitude,

$ZM = 90° - h =$ the co-altitude,

$\angle PZM = a \qquad =$ the azimuth;

to find

$PM = 90° - d =$ the polar distance,

$\angle ZPM = t \qquad =$ the hour angle.

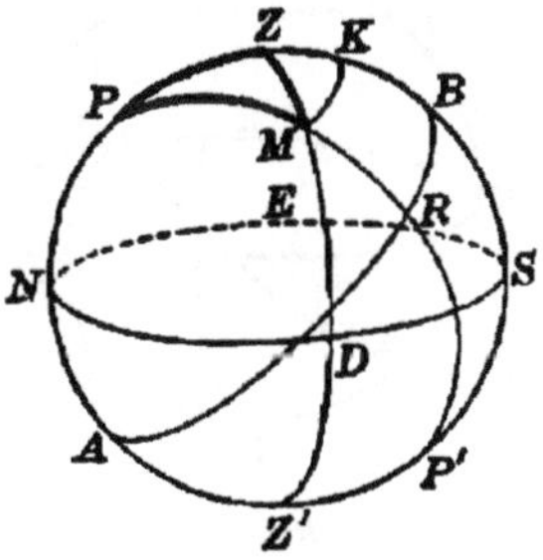

Draw $MK \perp$ to NZS.

FIG. 117

Let $\qquad\qquad ZK = m.$

Then, if $\quad a < 90°,\ PK = 90° - (l + m)$;

and if $\qquad a > 90°,\ PK = 90° - (l - m)$.

By Napier's Rules,

$$\cos a = \pm \tan m \tan h, \qquad\qquad (1)$$
$$\sin d = \cos PK \cos MK, \qquad\qquad (2)$$
$$\sin h = \cos m \cos MK. \qquad\qquad (3)$$

From (1), $\qquad \tan m = \pm \cot h \cos a. \qquad\qquad$ **(A)**

From (3), $\qquad \cos MK = \sin h \sec m. \qquad\qquad$ (4)

From (2), $\qquad \sin d = \cos [90° - (l \pm m)] \cos MK.$

$\qquad\qquad \therefore \sin d = \sin (l \pm m) \cos MK. \qquad\qquad$ (5)

Substitute in (5) the value of $\cos MK$.

Then, $\qquad\qquad \sin d = \sin (l \pm m) \sin h \sec m. \qquad$ **(B)**

In equations (**A**) and (**B**) the $-$ sign is to be used if $a > 90°$. The hour angle may then be found by means of [44], p. 158, whence $\qquad\qquad \sin t = \sin a \cos h \sec d.$

SECTION LXX

PROBLEM

To find the hour angle of a heavenly body when its declination, its altitude, and the latitude of the place are known.

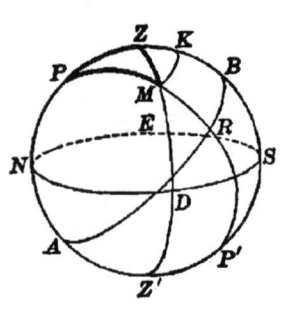

Fig. 118

In the triangle ZPM (Fig. 118),

given $PZ = 90° - l,$

$PM = 90° - d = p,$

$ZM = 90° - h;$

required

$\angle ZPM = t.$

If, in the first formula of [47], p. 161,

$$\sin \tfrac{1}{2} A = \sqrt{\sin(s-b)\sin(s-c)\csc b \csc c},$$

we put $A = t$, $a = 90° - h$, $b = p$, $c = 90° - l$,

and $2s = a + b + c.$

Then, $2s = 90° - h + p + 90° - l,$

or $2s = 180° - l + p - h;$

whence, $s = 90° - \tfrac{1}{2}l + \tfrac{1}{2}p - \tfrac{1}{2}h,$

$$s - b = 90° - \tfrac{1}{2}(l + p + h),$$

$$s - c = \tfrac{1}{2}(l + p - h);$$

and the formula becomes

$\sin \tfrac{1}{2} t$

$$= \pm \sqrt{\sin\left[90° - \tfrac{1}{2}(l+p+h)\right]\sin \tfrac{1}{2}(l+p-h)\csc p \csc (90° - l)}$$

$$= \pm \sqrt{\cos \tfrac{1}{2}(l+p+h)\sin \tfrac{1}{2}(l+p-h)\csc p \sec l,}$$

in which the — sign is to be taken when the body is east of the meridian.

If the body is the sun, how can the local time be found when the hour angle has been computed (Sect. LXVIII, p. 189)?

SECTION LXXI

PROBLEM

To find the altitude and the azimuth of a celestial body when its declination, its hour angle, and the latitude of the place are known.

In the triangle ZPM (Fig. **118**),

given
$$PZ = 90° - l,$$
$$PM = 90° - d = p,$$
$$\angle ZPM = t;$$

required
$$ZM = 90° - h,$$
$$\angle ZPM = a.$$

Draw $MK \perp$ to NZS and let $PK = m$.

Then, if $a < 90°$, $ZK = 90° - (l + m)$;

and if $a > 90°$, $ZK = (l + m) - 90°$.

By Napier's Rules,

$$\cos t = \tan m \tan d, \tag{1}$$
$$\sin h = \sin (l + m) \cos MK, \tag{2}$$
$$\sin d = \cos MK \cos m, \tag{3}$$
$$\cos (l + m) = \cot a \tan MK, \tag{4}$$
$$\sin m = \cot t \tan MK. \tag{5}$$

From (1), $\tan m = \cot d \cos t.$ **(A)**

From (3), $\cos MK = \sin d \sec m.$ (6)

Substitute in (2) the value of $\cos MK$,

$$\sin h = \sin (l + m) \sin d \sec m. \tag{B}$$

From (5), $\tan MK = \sin m \tan t.$

Substitute in (4) the value of $\tan MK$,

$$\tan a = \sec (l + m) \sin m \tan t. \tag{C}$$

Equations **(A)**, **(B)**, and **(C)** are the equations required.
In **(C)** a is E. or W. to agree with the hour angle.

SECTION LXXII

PROBLEM

To find the latitude of the place when the altitude of a celestial body, its declination, and its hour angle are known.

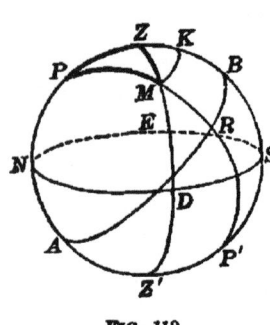

FIG. 119

In the triangle ZPM (Fig. 119),

given
$$ZM = 90° - h,$$
$$PM = 90° - d,$$
$$\angle ZPM = t\,;$$

required $PZ = 90° - l.$

Draw $MK \perp$ to $NZS.$

Let $PK = m, \quad ZK = n.$

Then, by Napier's Rules,

$$\cos t = \tan m \tan d,$$
$$\sin h = \cos n \cos MK,$$
$$\sin d = \cos m \cos MK\,;$$

whence,
$$\tan m = \cot d \cos t,$$
$$\cos n = \cos m \sin h \csc d.$$

It is evident from the figure that

$$l = 90° - (m \pm n),$$

in which the sign $+$ or the sign $-$ is to be taken according as the celestial body and the elevated pole are *on the same side* of the prime vertical or *on opposite sides.*

In fact, both the values of l shown above may be possible for the same altitude and the same hour angle; but, unless n is very small, the two values will differ largely from each other, so that the observer has no difficulty in deciding which of them should be taken.

SECTION LXXIII

PROBLEM

Given the declination, the right ascension of a star, and the obliquity of the ecliptic, to find the latitude and the longitude of the star.

Let M (Fig. 121) be the star, P the pole of the celestial equator, and Q the pole of the ecliptic.

Then, in the triangle PMQ,

given

$$PQ = e = 23° 27',$$

$$PM = 90° - d,$$

$$\angle MPQ = 90° + r;$$

required

$$QM = 90° - u,$$

and

$$\angle PQM = 90° - v;$$

where

$r =$ the right ascension $= VR$,

$u =$ the latitude of M $= MT$,

and

$v =$ the longitude of $M = VT$.

Fig. 120

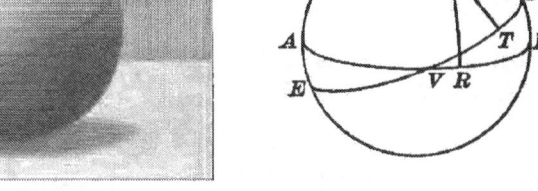

Fig. 121

Here, two sides and the included angle are given. Draw $MH \perp$ to PQ, and meeting it produced at H. Let $PH = n$,

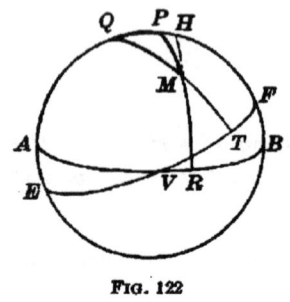

Fig. 122

By Napier's Rules,

$$\sin r = \tan n \tan d, \qquad (1)$$

$$\sin u = \cos(e + n) \cos MH, \quad (2)$$

$$\sin d = \cos n \cos MH, \qquad (3)$$

$$\sin(e+n) = \tan v \tan MH, \qquad (4)$$

$$\sin n = \tan r \tan MH. \qquad (5)$$

From (1),

$$\tan n = \cot d \sin r. \qquad \textbf{(A)}$$

From (3), $\cos MH = \sin d \sec n.$

Substitute in (2) the value of $\cos MH$,

$$\sin u = \cos(e + n)\sin d \sec n. \qquad \textbf{(B)}$$

From (4), $\tan v = \sin(e + n)\cot MH. \qquad (6)$

From (5), $\cot MH = \tan r \csc n.$

Substitute in (6) the value of $\cot MH$,

$$\tan v = \sin(e + n)\tan r \csc n. \qquad \textbf{(C)}$$

Equations **(A)**, **(B)**, and **(C)** determine u and v.

To avoid obtaining u from its sine we may proceed as follows:

From equations **(B)** and **(C)** we have, by division,

$$\frac{\sin u}{\tan v} = \frac{\sin d \cos(e + n) \sec n}{\tan r \sin(e + n) \csc n}.$$

$$\therefore \sin u = \tan v \sin d \cot r \cot(e + n)\tan n.$$

By taking MH as middle part, successively, in the triangles MQH and MPH, we obtain

$$\cos MH = \cos u \cos v,$$

and $\cos MH = \cos d \cos r.$

$$\therefore \cos u \cos v = \cos d \cos r.$$

$$\therefore \cos u = \sec v \cos d \cos r.$$

From these values of $\sin u$ and $\cos u$ we obtain, by division,

$$\frac{\sin u}{\cos u} = \frac{\tan v \cot (e + n) \sin d \cot r \tan n}{\sec v \cos d \cos r}.$$

$$\therefore \tan u = \sin v \cot (e + n) \tan d \csc r \tan n.$$

From the relation

$$\sin r = \tan n \tan d,$$

it follows that, dividing by $\sin r$,

$$\tan d \csc r \tan n = 1.$$

Therefore, $\tan u = \sin v \cot (e + n)$,

a formula by which u can be easily found after v has been computed.

EXERCISE XLIV

1. Find the dihedral angle made by adjacent lateral faces of a regular ten-sided pyramid; given the angle $V = 18°$, made at the vertex by two adjacent lateral edges.

2. Through the foot of a rod which makes the angle A with a plane a straight line is drawn in the plane. This line makes the angle B with the projection of the rod upon the plane. What angle does this line make with the rod?

3. Find the volume V of an oblique parallelopipedon; given the three unequal edges a, b, c, and the three angles l, m, n, which the edges make with one another.

4. The continent of Asia has nearly the shape of an equilateral triangle, the vertices being the East Cape, Cape Romania, and the Promontory of Baba. Assuming each side of this triangle to be 4800 geographical miles, and the earth's radius to be 3440 geographical miles, find the area of the triangle: (i) regarded as a plane triangle; (ii) regarded as a spherical triangle.

5. A ship sails from a harbor in latitude l and keeps on the arc of a great circle. Her *course* (or angle between the direction in which she sails and the meridian) at starting is a. Find where she will cross the equator, her course at the equator, and the distance she has sailed.

6. Two places have the same latitude l, and the distance between the places, measured on an arc of a great circle, is d. How much greater is the arc of the parallel of latitude between the places than the arc of the great circle? Compute the results for $l = 45°$, $d = 90°$.

7. The distance d between two places and the latitudes l and l' of the places are known. Find the difference between their longitudes.

8. Given the latitudes and longitudes of three places on the earth's surface, and also the radius of the earth; show how to find the area of the spherical triangle formed by arcs of great circles passing through the three places.

9. The distance between Paris and Berlin (the arc of a great circle) is equal to 472 geographical miles. The latitude of Paris is $48° 50' 13''$; that of Berlin, $52° 30' 16''$. When it is noon at Paris what time is it at Berlin?

NOTE. Owing to the apparent motion of the sun, the local time over the earth's surface at any instant varies at the rate of one hour for 15° of longitude; and the more *easterly* the place, the *later* the local time.

10. Given the altitude of the pole 45°, and the azimuth of a star on the horizon 45°; find the polar distance of the star.

11. Given the latitude l of the observer, and the declination d of the sun; find the local time (apparent solar time) of sunrise and sunset, and also the azimuth of the sun at these times (refraction being neglected). When and where does the sun rise on the longest day of the year (at which time $d = + 23° 27'$) in Boston ($l = 42° 21'$), and what is the length of the day from sunrise to sunset? Also, find when and where the sun rises

in Boston on the shortest day of the year (when $d = -23° 27'$), and the length of this day.

12. When is the solution of the problem in Example 11 impossible, and for what places is the solution impossible?

13. Given the latitude of a place and the sun's declination; find his altitude and azimuth at 6 o'clock A.M. (neglecting refraction). Compute the results for the longest day of the year at Munich ($l = 48° 9'$).

14. How does the altitude of the sun at 6 A.M. *on a given day* change as we go from the equator to the pole? At what time of the year is it a maximum *at a given place?* (Given $\sin h = \sin l \sin d$.)

15. Given the latitude of a place north of the equator, and the declination of the sun; find the time of day when the sun bears due east and due west. Compute the results for the longest day at St. Petersburg ($l = 59° 56'$).

16. Apply the general result in Example 15 ($\cos t = \cot l \tan d$) to the case when the days and nights are equal in length (that is, when $d = 0°$). Why can the sun in summer never be due east before 6 A.M., or due west after 6 P.M.? How does the time of bearing due east and due west change with the declination of the sun? Apply the general result to the cases where $l < d$ and $l = d$. What is it at the north pole?

17. Given the sun's declination and his altitude when he bears due east; find the latitude of the observer.

18. At a point O in a horizontal plane MN a staff OA is fixed so that its angle of inclination AOB with the plane is equal to the latitude of the place, 51° 30' N., and the direction OB is due north. What angle will OB make with the shadow of OA on the plane, at 1 P.M., when the sun is on the equinoctial?

19. What is the direction of a wall in latitude 52° 30' N. which casts no shadow at 6 A.M. on the longest day of the year?

20. Find the latitude of the place at which the sun rises exactly in the northeast on the longest day of the year.

21. Find the latitude of the place at which the sun sets at 10 o'clock on the longest day.

22. To what does the general formula for the hour angle, in Sect. LXX, reduce when (i) $h = 0°$, (ii) $l = 0°$ and $d = 0°$, (iii) l or $d = 90°$?

23. What does the general formula for the azimuth of a celestial body, in Sect. LXXI, become when $t = 90° = 6$ hours?

24. Show that the formulas of Sect. LXXII, if $t = 90°$, lead to the equation $\sin l = \sin h \csc d$; and that if $d = 0°$, they lead to the equation $\cos l = \sin h \sec t$.

25. Given the latitude of the place of observation $52° 30' 16''$, the declination of a star $38°$, its hour angle $28° 17' 15''$; find the altitude of the star.

26. Given the latitude of the place of observation $51° 19' 20''$, the polar distance of a star $67° 59' 5''$, its hour angle $15° 8' 12''$; find the altitude and the azimuth of the star.

27. Given the declination of a star $7° 54'$, its altitude $22° 45' 12''$, its azimuth $129° 45' 37''$; find the hour angle of the star and the latitude of the observer.

28. Given $e = 23° 27'$ and the longitude v of the sun; find the declination d and the right ascension r.

29. Given $e = 23° 27'$, the latitude of a star $51°$, its longitude $315°$; find its declination and its right ascension.

30. Given the latitude of the observer $44° 50'$, the azimuth of a star $138° 58'$, its hour angle $20°$; find its declination.

31. Given the latitude of the place of observation $51° 31' 48''$, the altitude of the sun west of the meridian $35° 14' 27''$, its declination $+ 21° 27'$; find the local apparent time.

32. Given the latitude of a place l, the polar distance p of a star, and its altitude h; find its azimuth a.

FORMULAS

PLANE TRIGONOMETRY

1. $\sin^2 A + \cos^2 A = 1.$

2. $\tan A = \dfrac{\sin A}{\cos A}.$

3. $\begin{cases} \sin A \times \csc A = 1. \\ \cos A \times \sec A = 1. \\ \tan A \times \cot A = 1. \end{cases}$

4. $\sin(x + y) = \sin x \cos y + \cos x \sin y.$

5. $\cos(x + y) = \cos x \cos y - \sin x \sin y.$

6. $\tan(x + y) = \dfrac{\tan x + \tan y}{1 - \tan x \tan y}.$

7. $\cot(x + y) = \dfrac{\cot x \cot y - 1}{\cot y + \cot x}.$

8. $\sin(x - y) = \sin x \cos y - \cos x \sin y.$

9. $\cos(x - y) = \cos x \cos y + \sin x \sin y.$

10. $\tan(x - y) = \dfrac{\tan x - \tan y}{1 + \tan x \tan y}.$

11. $\cot(x - y) = \dfrac{\cot x \cot y + 1}{\cot y - \cot x}.$

12. $\sin 2x = 2 \sin x \cos x.$

13. $\cos 2x = \cos^2 x - \sin^2 x.$

14. $\tan 2x = \dfrac{2 \tan x}{1 - \tan^2 x}.$

202

15. $\cot 2x = \dfrac{\cot^2 x - 1}{2 \cot x}$.

16. $\sin \frac{1}{2} z = \pm \sqrt{\dfrac{1 - \cos z}{2}}$.

17. $\cos \frac{1}{2} z = \pm \sqrt{\dfrac{1 + \cos z}{2}}$.

18. $\tan \frac{1}{2} z = \pm \sqrt{\dfrac{1 - \cos z}{1 + \cos z}}$.

19. $\cot \frac{1}{2} z = \pm \sqrt{\dfrac{1 + \cos z}{1 - \cos z}}$.

20. $\sin A + \sin B = 2 \sin \frac{1}{2}(A + B) \cos \frac{1}{2}(A - B)$.

21. $\sin A - \sin B = 2 \cos \frac{1}{2}(A + B) \sin \frac{1}{2}(A - B)$.

22. $\cos A + \cos B = 2 \cos \frac{1}{2}(A + B) \cos \frac{1}{2}(A - B)$.

23. $\cos A - \cos B = - 2 \sin \frac{1}{2}(A + B) \sin \frac{1}{2}(A - B)$.

24. $\dfrac{\sin A + \sin B}{\sin A - \sin B} = \dfrac{\tan \frac{1}{2}(A + B)}{\tan \frac{1}{2}(A - B)}$.

25. $\dfrac{a}{b} = \dfrac{\sin A}{\sin B}$.

26. $a^2 = b^2 + c^2 - 2\,bc \cos A$.

27. $\dfrac{a - b}{a + b} = \dfrac{\tan \frac{1}{2}(A - B)}{\tan \frac{1}{2}(A + B)}$.

28. $\sin \frac{1}{2} A = \sqrt{\dfrac{(s - b)(s - c)}{bc}}$.

29. $\cos \frac{1}{2} A = \sqrt{\dfrac{s(s - a)}{bc}}$.

30. $\tan \frac{1}{2} A = \sqrt{\dfrac{(s-b)(s-c)}{s(s-a)}}.$

31. $\sqrt{\dfrac{(s-a)(s-b)(s-c)}{s}} = r.$

32. $\tan \frac{1}{2} A = \dfrac{r}{s-a}.$

33. $F = \frac{1}{2} ac \sin B.$

34. $F = \dfrac{a^2 \sin B \sin C}{2 \sin (B+C)}.$

35. $F = \sqrt{s(s-a)(s-b)(s-c)}.$

36. $F = \dfrac{abc}{4R}.$

37. $F = \frac{1}{2} r (a+b+c) = rs.$

SPHERICAL TRIGONOMETRY

38. $\cos c = \cos a \cos b.$

39. $\begin{cases} \sin a = \sin c \sin A. \\ \sin b = \sin c \sin B. \end{cases}$

40. $\begin{cases} \cos A = \tan b \cot c. \\ \cos B = \tan a \cot c. \end{cases}$

41. $\begin{cases} \cos A = \cos a \sin B. \\ \cos B = \cos b \sin A. \end{cases}$

42. $\begin{cases} \sin b = \tan a \cot A. \\ \sin a = \tan b \cot B. \end{cases}$

43. $\cos c = \cot A \cot B.$

44. $\begin{cases} \sin a \sin B = \sin b \sin A. \\ \sin a \sin C = \sin c \sin A. \\ \sin b \sin C = \sin c \sin B. \end{cases}$

45. $\begin{cases} \cos a = \cos b \cos c + \sin b \sin c \cos A. \\ \cos b = \cos a \cos c + \sin a \sin c \cos B. \\ \cos c = \cos a \cos b + \sin a \sin b \cos C. \end{cases}$

46. $\begin{cases} \cos A = -\cos B \cos C + \sin B \sin C \cos a. \\ \cos B = -\cos A \cos C + \sin A \sin C \cos b. \\ \cos C = -\cos A \cos B + \sin A \sin B \cos c. \end{cases}$

47. $\begin{cases} \sin \tfrac{1}{2} A = \sqrt{\sin (s-b) \sin (s-c) \csc b \csc c}. \\ \cos \tfrac{1}{2} A = \sqrt{\sin s \sin (s-a) \csc b \csc c}. \\ \tan \tfrac{1}{2} A = \sqrt{\csc s \csc (s-a) \sin (s-b) \sin (s-c)}. \end{cases}$

48. $\begin{cases} \sin \tfrac{1}{2} a = \sqrt{-\cos S \cos (S-A) \csc B \csc C}. \\ \cos \tfrac{1}{2} a = \sqrt{\cos (S-B) \cos (S-C) \csc B \csc C}. \\ \tan \tfrac{1}{2} a = \sqrt{-\cos S \cos (S-A) \sec (S-B) \sec (S-C)}. \end{cases}$

49. $\begin{cases} \cos \tfrac{1}{2}(A+B) \cos \tfrac{1}{2} c = \cos \tfrac{1}{2}(a+b) \sin \tfrac{1}{2} C. \\ \sin \tfrac{1}{2}(A+B) \cos \tfrac{1}{2} c = \cos \tfrac{1}{2}(a-b) \cos \tfrac{1}{2} C. \\ \cos \tfrac{1}{2}(A-B) \sin \tfrac{1}{2} c = \sin \tfrac{1}{2}(a+b) \sin \tfrac{1}{2} C. \\ \sin \tfrac{1}{2}(A-B) \sin \tfrac{1}{2} c = \sin \tfrac{1}{2}(a-b) \cos \tfrac{1}{2} C. \end{cases}$

50. $\begin{cases} \tan \tfrac{1}{2}(A+B) = \dfrac{\cos \tfrac{1}{2}(a-b)}{\cos \tfrac{1}{2}(a+b)} \cot \tfrac{1}{2} C. \\[2mm] \tan \tfrac{1}{2}(A-B) = \dfrac{\sin \tfrac{1}{2}(a-b)}{\sin \tfrac{1}{2}(a+b)} \cot \tfrac{1}{2} C. \\[2mm] \tan \tfrac{1}{2}(a+b) = \dfrac{\cos \tfrac{1}{2}(A-B)}{\cos \tfrac{1}{2}(A+B)} \tan \tfrac{1}{2} c. \\[2mm] \tan \tfrac{1}{2}(a-b) = \dfrac{\sin \tfrac{1}{2}(A-B)}{\sin \tfrac{1}{2}(A+B)} \tan \tfrac{1}{2} c. \end{cases}$

51. $F = \dfrac{\pi R^2 E}{180}.$

52. $\tan^2 \tfrac{1}{4} E = \tan \tfrac{1}{2} s \tan \tfrac{1}{2}(s-a) \tan \tfrac{1}{2}(s-b) \tan \tfrac{1}{2}(s-c).$

The following diagram shows Prof. Blakslee's construction by which the direction ratios for plane right triangles give directly from a figure the analogies for a right trihedral or for a right spherical triangle :

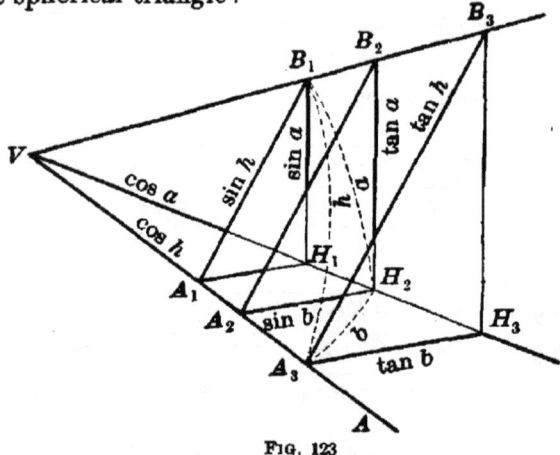

FIG. 123

The construction consists of two parts.

1. Lay off from the vertex V a unit's distance on each edge.

2. Pass through the three extremities of these distances three planes perpendicular to one of the edges, as VA. Now these three parallel planes will cut out three similar right triangles. The first being constructed in either of the two usual ways, the construction of the others is evident.

Since the plane angles A_1, A_2, A_3 all equal the dihedral A, and the nine right triangles in the three faces give the values in the figure, we have :

(1) $\sin A = \sin a : \sin h$; similarly, $\sin B = \sin b : \sin h$.

(2) $\cos A = \tan b : \tan h$; similarly, $\cos B = \tan a : \tan h$.

(3) $\tan A = \tan a : \sin b$; similarly, $\tan B = \tan b : \sin a$.

(4) $\cos h = \cos a \cos b$; by (3), $\cos h = \cot A \cot B$.

(5) $\sin A = \cos B : \cos b$; $\sin B = \cos A : \cos a$.

Note. If a sphere of unit radius is described about V as a centre, the three faces will cut out a right spherical triangle, having the sides a, b, and h, and angles A, B, and H. The above formulas are thus seen to be the analogies of:

(1) $\sin A = a : h$; $\sin B = b : h$.

(2) $\cos A = b : h$; $\cos B = a : h$.

(3) $\tan A = a : b$; $\tan B = b : a$.

(4) $\quad h^2 = a^2 + b^2$; $1 = \sin^2 + \cos^2$; $1 = \cot A \cot B$.

(5) $\sin A = \cos B$; $\sin B = \cos A$.

Napier's Rules give only the following, which follow from the analogies as numbered:

$$\text{By} \left.\begin{cases} \sin a = \sin A \sin h = \tan b \cot B \\ \sin b = \sin B \sin h = \tan a \cot A \end{cases}\right\} \text{ (3)}$$
(1)

$$\text{(5)} \left\{\begin{array}{l} \cos A = \sin B \cos a = \tan b \cot h \\ \cos B = \sin A \cos b = \tan a \cot h \end{array}\right\} \text{ (2)}$$

$$\text{(4)} \ \{ \cos h = \cos a \cos b = \cot A \cot B \} \text{ (4)}$$

GAUSS'S EQUATIONS

$$\cos \tfrac{1}{2}(A + B)\cos \tfrac{1}{2}c = \cos \tfrac{1}{2}(a + b)\sin \tfrac{1}{2}C.$$

$$\sin \tfrac{1}{2}(A + B)\cos \tfrac{1}{2}c = \cos \tfrac{1}{2}(a - b)\cos \tfrac{1}{2}C.$$

$$\cos \tfrac{1}{2}(A - B)\sin \tfrac{1}{2}c = \sin \tfrac{1}{2}(a + b)\sin \tfrac{1}{2}C.$$

$$\sin \tfrac{1}{2}(A - B)\sin \tfrac{1}{2}c = \sin \tfrac{1}{2}(a - b)\cos \tfrac{1}{2}C.$$

ANSWERS

PLANE TRIGONOMETRY

Exercise I. Page 2

1. $\frac{1}{3}\pi$; $\frac{1}{4}\pi$; $\frac{5}{8}\pi$; $1\frac{1}{3}\pi$; $\frac{7}{18}\pi$; $\frac{11}{16}\pi$; $\frac{5}{24}\pi$.

2. $120°$; $135°$; $112° \, 30'$; $168° \, 45'$; $84°$.

3. 0.017453; 0.0002909.

4. $206,265''$.

5. $\frac{1}{4}\pi$; $\frac{1}{3}\pi$.

6. $11° \, 27' \, 33''$.

7. $14° \, 27' \, 28''$.

8. 69.166 miles.

9. 57 feet 8.55 inches.

10. 3 hours 49 minutes 11 seconds.

11. 9 feet 2 inches.

12. $\frac{7}{165}$ seconds.

Exercise II. Page 5

1. $\sin B = \dfrac{b}{c}$; $\cos B = \dfrac{a}{c}$; $\tan B = \dfrac{b}{a}$; $\cot B = \dfrac{a}{b}$; $\sec B = \dfrac{c}{a}$; $\csc B = \dfrac{c}{b}$.

3. (i) $\sin = \frac{4}{5}$, $\cos = \frac{3}{5}$, $\tan = \frac{4}{3}$, $\cot = \frac{3}{4}$, $\sec = \frac{5}{3}$, $\csc = \frac{5}{4}$;

(ii) $\sin = \frac{5}{13}$, $\cos = \frac{12}{13}$, $\tan = \frac{5}{12}$, $\cot = \frac{12}{5}$, $\sec = \frac{13}{12}$, $\csc = \frac{13}{5}$;

(iii) $\sin = \frac{8}{17}$, $\cos = \frac{15}{17}$, $\tan = \frac{8}{15}$, $\cot = \frac{15}{8}$, $\sec = \frac{17}{15}$, $\csc = \frac{17}{8}$;

(iv) $\sin = \frac{9}{41}$, $\cos = \frac{40}{41}$, $\tan = \frac{9}{40}$, $\cot = \frac{40}{9}$, $\sec = \frac{41}{40}$, $\csc = \frac{41}{9}$;

(v) $\sin = \frac{15}{17}$, $\cos = \frac{8}{17}$, $\tan = \frac{15}{8}$, $\cot = \frac{8}{15}$, $\sec = \frac{17}{8}$, $\csc = \frac{17}{15}$;

(vi) $\sin = \frac{119}{169}$, $\cos = \frac{120}{169}$, $\tan = \frac{119}{120}$, $\cot = \frac{120}{119}$,

$\sec = \frac{169}{120}$, $\csc = \frac{169}{119}$.

4. The required condition is that $a^2 + b^2 = c^2$. It is.

5. (i) $\sin = \dfrac{2mn}{m^2 + n^2}$, $\cos = \dfrac{m^2 - n^2}{m^2 + n^2}$, $\tan = \dfrac{2mn}{m^2 - n^2}$,

$\cot = \dfrac{m^2 - n^2}{2mn}$, $\sec = \dfrac{m^2 + n^2}{m^2 - n^2}$, $\csc = \dfrac{m^2 + n^2}{2mn}$;

(ii) $\sin = \dfrac{2xy}{x^2 + y^2}$, $\cos = \dfrac{x^2 - y^2}{x^2 + y^2}$, $\tan = \dfrac{2xy}{x^2 - y^2}$,

$\cot = \dfrac{x^2 - y^2}{2xy}$, $\sec = \dfrac{x^2 + y^2}{x^2 - y^2}$, $\csc = \dfrac{x^2 + y^2}{2xy}$;

(iii) $\sin = \dfrac{q}{s}$, $\cos = \dfrac{q}{p}$, $\tan = \dfrac{p}{s}$, $\cot = \dfrac{s}{p}$, $\sec = \dfrac{p}{q}$, $\csc = \dfrac{s}{q}$;

(iv) $\sin = \dfrac{ms}{qr}$, $\cos = \dfrac{mpv}{nqr}$, $\tan = \dfrac{ns}{pv}$,

$\cot = \dfrac{pv}{ns}$, $\sec = \dfrac{nqr}{mpv}$, $\csc = \dfrac{qr}{ms}$.

7. In (iii) $p^2q^2 + q^2s^2 = p^2s^2$; in (iv) $m^2n^2s^2 + m^2p^2v^2 = n^2q^2r^2$.

8. $\sin A = \tfrac{24}{145} = \cos B$; $\cos A = \tfrac{143}{145} = \sin B$; $\tan A = \tfrac{24}{143} = \cot B$;
$\cot A = \tfrac{143}{24} = \tan B$; $\sec A = \tfrac{145}{143} = \csc B$; $\csc A = \tfrac{145}{24} = \sec B$.

9. $\sin A = \tfrac{24}{25} = \cos B$; $\cos A = \tfrac{7}{25} = \sin B$; $\tan A = \tfrac{24}{7} = \cot B$;
$\cot A = \tfrac{7}{24} = \tan B$; $\sec A = \tfrac{25}{7} = \csc B$; $\csc A = \tfrac{25}{24} = \sec B$.

10. $\sin A = \tfrac{168}{193} = \cos B$; $\cos A = \tfrac{95}{193} = \sin B$; $\tan A = \tfrac{168}{95} = \cot B$;
$\cot A = \tfrac{95}{168} = \tan B$; $\sec A = \tfrac{193}{95} = \csc B$; $\csc A = \tfrac{193}{168} = \sec B$.

11. $\sin A = \dfrac{\sqrt{p^2+q^2}}{p+q} = \cos B$; $\cos A = \dfrac{\sqrt{2pq}}{p+q} = \sin B$;

$\tan A = \dfrac{\sqrt{p^2+q^2}}{\sqrt{2pq}} = \cot B$; $\cot A = \dfrac{\sqrt{2pq}}{\sqrt{p^2+q^2}} = \tan B$;

$\sec A = \dfrac{p+q}{\sqrt{2pq}} = \csc B$; $\csc A = \dfrac{p+q}{\sqrt{p^2+q^2}} = \sec B$.

12. $\sin A = \dfrac{\sqrt{p^2+pq}}{p+q} = \cos B$; $\cos A = \dfrac{\sqrt{q^2+pq}}{p+q} = \sin B$;

$\tan A = \sqrt{\dfrac{p}{q}} = \cot B$; $\cot A = \sqrt{\dfrac{q}{p}} = \tan B$;

$\sec A = \dfrac{p+q}{\sqrt{q^2+pq}} = \csc B$; $\csc A = \dfrac{p+q}{\sqrt{p^2+pq}} = \sec B$.

13. $\sin A = \dfrac{p-q}{p+q} = \cos B$; $\cos A = \dfrac{2\sqrt{pq}}{p+q} = \sin B$;

$\tan A = \dfrac{p-q}{2\sqrt{pq}} = \cot B$; $\cot A = \dfrac{2\sqrt{pq}}{p-q} = \tan B$;

$\sec A = \dfrac{p+q}{2\sqrt{pq}} = \csc B$; $\csc A = \dfrac{p+q}{p-q} = \sec B$.

14. $\sin A = \tfrac{2}{3}\sqrt{5}$; $\cos A = \tfrac{1}{3}\sqrt{5}$; $\tan A = 2$; $\cot A = \tfrac{1}{2}$;
$\sec A = \sqrt{5}$; $\csc A = \tfrac{1}{2}\sqrt{5}$.

15. $\sin A = \tfrac{1}{3}$; $\cos A = \tfrac{1}{3}\sqrt{5}$; $\tan A = \tfrac{2}{3}\sqrt{5}$; $\cot A = \tfrac{1}{2}\sqrt{5}$;
$\sec A = \tfrac{3}{5}\sqrt{5}$; $\csc A = \tfrac{3}{2}$.

16. $\sin A = \frac{1}{7}(5 + \sqrt{7})$;
    $\tan A = \frac{1}{9}(16 + 5\sqrt{7})$;
    $\sec A = \frac{2}{7}(5 + \sqrt{7})$;

    $\cos A = \frac{1}{7}(5 - \sqrt{7})$;
    $\cot A = \frac{1}{9}(16 - 5\sqrt{7})$;
    $\csc A = \frac{2}{7}(5 - \sqrt{7})$.

17. $\sin A = \frac{1}{6}(\sqrt{31} + 1)$;
    $\tan A = \frac{1}{15}(16 + \sqrt{31})$;
    $\sec A = \frac{4}{15}(\sqrt{31} + 1)$;

    $\cos A = \frac{1}{6}(\sqrt{31} - 1)$;
    $\cot A = \frac{1}{15}(16 - \sqrt{31})$;
    $\csc A = \frac{4}{15}(\sqrt{31} - 1)$.

18. $a = 12.3$.   20. $a = 9$.   22. $c = 40$.

19. $b = 1.54$.   21. $b = 68$.   23. $c = 229.62$.

24. Construct a rt. $\triangle$ with legs equal to 3 and 2, respectively; then construct a similar $\triangle$ with hypotenuse equal to 6.

25. $a = 1.5$ miles; $b = 2$ miles.

30. $a = 0.342$, $b = 0.940$; $a = 1.368$, $b = 3.760$.   31. 142.926 yards.

Exercise III. Page 9

5. Through A (Fig. 3) draw a tangent, and take AT equal to 3; the angle AOT is the required angle.

6. From O (Fig. 3) as a centre, with a radius equal to 2, describe an arc cutting at S the tangent drawn through B; the angle AOS is the required angle.

7. In Fig. 3, take OM equal to $\frac{1}{2}$, and erect $MP \perp OA$, intersecting the circumference at P; the angle POM is the required angle.

8. Since $\sin x = \cos x$, $OM = PM$ (Fig. 3), and $x = 45°$; hence, construct x equal to $45°$.

9. Construct a rt. $\triangle$ with one leg equal to twice the other; the angle opposite the longer leg is the required angle.

10. Divide OA (Fig. 3) into four equal parts; at the first point of division from O erect a perpendicular meeting the circumference at some point P. Draw OP; the angle AOP is the required angle.

12. $x = 18°$.   21. $r \sin x$.   22. $a = mc$; $b = nc$.

Exercise IV. Page 12

1. $\cos 60°$; $\sin 45°$; $\cot 1°$; $\tan 75°$;
   $\sec 71° 50'$; $\sin 52° 36'$; $\tan 7° 41'$; $\sec 35° 14'$.

2. $\cos 30°$; $\sin 15°$; $\cot 83°$; $\tan 6°$;
   $\sec 20° 58'$; $\sin 4° 21'$; $\tan 0° 1'$; $\sec 44° 59'$.

3. $\frac{1}{4}\sqrt{3}$.	**6.** 30°.	**9.** 22° 30′.	**12.** $\dfrac{90°}{n+1}$.
4. 45°.	**7.** 90°.	**10.** 18°.	
5. 30°.	**8.** 60°.	**11.** 10°.	

Exercise VI. Page 16

1. $\cos A = \frac{5}{13}$; $\tan A = \frac{12}{5}$; $\cot A = \frac{5}{12}$; $\sec A = \frac{13}{5}$; $\csc A = \frac{13}{12}$.

2. $\cos A = 0.6$; $\tan A = 1.3333$; $\cot A = 0.75$; $\sec A = 1.6667$; $\csc A = 1.25$.

3. $\sin A = \frac{15}{17}$; $\tan A = \frac{15}{8}$; $\cot A = \frac{8}{15}$; $\sec A = \frac{17}{8}$; $\csc A = \frac{17}{15}$.

4. $\sin A = 0.96$; $\tan A = 3.4286$; $\cot A = 0.2917$; $\sec A = 3.5714$; $\csc A = 1.0417$.

5. $\sin A = 0.8$; $\cos A = 0.6$; $\cot A = 0.75$; $\sec A = 1.6667$; $\csc A = 1.25$.

6. $\sin A = \frac{1}{2}\sqrt{2}$; $\cos A = \frac{1}{2}\sqrt{2}$; $\tan A = 1$; $\sec A = \sqrt{2}$; $\csc A = \sqrt{2}$.

7. $\sin A = 0.90$; $\cos A = 0.45$; $\tan A = 2$; $\sec A = 2.22$; $\csc A = 1.11$.

8. $\sin A = \frac{1}{2}\sqrt{3}$; $\cos A = \frac{1}{2}$; $\tan A = \sqrt{3}$; $\cot A = \frac{1}{3}\sqrt{3}$; $\csc A = \frac{2}{3}\sqrt{3}$.

9. $\sin A = \frac{1}{2}\sqrt{2}$; $\cos A = \frac{1}{2}\sqrt{2}$; $\tan A = 1$; $\cot A = 1$; $\sec A = \sqrt{2}$.

10. $\cos A = \sqrt{1-m^2}$; $\tan A = \dfrac{m}{1-m^2}\sqrt{1-m^2}$; $\cot A = \dfrac{1}{m}\sqrt{1-m^2}$;

$\sec A = \dfrac{1}{\sqrt{1-m^2}}$; $\csc A = \dfrac{1}{m}$.

11. $\cos A = \dfrac{1-m^2}{1+m^2}$; $\tan A = \dfrac{2m}{1-m^2}$; $\cot A = \dfrac{1-m^2}{2m}$;

$\sec A = \dfrac{1+m^2}{1-m^2}$; $\csc A = \dfrac{1+m^2}{2m}$.

12. $\sin A = \dfrac{m^2-n^2}{m^2+n^2}$; $\tan A = \dfrac{m^2-n^2}{2mn}$; $\cot A = \dfrac{2mn}{m^2-n^2}$;

$\sec A = \dfrac{m^2+n^2}{2mn}$; $\csc A = \dfrac{m^2+n^2}{m^2-n^2}$.

13. $\sin = \frac{1}{2}\sqrt{2}$; $\cos = \frac{1}{2}\sqrt{2}$; $\cot = 1$; $\sec = \sqrt{2}$; $\csc = \sqrt{2}$.

14. $\cos = \frac{1}{2}\sqrt{3}$; $\tan = \frac{1}{3}\sqrt{3}$; $\cot = \sqrt{3}$; $\sec = \frac{2}{3}\sqrt{3}$; $\csc = 2$.

15. $\sin = \frac{1}{2}\sqrt{3}$; $\cos = \frac{1}{2}$; $\tan = \sqrt{3}$; $\cot = \frac{1}{3}\sqrt{3}$; $\sec = 2$.

16. $\sin = \frac{1}{2}\sqrt{2-\sqrt{3}}$; $\cos = \frac{1}{2}\sqrt{2+\sqrt{3}}$; $\cot = 2+\sqrt{3}$;

$\sec = 2(2-\sqrt{3})\sqrt{2+\sqrt{3}}$; $\csc = 2(2+\sqrt{3})\sqrt{2-\sqrt{3}}$.

17. $\sin = \frac{1}{2}\sqrt{2-\sqrt{2}}$; $\cos = \frac{1}{2}\sqrt{2+\sqrt{2}}$; $\tan = \sqrt{2}-1$;

$\sec = (2-\sqrt{2})\sqrt{2+\sqrt{2}}$; $\csc = (2+\sqrt{2})\sqrt{2-\sqrt{2}}$.

18. $\cos = 1$; $\tan = 0$; $\cot = \infty$; $\sec = 1$; $\csc = \infty$.

19. $\cos = 0$; $\tan = \infty$; $\cot = 0$; $\sec = \infty$; $\csc = 1$.

20. $\sin = 1$; $\cos = 0$; $\cot = 0$; $\sec = \infty$; $\csc = 1$.

21. $\cos A = \sqrt{1-\sin^2 A}$; $\tan A = \dfrac{\sin A}{\sqrt{1-\sin^2 A}}$; $\cot A = \dfrac{\sqrt{1-\sin^2 A}}{\sin A}$;

$\sec A = \dfrac{1}{\sqrt{1-\sin^2 A}}$; $\csc A = \dfrac{1}{\sin A}$.

22. $\sin A = \sqrt{1-\cos^2 A}$; $\tan A = \dfrac{\sqrt{1-\cos^2 A}}{\cos A}$; $\cot A = \dfrac{\cos A}{\sqrt{1-\cos^2 A}}$;

$\sec A = \dfrac{1}{\cos A}$; $\csc A = \dfrac{1}{\sqrt{1-\cos^2 A}}$

23. $\sin A = \dfrac{\tan A}{\sqrt{1+\tan^2 A}}$; $\cos A = \dfrac{1}{\sqrt{1+\tan^2 A}}$; $\cot A = \dfrac{1}{\tan A}$;

$\sec A = \sqrt{1+\tan^2 A}$; $\csc A = \dfrac{\sqrt{1+\tan^2 A}}{\tan A}$.

24. $\sin A = \dfrac{1}{\sqrt{1+\cot^2 A}}$; $\cos A = \dfrac{\cot A}{\sqrt{1+\cot^2 A}}$; $\tan A = \dfrac{1}{\cot A}$;

$\sec A = \dfrac{\sqrt{1+\cot^2 A}}{\cot A}$; $\csc A = \sqrt{1+\cot^2 A}$.

25. $\sin A = \frac{1}{3}\sqrt{5}$; $\cos A = \frac{2}{3}\sqrt{5}$.   **27.** $\sin A = \frac{9}{41}$; $\cos A = \frac{40}{41}$.

26. $\sin A = \frac{1}{4}\sqrt{15}$; $\tan A = \sqrt{15}$.   **28.** $\dfrac{1-3\cos^2 A + 3\cos^4 A}{\cos^2 A - \cos^4 A}$.

Exercise VII. Page 18

1. $x = 45°$.   **6.** $x = 45°$.   **11.** $x = 30°$.   **16.** $x = 45°$.

2. $x = 30°$.   **7.** $x = 45°$.   **12.** $x = 45°$.   **17.** $x = 60°$.

3. $x = 0°$, or $60°$.   **8.** $x = 45°$.   **13.** $x = 0°$, or $60°$.

4. $x = 45°$.   **9.** $x = 60°$.   **14.** $x = 30°$.

5. $x = 60°$.   **10.** $x = 60°$.   **15.** $x = 30°$, or $45°$.

Exercise VIII. Page 24

1. $c = \dfrac{b}{\cos A}$.   **2.** $c = \dfrac{a}{\sin A}$.   **3.** $b = c \cos A$.   **4.** $c = \dfrac{a}{\sin A}$.

5. $A = 90° - B$; $a = c \cos B$; $b = c \sin B$.

6. $A = 90° - B$; $a = b \cot B$; $c = \dfrac{b}{\sin B}$.

7. $A = 90° - B$; $b = a \tan B$; $c = \dfrac{a}{\cos B}$.

8. $\cos A = \dfrac{b}{c}$; $B = 90° - A$; $a = \sqrt{(c + b)(c - b)}$.

Exercise IX. Page 28

31. $c = 7.8112$; $A = 39° 48'$; $B = 50° 12'$; $F = 15$.

32. $b = 69.997$; $A = 30' 12''$; $B = 89° 29' 48''$; $F = 21.525$.

33. $a = 1.1886$; $A = 43° 20'$; $B = 46° 40'$; $F = 0.74876$.

34. $b = 21.249$; $c = 22.372$; $B = 71° 46'$; $F = 74.372$.

35. $a = 6.6882$; $c = 13.738$; $B = 60° 52'$; $F = 40.129$.

36. $a = 63.859$; $b = 23.369$; $B = 20° 6'$; $F = 746.15$.

37. $a = 19.40$; $b = 18.778$; $A = 45° 56'$; $F = 182.15$.

38. $b = 58.719$; $c = 71.377$; $A = 41° 11'$; $F = 1262.4$.

39. $a = 12.981$; $c = 15.796$; $A = 55° 16'$; $F = 58.416$.

40. $a = 0.58046$; $b = 8.442$; $A = 3° 56'$; $F = 2.4501$.

41. $F = \frac{1}{2} c^2 \sin A \cos A$.   **43.** $F = \frac{1}{2} b^2 \tan A$.

42. $F = \frac{1}{2} a^2 \cot A$.   **44.** $F = \frac{1}{2} a \sqrt{c^2 - a^2}$.

45. $b = 11.6$; $c = 15.315$; $A = 40° 45' 48''$; $B = 49° 14' 12''$.

46. $a = 7.2$; $c = 8.7658$; $A = 55° 13' 20''$; $B = 34° 46' 40''$.

47. $a = 3.6474$; $b = 6.58$; $c = 7.5233$; $B = 61°$.

48. $a = 10.283$; $b = 19.449$; $A = 27° 52'$; $B = 62° 8'$.

49. $19° 28' 17''$ and $70° 31' 43''$.   **52.** $36° 52' 12''$ and $53° 7' 48''$.

50. 3 and 5.1961.   **53.** 212.1 feet.

51. $a = c \cos \dfrac{90°}{n + 1}$;   **54.** 732.22 feet.

$\qquad b = c \sin \dfrac{90°}{n + 1}$.   **55.** 3270 feet.

          56. 37.3 feet.

57. 1° 25′ 56″. **58.** 59° 44′ 35″. **59.** 95.84 feet.

60. 7.0712 miles in each direction.

61. 20.88 feet. **63.** 685.9 feet. **65.** 140 feet.

62. 56.65 feet. **64.** 136.6 feet. **66.** 84.74 feet.

Exercise X. Page 33

1. $C = 2(90° - A)$; $c = 2a\cos A$; $h = a\sin A$.

2. $A = \frac{1}{2}(180° - C)$; $c = 2a\cos A$; $h = a\sin A$.

3. $C = 2(90° - A)$; $a = \dfrac{c}{2\cos A}$; $h = a\sin A$.

4. $A = \frac{1}{2}(180° - C)$; $a = \dfrac{c}{2\cos A}$; $h = a\sin A$.

5. $C = 2(90° - A)$; $a = \dfrac{h}{\sin A}$; $c = 2a\cos A$.

6. $A = \frac{1}{2}(180° - C)$; $a = \dfrac{h}{\sin A}$; $c = 2a\cos A$.

7. $\sin A = \dfrac{h}{a}$; $C = 2(90° - A)$; $c = 2a\cos A$.

8. $\tan A = \dfrac{2h}{c}$; $C = 2(90° - A)$; $a = \dfrac{h}{\sin A}$.

9. $A = 67° 22′ 50″$; $C = 45° 14′ 20″$; $h = 13.2$.

10. $c = 0.21943$; $h = 0.27384$; $F = 0.03004$.

11. $a = 2.055$; $h = 1.6852$; $F = 1.9819$.

12. $a = 7.706$; $c = 3.6676$; $F = 13.725$.

13. $A = 79° 36′ 30″$; $C = 20° 47′$; $c = 2.4206$.

14. $A = 77° 19′ 11″$; $C = 25° 21′ 38″$; $a = 20.5$.

15. $A = 25° 27′ 47″$; $C = 129° 4′ 26″$; $a = 81.41$; $h = 35$.

16. $A = 81° 12′ 9″$; $C = 17° 35′ 42″$; $a = 17$; $c = 5.2$.

17. $F = \frac{1}{4} c \sqrt{4a^2 - c^2}$. **22.** 0.76536.

18. $F = a^2 \sin \frac{1}{2} C \cos \frac{1}{2} C$. **23.** 94° 20′.

19. $F = a^2 \sin A \cos A$. **24.** 2.7261.

20. $F = h^2 \tan \frac{1}{2} C$. **25.** 38° 56′ 33″.

21. 28.284 feet; 4525.44 square feet. **26.** 37.699.

Exercise XI. Page 35

1. $r = 1.618$; $h = 1.5388$; $F = 7.694$.
2. $h = 0.9848$; $p = 6.2514$; $F = 3.0782$.
3. $h = 19.754$; $c = 6.257$; $F = 1236$.
4. $r = 1.0824$; $c = 0.82842$; $F = 8.3137$.
5. $r = 2.5933$; $h = 2.4882$; $c = 1.4615$.
6. $r = 1.5994$; $h = 1.441$; $p = 9.716$.

7. 0.61803.
8. 0.64984.
9. 0.51764.
10. $b = \dfrac{c}{2 \cos \dfrac{90°}{n}}$.

11. 0.2238.
12. 0.310.
13. 0.82842.
14. 94.63.
15. 414.97.

16. 11.636.
17. 99.640.
18. 1.0235.
19. 0.635.

Exercise XII. Page 45

5. Two angles; one in Quadrant I, one in Quadrant II.

6. Four values; two in Quadrant I, two in Quadrant IV.

7. x may have two values in the first case, and one value in each of the other cases.

8. If $\cos x = -\frac{2}{3}$, x is between 90° and 270°; if $\cot x = 4$, x is between 0° and 90° or between 180° and 270°; if $\sec x = 80$, x is between 0° and 90° or between 270° and 360°; if $\csc x = -3$, x is between 180° and 360°.

9. In Quadrant III; in Quadrant II; in Quadrant III.

10. 40 angles; 20 positive and 20 negative.

11. $+$, when x is known to be in Quadrant I or IV; $-$, when x is known to be in Quadrant II or III.

12. $\sin x = +\frac{1}{2}\sqrt{2}$; $\tan x = -1$; $\cot x = -1$; $\sec x = -\sqrt{2}$; $\csc x = +\sqrt{2}$.

13. $\sin x = -\frac{1}{2}\sqrt{3}$; $\cos x = -\frac{1}{2}$; $\cot x = +\frac{1}{3}\sqrt{3}$; $\sec x = -2$; $\csc x = -\frac{2}{3}\sqrt{3}$.

14. $\sin x = -\frac{4}{7}\sqrt{3}$; $\cos x = \frac{1}{7}$; $\tan x = -4\sqrt{3}$; $\cot x = -\frac{1}{12}\sqrt{3}$; $\csc x = -\frac{7}{12}\sqrt{3}$.

15. $\sin x = \pm\frac{1}{10}\sqrt{10}$; $\cos x = \mp\frac{3}{10}\sqrt{10}$; $\tan x = -\frac{1}{3}$; $\sec x = \mp\frac{1}{3}\sqrt{10}$; $\csc x = \pm\sqrt{10}$.

16. The cosine, the tangent, the cotangent, and the secant are negative when the angle is obtuse.

18. $\sin 90° = 1$, $\cos 90° = 0$, $\cot 90° = 0$, $\sec 90° = \infty$, $\csc 90° = 1$;
$\sin 180° = 0$, $\tan 180° = 0$, $\cot 180° = \infty$, $\sec 180° = -1$, $\csc 180° = \infty$;
$\sin 270° = -1$, $\cos 270° = 0$, $\tan 270° = \infty$, $\sec 270° = \infty$. $\csc 270° = -1$;
$\sin 360° = 0$, $\cos 360° = 1$, $\tan 360° = 0$, $\cot 360° = \infty$, $\sec 360° = 1$.

19. $\sin 450° = 1$; $\tan 540° = 0$; $\cos 630° = 0$; $\cot 720° = \infty$; $\sin 810° = 1$; $\csc 900° = \infty$.

20. 0.　　　　**21.** 0.　　　　**22.** 0.　　　　**23.** $a^2 - b^2 + 4ab$.

Exercise XIII.　Page 51

1. $-\cos 20°$.

2. $\sin 8°$.

3. $-\sin 10°$.

4. $-\cot 35°$.

5. $-\tan 1°$.

6. $-\csc 20°$.

7. $\csc 23°$.

8. $-\sin 24°$.

9. $\cos 1°$.

10. $-\cot 30°$.

11. $\tan 6°$.

12. $-\csc 26°$.

13. $-\sec 1°$.

14. $\sin 16° 11'$.

15. $-\cos 15° 38'$.

16. $\cot 0° 45'$.

17. $-\cot 40° 43'$.

18. $\csc 29° 45'$.

19. $\sec 2° 25'$.

20. $\sin(-75°) = -\cos 15°$;
$\cos(-75°) = \sin 15°$;
$\tan(-75°) = -\cot 15°$;
$\cot(-75°) = -\tan 15°$.

21. $\sin(-127°) = -\cos 37°$;
$\cos(-127°) = -\sin 37°$;
$\tan(-127°) = \cot 37°$;
$\cot(-127°) = \tan 37°$.

22. $\sin(-200°) = \sin 20°$;
$\cos(-200°) = -\cos 20°$;
$\tan(-200°) = -\tan 20°$;
$\cot(-200°) = -\cot 20°$.

23. $\sin(-345°) = \sin 15°$;
$\cos(-345°) = \cos 15°$;
$\tan(-345°) = \tan 15°$;
$\cot(-345°) = \cot 15°$.

24. $\sin(-52° 37') = -\cos 37° 23'$;
$\cos(-52° 37') = \sin 37° 23'$;
$\tan(-52° 37') = -\cot 37° 23'$;
$\cot(-52° 37') = -\tan 37° 23'$.

25. $\sin(-196° 54') = \sin 16° 54'$;
$\cos(-196° 54') = -\cos 16° 54'$;
$\tan(-196° 54') = -\tan 16° 54'$;
$\cot(-196° 54') = -\cot 16° 54'$.

26. $\sin 120° = +\frac{1}{2}\sqrt{3}$; $\cos 120° = -\frac{1}{2}$; $\tan 120° = -\sqrt{3}$; $\cot 120° = -\frac{1}{3}\sqrt{3}$.

27. $\sin 135° = +\frac{1}{2}\sqrt{2}$; $\cos 135° = -\frac{1}{2}\sqrt{2}$; $\tan 135° = -1$; $\cot 135° = -1$.

28. $\sin 150° = \frac{1}{2}$; $\cos 150° = -\frac{1}{2}\sqrt{3}$; $\tan 150° = -\frac{1}{3}\sqrt{3}$; $\cot 150° = -\sqrt{3}$.

29. $\sin 210° = -\frac{1}{2}$; $\cos 210° = -\frac{1}{2}\sqrt{3}$; $\tan 210° = +\frac{1}{3}\sqrt{3}$; $\cot 210° = +\sqrt{3}$.

30. $\sin 225° = -\frac{1}{2}\sqrt{2}$; $\cos 225° = -\frac{1}{2}\sqrt{2}$; $\tan 225° = 1$; $\cot 225° = 1$.

31. $\sin 240° = -\frac{1}{2}\sqrt{3}$; $\cos 240° = -\frac{1}{2}$; $\tan 240° = +\sqrt{3}$; $\cot 240° = +\frac{1}{3}\sqrt{3}$.

32. $\sin 300° = -\frac{1}{2}\sqrt{3}$; $\cos 300° = \frac{1}{2}$; $\tan 300° = -\sqrt{3}$; $\cot 300° = -\frac{1}{3}\sqrt{3}$.

33. $\sin(-30°) = -\frac{1}{2}$; $\cos(-30°) = +\frac{1}{2}\sqrt{3}$; $\tan(-30°) = -\frac{1}{3}\sqrt{3}$; $\cot(-30°) = -\sqrt{3}$.

34. $\sin(-225°) = +\frac{1}{2}\sqrt{2}$; $\cos(-225°) = -\frac{1}{2}\sqrt{2}$; $\tan(-225°) = -1$; $\cot(-225°) = -1$.

35. $\cos x = -\frac{1}{2}\sqrt{2}$; $\tan x = 1$; $\cot x = 1$; $x = 225°$.

36. $\sin x = \frac{1}{2}$; $\cos x = -\frac{1}{2}\sqrt{3}$; $\tan x = -\frac{1}{3}\sqrt{3}$; $x = 150°$.

37. $\sin 3540° = -\frac{1}{2}\sqrt{3}$; $\cos 3540° = \frac{1}{2}$; $\tan 3540° = -\sqrt{3}$; $\cot 3540° = -\frac{1}{3}\sqrt{3}$.

38. $210°$ and $330°$; $120°$ and $300°$.

39. $135°$, $225°$, and $-225°$; $150°$ and $-30°$.

40. $30°$, $150°$, $390°$, and $510°$.

41. $\sin 168°$; $\cos 334°$; $\tan 225°$; $\cot 252°$; $\sin 349°$; $\cos 240°$; $\tan 64°$; $\cot 177°$.

42. 0.8480.   **43.** -1.9522.   **44.** $(a-b)\sin x$.

45. $m \sin x \cos x$.   **48.** 0.

46. $(a-b)\cot x - (a+b)\tan x$.   **49.** $\cos x \sin y - \sin x \cos y$.

47. $a^2 + b^2 + 2ab \cos x$.   **50.** $\tan x$.

51. Positive between $x = 0°$ and $x = 135°$, and between $x = 315°$ and $x = 360°$; negative between $x = 135°$ and $x = 315°$.

52. Positive between $x = 45°$ and $x = 225°$; negative between $x = 0°$ and $x = 45°$, and between $x = 225°$ and $x = 360°$.

53. $\sin(x - 90°) = -\cos x$; $\cos(x - 90°) = \sin x$; $\tan(x - 90°) = -\cot x$; $\cot(x - 90°) = -\tan x$.

54. $\sin(x - 180°) = -\sin x$; $\cos(x - 180°) = -\cos x$; $\tan(x - 180°) = \tan x$; $\cot(x - 180°) = \cot x$.

Exercise XIV. Page 60

1. $\sin(x+y) = \frac{6}{5}\frac{3}{6}$; $\cos(x+y) = \frac{3}{6}\frac{3}{6}$.

2. $\sin(90° - y) = \cos y$; $\cos(90° - y) = \sin y$.

3. $\sin(90° + y) = \cos y$; $\cos(90° + y) = -\sin y$;
   $\tan(90° + y) = -\cot y$; $\cot(90° + y) = -\tan y$.

4. $\sin(180° - y) = \sin y$; $\cos(180° - y) = -\cos y$;
   $\tan(180° - y) = -\tan y$; $\cot(180° - y) = -\cot y$.

5. $\sin(180° + y) = -\sin y$; $\cos(180° + y) = -\cos y$;
   $\tan(180° + y) = \tan y$; $\cot(180° + y) = \cot y$.

6. $\sin(270° - y) = -\cos y$; $\cos(270° - y) = -\sin y$;
   $\tan(270° - y) = \cot y$; $\cot(270° - y) = \tan y$.

7. $\sin(270° + y) = -\cos y$; $\cos(270° + y) = \sin y$;
   $\tan(270° + y) = -\cot y$; $\cot(270° + y) = -\tan y$.

8. $\sin(360° - y) = -\sin y$; $\cos(360° - y) = \cos y$;
   $\tan(360° - y) = -\tan y$; $\cot(360° - y) = -\cot y$.

9. $\sin(360° + y) = \sin y$; $\cos(360° + y) = \cos y$;
   $\tan(360° + y) = \tan y$; $\cot(360° + y) = \cot y$.

10. $\sin(x - 90°) = -\cos x$; $\cos(x - 90°) = \sin x$;
    $\tan(x - 90°) = -\cot x$; $\cot(x - 90°) = -\tan x$.

11. $\sin(x - 180°) = -\sin x$; $\cos(x - 180°) = -\cos x$;
    $\tan(x - 180°) = \tan x$; $\cot(x - 180°) = \cot x$.

12. $\sin(x - 270°) = \cos x$; $\cos(x - 270°) = -\sin x$;
    $\tan(x - 270°) = -\cot x$; $\cot(x - 270°) = -\tan x$.

13. $\sin(-y) = -\sin y$; $\cos(-y) = \cos y$;
    $\tan(-y) = -\tan y$; $\cot(-y) = -\cot y$.

14. $\sin(45° - y) = \frac{1}{2}\sqrt{2}(\cos y - \sin y)$; $\cos(45° - y) = \frac{1}{2}\sqrt{2}(\cos y + \sin y)$;
    $\tan(45° - y) = \dfrac{1 - \tan y}{1 + \tan y}$; $\cot(45° - y) = \dfrac{\cot y + 1}{\cot y - 1}$.

15. $\sin(45° + y) = \frac{1}{2}\sqrt{2}(\cos y + \sin y)$; $\cos(45° + y) = \frac{1}{2}\sqrt{2}(\cos y - \sin y)$;
    $\tan(45° + y) = \dfrac{1 + \tan y}{1 - \tan y}$; $\cot(45° + y) = \dfrac{\cot y - 1}{\cot y + 1}$.

16. $\sin(30° + y) = \frac{1}{2}(\cos y + \sqrt{3}\sin y)$; $\cos(30° + y) = \frac{1}{2}(\sqrt{3}\cos y - \sin y)$;
    $\tan(30° + y) = \dfrac{\frac{1}{3}\sqrt{3} + \tan y}{1 - \frac{1}{3}\sqrt{3}\tan y}$; $\cot(30° + y) = \dfrac{\sqrt{3}\cot y - 1}{\cot y + \sqrt{3}}$.

17. $\sin(60° - y) = \frac{1}{2}(\sqrt{3}\cos y - \sin y)$; $\cos(60° - y) = \frac{1}{2}(\cos y + \sqrt{3}\sin y)$;

$\tan(60° - y) = \dfrac{\sqrt{3} - \tan y}{1 + \sqrt{3}\tan y}$; $\cot(60° - y) = \dfrac{\frac{1}{2}\sqrt{3}\cot y + 1}{\cot y - \frac{1}{2}\sqrt{3}}$.

18. $3\sin x - 4\sin^3 x$.   **19.** $4\cos^3 x - 3\cos x$.   **20.** 0.   **21.** $\frac{1}{2}\sqrt{3}$.

22. $\sin\frac{1}{2}x = \sqrt{\dfrac{1 - 0.4\sqrt{6}}{2}} = 0.10051$; $\cos\frac{1}{2}x = \sqrt{\dfrac{1 + 0.4\sqrt{6}}{2}} = 0.99493$.

23. $\cos 2x = -\frac{1}{2}$, $\tan 2x = -\sqrt{3}$.

24. $\sin 22\frac{1}{2}° = \frac{1}{2}\sqrt{2 - \sqrt{2}} = 0.3827$; $\cos 22\frac{1}{2}° = \frac{1}{2}\sqrt{2 + \sqrt{2}} = 0.9239$;
$\tan 22\frac{1}{2}° = \sqrt{2} - 1 = 0.4142$; $\cot 22\frac{1}{2}° = \sqrt{2} + 1 = 2.4142$.

25. $\sin 15° = \frac{1}{2}\sqrt{2 - \sqrt{3}} = 0.2588$; $\cos 15° = \frac{1}{2}\sqrt{2 + \sqrt{3}} = 0.9659$;
$\tan 15° = 2 - \sqrt{3} = 0.2679$; $\cot 15° = 2 + \sqrt{3} = 3.7321$.

34. $\sin A + \sin B + \sin C = \sin A + \sin B + \sin[180° - (A + B)]$
              $= \sin A + \sin B + \sin(A + B)$

By [20] and [12],
           $= 2\sin\frac{1}{2}(A + B)\cos\frac{1}{2}(A - B) + 2\sin\frac{1}{2}(A + B)\cos\frac{1}{2}(A + B)$
           $= 2\sin\frac{1}{2}(A + B)[\cos\frac{1}{2}(A - B) + \cos\frac{1}{2}(A + B)]$

By [22],
           $= 2\sin\frac{1}{2}(A + B)(2\cos\frac{1}{2}A\cos\frac{1}{2}B)$
           $= 4\sin\frac{1}{2}(A + B)\cos\frac{1}{2}A\cos\frac{1}{2}B$.

But $\cos\frac{1}{2}C = \cos[90° - \frac{1}{2}(A + B)] = \sin\frac{1}{2}(A + B)$.
     $\therefore \sin A + \sin B + \sin C = 4\cos\frac{1}{2}A\cos\frac{1}{2}B\cos\frac{1}{2}C$.

35. Proof similar to that for 34.

38. $\dfrac{2}{\sin 2x}$.

39. $2\cot 2x$.

40. $\dfrac{\cos(x - y)}{\sin x \cos y}$.

41. $\dfrac{\cos(x + y)}{\sin x \cos y}$.

42. $\tan^2 x$.

43. $\dfrac{\cos(x - y)}{\cos x \cos y}$.

44. $\dfrac{\cos(x + y)}{\cos x \cos y}$.

45. $\dfrac{\cos(x - y)}{\sin x \sin y}$.

46. $\dfrac{\cos(x + y)}{\sin x \sin y}$.

47. $\tan x \tan y$.

Exercise XV. Page 63

1. $\sin^{-1}\frac{1}{2}\sqrt{3} = 60° + 2n\pi$ or $120° + 2n\pi$;
     $\tan^{-1}\frac{1}{2}\sqrt{3} = 30° + 2n\pi$ or $210° + 2n\pi$;
     $\text{vers}^{-1}\frac{1}{2} = 60° + 2n\pi$ or $300° + 2n\pi$;

$$\cos^{-1}(-\tfrac{1}{2}\sqrt{2}) = 135° + 2\,n\pi \text{ or } 225° + 2\,n\pi;$$
$$\csc^{-1}\sqrt{2} = 45° + 2\,n\pi \text{ or } 135° + 2\,n\pi;$$
$$\tan^{-1}\infty = 90° + 2\,n\pi \text{ or } 270° + 2\,n\pi;$$
$$\sec^{-1}2 = 60° + 2\,n\pi \text{ or } 300° + 2\,n\pi;$$
$$\cos^{-1}(-\tfrac{1}{2}\sqrt{3}) = 150° + 2\,n\pi \text{ or } 210° + 2\,n\pi.$$

4. $\tfrac{1}{2}\sqrt{2}$.   **10.** $\pm\tfrac{5}{13}$.   **12.** $\pm\tfrac{1}{2}\sqrt{2}$.

8. $0°, 90°, 180°$.   **11.** $\pm\tfrac{7}{24}$.   **13.** $z = 0$ or $\pm\tfrac{1}{2}\sqrt{3}$.

Exercise XVI. Page 67

1. If, for instance, $C = 90°$, [25] becomes $\dfrac{a}{c} = \sin A$.

3. $a^2 = b^2 + c^2$; $a^2 = b^2 + c^2 - 2\,bc$; $a^2 = b^2 + c^2 + 2\,bc$; a right triangle; a straight line; a straight line.

4. $b = a\cos C + c\cos A$; $a = b\cos C + c\cos B$; $c = b\cos A$.

6. $90°$.

7. (i) $\dfrac{a - b}{a + b} = \tan(A - 45°)$; a right triangle.

    (ii) $a + b = (a - b)(2 + \sqrt{3})$; an isosceles triangle with the angles $30°$, $30°$, $120°$.

Exercise XVII. Page 69

9. 300 yards.

10. $AB = 59.564$ miles; $AC = 54.285$ miles.

11. 4.6064 miles; 4.4494 miles; 8.7733 miles.

12. 4.1501 and 8.67.

13. 6.1433 miles and 8.7918 miles.

14. 8 and 5.4723.

15. $a = 5$; $c = 9.6593$.

16. $a = 7$; $b = 8.573$.

17. Sides, 600 feet and 1039.2 feet; altitude, 519.6 feet.

18. $855 : 1607$.

19. 5.438 and 6.857.

20. 15.588.

Exercise XVIII. Page 74

1. Two; one; no solution; one; two; no solution; one.

11. 420.   **12.** 124.617.

Exercise XIX. Page 78

11. 6.
12. 10.392.
14. 8.9212.

15. 25.
16. 3800 yards.
17. 729.67 yards.

18. 10.266 miles.
19. 5.0032 and 2.3385.
20. 26° 0′ 10″ and 14° 5′ 50″.

21. 430.85 yards.

Exercise XX. Page 83

11. $A = 36° 52′ 12″$; $B = 53° 7′ 48″$; $C = 90°$.
12. $A = B = 33° 33′ 27″$; $C = 112° 53′ 6″$.
13. $A = B = C = 60°$.
14. $A = 28° 57′ 18″$; $B = 46° 34′ 6″$; $C = 104° 28′ 36″$.
15. $A = 45°$; $B = 120°$; $C = 15°$.   21. 54.516 miles.
16. $A = 45°$; $B = 60°$; $C = 75°$.   22. 84° 14′ 34″.
17. 4° 23′ 2″ W. of N., or W. of S.   23. 54° 48′ 54″.
18. 60°.   24. 105°; 15°; 60°.
20. 0.88877.   25. 12.434 inches.

Exercise XXI. Page 87

1. 4,333,600.
2. 365.68.
3. 13,260.
4. 8160.
5. 240.

6. 26,208.
7. 15,540.
8. 29,450 or 6982.8.
9. 17.3206.
10. 10.392.

11. 0.19975.
12. $F = ab \sin A$.
13. $F = \frac{1}{2}(a^2 - b^2)\tan A$.
14. 2,421,000.
15. 30°; 30°; 120°.

Exercise XXII. Page 88

1. 21.166 miles; 24.966 miles.   4. 30°.
2. 6.3399 miles.   5. 20 feet.
3. 119.29 feet.   6. 2.6247 or 21.4587.

Exercise XXIII. Page 90

1. 106.70 feet;
   142.86 feet.
2. 1023.9 feet.

3. 37° 34′ 5″.
4. 238,410 miles.
5. 861,860 miles.

6. 2922.4 miles.
7. 60°.
8. 8.2068.

9. 6.6081.

10. 199.56 feet.

11. 43.107 feet.

12. 45 feet.

13. 26° 34′.

14. 78.367 feet.

15. 75 feet.

16. 1.4446 miles.

17. 7912.8 miles.

18. 56.649 feet.

19. 69.282 feet.

20. 260.21 feet; 3690.3 feet.

21. 1.3438 miles.

22. 235.81 yards.

26. 8.0076 inches.

29. 460.46 feet.

30. 88.936 feet.

31. 13.657 miles per hour.

33. 56.564 feet.

34. 51.595 feet.

35. 101.892 feet.

37. N. 76° 56′ E.; 13.938 miles per hour.

38. 442.11 yards.

39. 255.78 feet.

40. 3121.1 feet; 3633.5 feet.

41. 529.49 feet.

42. 41.411 feet.

43. 234.51 feet.

44. 25.433 miles.

45. 294.69 feet.

46. 12,492.6 feet.

47. 6.3397 miles.

48. 210.44 feet.

50. 757.50 feet.

51. 520.01 yards.

52. 1366.4 feet.

53. 658.36 pounds; 22° 23′ 47″ with first force.

54. 88.326 pounds; 45° 37′ 16″ with known force.

57. 536.28; 500.16.

58. 345.48 feet.

59. 345.46 yards.

60. 61.23 feet.

62. 307.77 yards.

63. 19.8; 35.7; 44.5.

64. 45°, 135°, 225°, or 315°.

65. $\cos A = \dfrac{-m \pm \sqrt{m^2 + 4\,(n+1)}}{2}$.

66. $\sin A = \sqrt{\dfrac{m^2 - n^2}{1 - n^2}}$; $\cos B = \dfrac{n}{m}\sqrt{\dfrac{1 - m^2}{1 - n^2}}$.

67. 60°, 120°, 240°, or 300°.

68. 0°, 60°, 180°, or 300°.

69. 0°, 30°, 150°, 180°, 210°, 330°.

71. $r = \dfrac{a}{2}\cot\dfrac{180°}{n}$; $R = \dfrac{a}{2}\csc\dfrac{180°}{n}$.

72. $F = \frac{1}{2}\,bc \sin A$.

73. $F = \frac{1}{2}c^2 \sin A \sin B \csc (A + B)$.

74. $F = \sqrt{s\,(s - a)\,(s - b)\,(s - c)}$.

76. 199 acres 8 square chains.

77. 210 acres 9.1 square chains.

78. 12 acres 9.78 square chains.

79. 3 acres 0.392 square chains.

80. 12 acres 3.45 square chains.

81. 4 acres 6.634 square chains.

82. 14 acres 5.54 square chains.

83. 61 acres 4.97 square chains.

84. 4 acres 6.633 square chains.

85. 13.93 chains; 23.21 chains; 32.50 chains.

86. 9 acres 0.055 square chains.

88. 876.34.

89. 1229.5.

91. 1075.3.

92. 2660.4.

93. 16,281.

94. 435.76 square feet.

95. 49,088 square feet.

96. 749.95 square feet.

97. 422.38 square feet.

98. 1884.95 square feet.

99. 26.88.

102. 6.

107. 6.

108. 6086.4 feet.

109. 5° 25′ 6″ S.; 457.49 miles.

110. 460.79 miles; 383.13 miles.

111. 228.98 miles; 11° 39′ 6″ S.

112. S. 56° 7′ 32″ E.; 202.58 miles.

113. N. 17° 25′ 22″ W.; 37° 46′ 13″ N.

114. 244.35 miles; S. 56° 10′ 49″ E.

115. 359.87 miles.

117. Long. 68° 54′ 39″ W.

118. 103.57 miles.

119. 33° 18′ 22″ N.; 36° 23′ 53″ W.

120. N. 28° 47′ 26″ E.; 1292.8 miles.

121. S. 50° 39′ 44″ W.; 260.84 miles; 20° 9′ 30″ W.

122. 38° 20′ 34″ N.; 55° 12′ 4″ W.

123. 171.14 miles; 32° 43′ 38″ W.

124. N. 36° 52′ 12″ W.; 36° 7′ 37″ W.

125. 173.18 miles; 51° 16′ 16″ S.; 34° 12′ 43″ E.

126. S. 50° 57′ 48″ E.; 47° 14′ 35″ N.; 20° 48′ 37″ W.

127. N. 53° 20′ 21″ E., 16° 6′ 57″ W.; or N. 53° 20′ 21″ W., 25° 53′ 3″ W.

128. N. 47° 42′ 33″ E., 19° 27′ 22″ N., 121° 50′ 34″ E.; or N. 47° 42′ 33″ W., 19° 27′ 22″ N., 116° 9′ 26″ E.; or S. 47° 42′ 33″ E., 14° 32′ 38″ N., 121° 48′ 20″ E.; or S. 47° 42′ 33″ W., 14° 32′ 38″ N., 116° 11′ 40″ E.

129. 359.82 miles; 359.73 miles; 359.50 miles.

130. 35° 49′ 10″ S., 22° 2′ 44″ W.; N. 61° 42′ W.; 183.16 miles.

131. 42° 15′ 29″ N., 69° 5′ 11″ W.; N. 72° 32′ 40″ E.; 44.939 miles.

132. 32° 58′ 34″ S., 13° 1′ 53″ E.; N. 72° 3′ 43″ W.; 287.16 miles.

Exercise XXIV. Page 107

(The solutions here given are for angles less than 360°.)

79. $\sin \frac{1}{2}x = \pm \frac{1}{2}\sqrt{5}$; $\cos \frac{1}{2}x = \pm \frac{2}{5}\sqrt{5}$.

80. $\pm \sqrt{5} - 2$.

81. $\pm \frac{1}{4}\sqrt{3}$.

82. $\pm \frac{2}{3}$ or $\pm \frac{2}{3}$.

83. $\pm 2\sqrt{2}$, $\pm \frac{1}{23}(9\sqrt{3} + 8\sqrt{2})$, or $\pm \frac{1}{23}(9\sqrt{3} - 8\sqrt{2})$.

84. $\pm \frac{1}{2}$.

85. $\frac{1}{2}(\sqrt{5} - 1)$; $\frac{1}{2}(\sqrt{5} + 1)$.

86. $(a^{\frac{2}{3}} + b^{\frac{2}{3}})^{\frac{3}{2}}$.

87. $\left(\frac{1 \pm m}{2}\right)^{\frac{1}{2}} (1 \mp 2m)$.

88. $\pm \frac{1}{2}\sqrt{2}$ or $\pm \frac{1}{2}\sqrt{3}$.

89. $\frac{1}{2}$.

90. $\frac{2}{3}$ or $-\frac{1}{3}$.

91. $\pm \dfrac{a + 1}{\sqrt{2a + 1}}$.

92. 4.

93. $\tan(x + y)$.

94. $\dfrac{\sin x}{\sin y}$.

95. $-\tan x$.

96. $\tan^{-1} \dfrac{2x}{1 - 2x^3}$.

97. 2.

98. $-\tan^2 x + \cot^2 x$.

99. $x = \frac{1}{4}\pi$ or $\frac{3}{4}\pi$.

100. $x = 90°$ or $270°$.

101. $x = 21° 28'$ or $158° 32'$.

102. $x = 0°$ or $90°$.

103. $x = 30°, 150°, 199° 28'$, or $340° 32'$.

104. $x = 51° 19', 180°$, or $308° 41'$.

105. $x = 0°, 120°, 180°$, or $240°$.

106. $x = 45°, 161° 34', 225°$, or $341° 34'$.

107. $\theta = 60°, 120°, 240°$, or $300°$.

108. $\theta = 26° 34'$ or $206° 34'$. **110.** $x = 45°$ or $135°$.

109. $x = 30°$ or $150°$. **111.** $x = 30°, 150°$, or $270°$.

112. $x = 35° 16'', 144° 44', 215° 16'$, or $324° 44'$.

113. $x = 75° 58'$ or $255° 58'$.

114. $\theta = 60°, 180°$, or $300°$. **116.** $x = 30°, 150°, 210°$, or $330°$.

115. $\theta = 90°$ or $143° 8'$. **117.** $x = 30°, 150°$, or $270°$.

118. $x = 26° 34', 90°, 206° 34'$, or $270°$.

119. $x = 45°, 135°, 225°$, or $315°$.

120. $x = 45°, 135°, 225°$, or $315°$.

121. $x = 15°, 75°, 135°, 195°, 255°$, or $315°$

122. $x = 45°, 135°, 225°$, or $315°$.

123. $x = 0°, 60°, 120°, 180°, 240°$, or $300°$.

124. $x = 27° 58', 135°, 242° 2'$, or $315°$.

125. $x = 0°, 45°, 180°$, or $225°$.

126. $x = 32° 46', 147° 14', 212° 46'$, or $327° 14'$.

127. $x = 0°, 45°, 90°, 180°, 225°$, or $270°$.

128. $x = 0°, 65° 42', 180°$, or $204° 18'$.

129. $x = 0°, 90°, 120°, 240°$, or $270°$.

130. $x = 0°, 36°, 72°, 108°, 144°, 180°, 216°, 252°, 288°$, or $324°$.

131. $x = 30°, 150°, 210°$, or $330°$.

132. $x = 60°$ or $240°$.

133. $x = 54° 44', 125° 16', 234° 44'$, or $305° 16'$.

134. $x = 105°$ or $345°$.

135. $x = \tan^{-1} \dfrac{a^2 - 1}{2a}$.

136. $x = \cos^{-1} \dfrac{-a \pm \sqrt{a^2 + 8a + 8}}{4}$.

137. $x = 135°$, $315°$, or $\frac{1}{2}\sin^{-1}(1-a)$.

138. $x = 30°$, $60°$, $120°$, $150°$, $210°$, $240°$, $300°$, or $330°$.

139. $x = 60°$, $90°$, $120°$, $240°$, $270°$, or $300°$.

140. $x = 60°$, $90°$, $120°$, $240°$, $270°$, or $300°$.

141. $x = 120°$.　　　　　**142.** $x = 14° 29'$, $30°$, $150°$, or $165° 31'$.

143. $x = 60°$, $90°$, $270°$, or $300°$.

144. $x = 0°$, $20°$, $100°$, $140°$, $180°$, $220°$, $260°$, or $340°$.

145. $x = 45°$, $90°$, $135°$, $225°$, $270°$, or $315°$.

146. $x = 30°$, $60°$, $90°$, $120°$, $150°$, $210°$, $240°$, $270°$, $300°$, or $330°$.

147. $x = 0°$, $45°$, $90°$, $180°$, $225°$, or $270°$.

148. $x = 30°$, $60°$, $120°$, $150°$, $210°$, $240°$, $300°$, or $330°$.

149. $x = 30°$, $90°$, $150°$, $210°$, $270°$, or $330°$.

150. $x = 0°$, $45°$, $180°$, or $225°$.

151. $x = 45°$, $60°$, $120°$, $135°$, $225°$, $240°$, $300°$, or $315°$.

152. $x = 0°$, $45°$, $135°$, $225°$, or $315°$.

153. $x = 30°$, $90°$, $150°$, $210°$, $270°$, or $330°$.

154. $x = 8°$ or $168°$.

155. $x = 40° 12'$, $139° 48'$, $220° 12'$, or $319° 48'$.

156. $x = 30°$ or $330°$.

157. $x = 60°$, $120°$, $240°$, or $300°$.

158. $x = 30°$, $60°$, $120°$, $150°$, $210°$, $240°$, $300°$, or $330°$.

159. $x = 53° 8'$, $126° 52'$, $233° 8'$, or $306° 52'$.

160. $x = 30°$.　　　　　**161.** $x = 22° 37'$ or $148° 8'$.

162. $x = 0°$, $20°$, $40°$, $60°$, $80°$, $100°$, $120°$, $140°$, $160°$, $180°$, $200°$, $220°$, $240°$, $260°$, $280°$, $300°$, $320°$, or $340°$.

163. $x = 22° 30'$, $45°$, $67° 30'$, $90°$, $112° 30'$, $135°$, $157° 30'$, $202° 30'$, $225°$, $247° 30'$, $270°$, $292° 30'$, $315°$, or $337° 30'$.

164. $x = 45°$ or $225°$.　　　　　**169.** $x = 1$.

165. $x = \pm 1$ or $\pm \frac{1}{2}\sqrt{21}$.　　　　　**170.** $x = 0$ or ± 1.

166. $x = \frac{1}{2}\sqrt{3}$ or $-\frac{1}{2}\sqrt{3}$.　　　　　**171.** $x = \pm \frac{1}{2}\sqrt{2}$.

167. $x = \frac{1}{2}\sqrt{3}$.　　　　　**172.** $x = \frac{1}{3}\sqrt{3}$.

168. $x = \frac{1}{2}$.　　　　　**173.** $\theta = 120°$ or $240°$.

174. $x = 60°$, $120°$, $240°$, or $300°$.

175. $x = 0°$, $45°$, $135°$, $225°$, or $315°$.

176. $x = 0°, 180°, 220° 39'$, or $319° 21'$.

177. $x = 0°, 60°, 120°, 180°, 240°$, or $300°$.

178. $\theta = 18°, 90°, 162°, 234°, 270°$, or $306°$.

179. $x = 0°, 30°, 90°, 150°, 180°, 210°, 270°$, or $330°$.

180. $x = 0°, 90°, 120°, 180°, 240°$, or $270°$.

181. $\theta = 0°, 74° 5', 127° 25', 180°, 232° 35'$, or $285° 55'$.

182. $x = 0°, 90°, 180°$, or $270°$.

183. $x = 0°, 45°, 90°, 180°, 225°$, or $270°$.

184. $x = 0°, 45°, 120°, 135°, 180°, 225°, 240°$, or $315°$.

185. $\theta = 10° 12', 34° 48', 190° 12',$ **187.** $\theta = 29° 19', 105° 41', 209° 19',$
or $214° 48'$.   or $285° 41'$.

186. $x = 90°$ or $270°$.   **188.** $x = 1$.

189. $x = \dfrac{b \sin \beta - a \cos \beta}{\sin(\beta - \alpha)}$; $y = \dfrac{a \cos \alpha - b \sin \alpha}{\sin(\beta - \alpha)}$.

190. $x = \tan^{-1}\dfrac{a}{b} + \cos^{-1}\frac{1}{2}\sqrt{a^2 + b^2}$; $y = \tan^{-1}\dfrac{a}{b} - \cos^{-1}\frac{1}{2}\sqrt{a^2 + b^2}$.

191. $\theta = \tan^{-1}\dfrac{a}{b}$; $r = \sqrt{a^2 + b^2}$.

192. $r = \dfrac{a}{\sin\left(\tan^{-1}\dfrac{a \cos \beta - b \sin \alpha}{a \sin \beta + b \cos \alpha} + \alpha\right)} = \dfrac{b}{\cos\left(\tan^{-1}\dfrac{a \cos \beta - b \sin \alpha}{a \sin \beta + b \cos \alpha} + \beta\right)}$;

$\theta = \tan^{-1}\dfrac{a \cos \beta - b \sin \alpha}{a \sin \beta + b \cos \alpha}$.

193. $r = \sqrt{a^2 + b^2 + c^2}$; $\theta = \tan^{-1}\dfrac{a}{b}$; $\phi = \tan^{-1}\dfrac{c}{\sqrt{a^2 + b^2}}$.

194. $x = 100$; $y = 200$.   **195.** $x = 76° 10'$; $y = 15° 30'$.

196. $r = 225.12$, $\theta = 24° 13'$; or $r = -225.12$, $\theta = 204° 13'$.

197. $r = 151$, $\theta = 42° 28'$; or $r = -151$, $\theta = 222° 28'$.

198. $r = 108$, $\phi = 120°$, $\theta = 330°$; $r = 108$, $\phi = 300°$, $\theta = 210°$;
$r = -108$, $\phi = 120°$, $\theta = 150°$; or $r = -108$, $\phi = 300°$, $\theta = 30°$.

199. $x = r \operatorname{vers}^{-1}\left(\dfrac{y}{r}\right) \mp \sqrt{2 ry - y^2}$.

Exercise XXV. Page 121

1. $\log_{10} 6 = 0.77815$; $\log_{10} 14 = 1.14613$; $\log_{10} 21 = 1.32222$;
$\log_{10} 4 = 0.60206$; $\log_{10} 12 = 1.07918$; $\log_{10} 5 = 0.69897$;
$\log_{10} \frac{1}{2} = \bar{1}.69897$; $\log_{10} \frac{1}{4} = \bar{1}.39794$; $\log_{10} \frac{7}{9} = \bar{1}.89086$;
$\log_{10} \frac{21}{20} = 0.02119$.

2. $\log_2 10 = 3.3219$; $\quad \log_2 5 = 2.3219$; $\quad \log_8 5 = 1.4650$;
$\log_7 \frac{1}{2} = -0.3562$; $\quad \log_5 \frac{2}{173} = -2.2620$.

3. $\log_e 2 = 0.69315$; $\quad \log_e 3 = 1.09862$; $\quad \log_e 5 = 1.60945$;
$\log_e 7 = 1.94593$; $\quad \log_e 8 = 2.07946$; $\quad \log_e 9 = 2.19724$;
$\log_e \frac{2}{3} = -0.40547$; $\quad \log_e \frac{4}{5} = -0.22315$; $\quad \log_e \frac{12}{5} = 0.25952$;
$\log_e \frac{7}{60} = -2.14845$.

4. $x = 1.54396$; $x = 0.83048$; $x = 0.42062$.

Exercise XXVI. Page 126

1. $\log_e 3 = 1.09861$. **2.** $\log_e 5 = 1.60944$. **3.** $\log_e 7 = 1.94591$.

4. $\log_e 10 = 2.3025850930$.

5. $\log_{10} 2 = 0.30103$; $\quad \log_{10} e = 0.43429$; $\quad \log_{10} 11 = 1.04139$.

Exercise XXVII. Page 128

1. $\sin 1' = 0.00029088820$; $\cos 1' = 0.99999995769$;
$\tan 1' = 0.000290888212$.

2. $\sin 2' = 0.000581776$. **3.** $\sin 1° = 0.0175$. **6.** $0° \, 40' \, 9''$.

Exercise XXVIII. Page 130

1. $\sin 6' = 0.0017453$; $\cos 6' = 0.9999985$.
2. $\sin 2° = 0.034902$; $\cos 2° = 0.999392$.
3. $\sin 3° = 0.052339$; $\cos 3° = 0.998632$.
4. $\sin 4° = 0.069760$; $\cos 4° = 0.997564$.
5. $\sin 5° = 0.087160$; $\cos 5° = 0.996193$.

Exercise XXIX. Page 135

1. The 6 sixth roots of -1 are:
$$\frac{\sqrt{3}+i}{2}, \; i, \; \frac{-\sqrt{3}+i}{2}, \; \frac{-\sqrt{3}-i}{2}, \; -i, \; \frac{\sqrt{3}-i}{2}.$$
The 6 sixth roots of $+1$ are:
$$1, \frac{1+\sqrt{-3}}{2}, \; \frac{-1+\sqrt{-3}}{2}, \; -1, \; \frac{-1-\sqrt{-3}}{2}, \; \frac{1-\sqrt{-3}}{2}.$$

2. $\dfrac{\sqrt{3}+i}{2}, \; \dfrac{-\sqrt{3}+i}{2}, \; -i.$

3. $\cos 67\frac{1}{2}°.+ i \sin 67\frac{1}{2}°$, $\cos 157\frac{1}{2}° + i \sin 157\frac{1}{2}°$, $\cos 247\frac{1}{2}° + i \sin 247\frac{1}{2}°$, $\cos 337\frac{1}{2}° + i \sin 337\frac{1}{2}°$.

4. $\sin 4\theta = 4 \cos^3 \theta \sin \theta - 4 \cos \theta \sin^3 \theta$;
$\cos 4\theta = \cos^4 \theta - 6 \cos^2 \theta \sin^2 \theta + \sin^4 \theta$.

Exercise XXX. Page 137

5. $\sec x = 1 + \dfrac{x^2}{2} + \dfrac{5 x^4}{24} + \dfrac{61 x^6}{720} + \cdots$

6. $x \cot x = 1 - \dfrac{x^2}{3} - \dfrac{x^4}{45} - \dfrac{2 x^6}{945} - \cdots$

7. $\sin 10° = 0.173648$; $\cos 10° = 0.984808$.

8. $\tan 15° = 0.267949$.

SPHERICAL TRIGONOMETRY

Exercise XXXI. Page 142

1. $110°$; $100°$; $80°$. 2. $140°$; $90°$; $55°$.

5. Multiply their measures in degrees by $\dfrac{\pi r}{180}$.

6. $\frac{1}{3} \pi$ feet; 2π feet; $\frac{2\frac{1}{3}}{} \pi$ feet.

Exercise XXXII. Page 146

3. (i) Either a or $b = 90°$; (ii) $A = 90°$ and $B = b$;
(iii) $A = 90°$ and $B = b$; (iv) $c = 90°$ and $b = B = 90°$.

Exercise XXXIII. Page 148

2. RULE I. The cosine of any middle part is equal to the product of the cotangents of the adjacent parts.
RULE II. The cosine of any middle part is equal to the product of the sines of the opposite parts.

Exercise XXXIV. Page 153

24. $A = 175° 57' 10''$; $B = 135° 42' 50''$; $C = 135° 34' 7''$.

25. $a = 104° 53' 2''$; $b = 133° 39' 48''$; $C = 104° 41' 39''$.

26. $a = 90°$; $b = B$; b and B are otherwise indeterminate.

27. $a = 60°$; $b = 90°$; $B = 90°$.

28. The triangle is impossible.

29. $b = 130° 41' 42''$; $c = 71° 27' 43''$; $A = 112° 57' 2''$.

30. $a = 26° 3' 51''$; $A = 85°$; $B = 65° 46'$.

31. The triangle is impossible.

32. $a = 60° 16' 17''$; $b = 29° 41' 4''$; $B = 38° 16' 54''$.

33. $b = 42° 10' 17''$; $c = 106° 37' 37''$; $A = 105° 41' 39''$.

34. $a = 113° 51' 5''$; $c = 105° 37' 54''$; $B = 50° 44' 19''$.

35. $a = 124° 10' 37''$; $b = 107° 7' 22''$; $c = 80° 28' 49''$.

Exercise XXXV. Page 157

1. $\cos A = \cot a \tan \tfrac{1}{2} b$; $\sin \tfrac{1}{2} B = \csc a \sin \tfrac{1}{2} b$; $\cos h = \cos a \sec \tfrac{1}{2} b$.

2. $\sin \tfrac{1}{2} A = \tfrac{1}{2} \sec \tfrac{1}{2} a$.

3. $\sin \tfrac{1}{2} A = \sec \tfrac{1}{2} a \cos \dfrac{180°}{n}$; $\sin R = \sin \tfrac{1}{2} a \csc \dfrac{180°}{n}$;

$\sin r = \tan \tfrac{1}{2} a \cot \dfrac{180°}{n}$.

4. Tetrahedron, $70° 31' 46''$; hexahedron, $90°$; octahedron, $109° 28' 14''$;
dodecahedron, $116° 33' 45''$; icosahedron $138° 11' 36''$.

5. $\cot \tfrac{1}{2} A = \sqrt{\cos a}$.

Exercise XXXVI. Page 160

1. (i) $\sin a \sin B = \sin b$, $\sin a \sin C = \sin c$;
  (ii) $\sin a = \sin b \sin A$, $\sin b \sin C = \sin c$;
 (iii) $\sin a = \sin c \sin A$, $\sin b = \sin c \sin B$;
 (iv) $\sin B = \sin b \sin A$, $\sin C = \sin c \sin A$;
  (v) $\sin a = \sin b$, $\sin c = \sin a \sin C = \sin b \sin C$;
 (vi) $\sin B = \sin A$, $\sin C = \sin c \sin A = \sin c \sin B$.

2. (i) $\cos a = \cos b \cos c$; (ii) $\cos b = \cos a \cos c$; (iii) $\cos c = \cos a \cos b$;
 (iv) $\cos a = \cos b \cos c$, $\cos b = \cos a \cos c$, $\cos c = \cos a \cos b$.

3. (i) $\cos a = \cos(b - c)$; (ii) $\cos a = \cos b \cos c$; (iii) $\cos a = \cos(b + c)$.

Exercise XXXVII. Page 167

1. (i) $\tan m = \tan b \cos A$, $\cos a = \cos b \sec m \cos(c - m)$;
 (ii) $\tan m = \tan c \cos B$, $\cos b = \cos c \sec m \cos(a - m)$.

Exercise XXXVIII. Page 169

1. (i) $\cot x = \tan B \csc a$, $\cos A = \cos B \csc x \sin (C - x)$;
   (ii) $\cot x = \tan C \csc b$, $\cos B = \cos C \csc x \sin (A - x)$.

Exercise XLIII. Page 180

4. $2.2298\ R^2$.
8. $1.1891\ R^2$.
12. $3.1416\ R^2$.
5. $1.4956\ R^2$ or $0.17085\ R^2$.
9. $0.7105\ R^2$.
13. $5.4206\ R^2$.
6. $0.95484\ R^2$.
10. $0.09301\ R^2$.
14. 2070.1 square miles.
7. $0.024832\ R^2$.
11. $2.8624\ R^2$.

Exercise XLIV. Page 197

1. $148°\ 42'$.
2. $\cos x = \cos A \cos B$.

3. Let w equal the inclination of the edge c to the plane of a and b.
   Then it is easily shown that $V = abc \sin l \sin w$. Now, conceive
   a sphere constructed having for centre the vertex of the trihedral
   angle whose edges are a, b, c. The spherical triangle, whose
   vertices are the points where a, b, c meet the surface of this
   sphere, has for its sides, l, m, n; and w is equal to the perpendicular arc to the side l from the opposite vertex. Let L, M, N
   denote the angles of this triangle. Then, by [39] and [47],

   $$\sin w = \sin m \sin N$$
   $$= 2 \sin m \sin \tfrac{1}{2} N \cos \tfrac{1}{2} N$$
   $$= \frac{2}{\sin l} \sqrt{\sin s \sin (s - l) \sin (s - m) \sin (s - n)},$$

   where
   $$s = \tfrac{1}{2}(l + m + n).$$
   $$\therefore V = 2\,abc \sqrt{\sin s \sin (s - l) \sin (s - m) \sin (s - n)}.$$

4. (i) 9,976,500 square miles; (ii) 13,316,560 square miles.

5. Let m equal the longitude of the point where the ship crosses the
   equator, B her course at the equator, d the distance sailed.

   Then,
   $$\tan m = \sin l \tan a,$$
   $$\cos B = \cos l \sin a,$$
   $$\cot d = \cot l \cos a.$$

6. Let k equal the arc of the parallel between the places, x the difference
   required.

   Then,
   $$x = 2 \cos l \sin^{-1} (\sin \tfrac{1}{2} d \sec l) - d.$$
   $$x = 90° (\sqrt{2} - 1).$$

7. $\tan \frac{1}{2}(m - m') = \sqrt{\sec s \sec (s - d) \sin (s - l) \sin (s - l')}$; where $2s = l + l' + d$, and m and m' are the longitudes of the places.

9. 44 minutes past 12 o'clock.   10. 60°.

11. $\cos t = -\tan d \tan l$; time of sunrise $= 12 - \dfrac{t}{15}$ o'clock A.M. ; time of sunset $= \dfrac{t}{15}$ o'clock P.M.; $\cos a = \sin d \sec l$. For longest day at Boston : time of sunrise, 4 hr. 26 min. 50 sec. A.M.; time of sunset, 7 hr. 33 min. 10 sec. P.M. Azimuth of sun at these times, 57° 25′ 15″ ; length of day, 15 hr. 6 min. 20 sec. ; for shortest day, times of sunrise and sunset are 7 hr. 33 min. 10 sec. A.M. and 4 hr. 26 min. 50 sec. P.M. ; azimuth of sun, 122° 34′ 45″ ; length of day, 8 hr. 53 min. 40 sec.

12. The problem is impossible when $\cot d < \tan l$; that is, for places in the frigid zones.

13. For the northern hemisphere and positive declination,
$\sin h = \sin l \sin d$, $\cot a = \cos l \tan d$; $h = 17° 14′ 35″$, $a = 73° 51′ 34″$ E.

14. The farther the place from the equator, the greater the sun's altitude at 6 A.M. in summer. At the equator it is 0°. At the north pole it is equal to the sun's declination. At a given place the sun's altitude at 6 A.M. is a maximum on the longest day of the year, and then $\sin h = \sin l \sin e$ (where $e = 23° 27′$).

15. $\cos t = \cot l \tan d$. Times of bearing due east and due west are $12 - \dfrac{t}{15}$ o'clock A.M. and $\dfrac{t}{15}$ o'clock P.M., respectively ; 6 hr. 58 min. 10 sec. A.M. and 5 hr. 1 min. 50 sec. P.M.

16. When the days and nights are equal, $d = 0°$, $\cos t = 0$, $t = 90°$; that is, sun is everywhere due east at 6 A.M., and due west at 6 P.M. Since l and d must both be less than 90°, $\cos t$ cannot be negative, therefore t cannot be greater than 90°. As d increases, t decreases ; that is, the times in question both approach noon.

If $l < d$, then $\cos t > 1$; therefore this case is impossible.

If $l = d$, then $\cos t = 1$, and $t = 0°$; that is, the times both coincide with noon. The explanation of this result is, that for $l = d$ the sun at noon is in the zenith, and south of the prime vertical at every other time.

If $l > d$, the diurnal circle of the sun and the prime vertical of the place meet in two points which separate farther and farther as l increases. At the pole the prime vertical is indeterminate; but near the pole, $t = 90°$, and the sun is always east at 6 A.M.

17. $\sin l = \sin d \csc h.$ **18.** $11° 50' 35''.$

19. The bearing of the wall, reckoned from the north point of the horizon, is given by the equation $\cot z = \cos l \tan d$; whence, for the given case, $x = 75° 12' 28''.$

20. $55° 45' 6'' $ N.   **21.** $63° 23' 41'' $ N. or S.

22. (i) $\cos t = -\tan l \cot p$; (ii) $t = z$; (iii) the result is indeterminate.

23. $\cot a = \cos l \tan d.$ **28.** $\sin d = \sin e \sin v$; $\tan r = \cos e \tan v.$

25. $h = 65° 37' 20''.$ **29.** $d = 32° 24' 12''$; $r = 301° 48' 17''.$

26. $h = 58° 25' 8''$; $a = 152° 28'.$ **30.** $d = 20° 48' 38''.$

27. $t = 45° 42'$; $l = 67° 58' 54''.$ **31.** 3 hr. 59 min. $27\frac{4}{15}$ sec. P.M.

32. $\cos \frac{1}{2} a = \sqrt{\cos \frac{1}{2}(l + h + p) \cos \frac{1}{2}(l + h - p) \sec l \sec h}.$

FIVE-PLACE

LOGARITHMIC AND TRIGONOMETRIC

TABLES

ARRANGED BY

G. A. WENTWORTH, A.M.

AND

G. A. HILL, A.M.

BOSTON, U.S.A., AND LONDON

PUBLISHED BY GINN & COMPANY

1903

INTRODUCTION.

1. If the natural numbers are regarded as powers of ten, the exponents of the powers are the Common or Briggs Logarithms of the numbers. If A and B denote natural numbers, a and b their logarithms, then $10^a = A$, $10^b = B$; or, written in logarithmic form,

$$\log A = a, \qquad \log B = b.$$

2. The logarithm of a product is found by adding the logarithms of its factors.

For, $\qquad A \times B = 10^a \times 10^b = 10^{a+b}$.

Therefore, $\qquad \log (A \times B) = a + b = \log A + \log B$.

3. The logarithm of a quotient is found by subtracting the logarithm of the divisor from that of the dividend.

For, $\qquad \dfrac{A}{B} = \dfrac{10^a}{10^b} = 10^{a-b}$.

Therefore, $\qquad \log \dfrac{A}{B} = a - b = \log A - \log B$.

4. The logarithm of a power of a number is found by multiplying the logarithm of the number by the exponent of the power.

For, $\qquad A^n = (10^a)^n = 10^{an}$.

Therefore, $\qquad \log A^n = an = n \log A$.

5. The logarithm of the root of a number is found by dividing the logarithm of the number by the index of the root.

For, $\qquad \sqrt[n]{A} = \sqrt[n]{10^a} = 10^{\frac{a}{n}}$.

Therefore, $\qquad \log \sqrt[n]{A} = \dfrac{a}{n} = \dfrac{\log A}{n}$.

6. The logarithms of 1, 10, 100, etc., and of 0.1, 0.01, 0.001, etc., are integral numbers. The logarithms of all other numbers are fractions.

For, $10^0 =$ 1, hence $\log 1 = 0$; $10^{-1} =$ 0.1, hence $\log 0.1 = -1$;
     $10^1 =$ 10, hence $\log 10 = 1$; $10^{-2} =$ 0.01, hence $\log 0.01 = -2$;
     $10^2 =$ 100, hence $\log 100 = 2$; $10^{-3} = 0.001$, hence $\log 0.001 = -3$;
     $10^3 = 1000$, hence $\log 1000 = 3$;   and so on.

If the number is between   1 and   10, the logarithm is between   0 and   1.
If the number is between   10 and   100, the logarithm is between   1 and   2.
If the number is between 100 and 1000, the logarithm is between   2 and   3.
If the number is between   1 and   0.1, the logarithm is between   0 and −1.
If the number is between   0.1 and   0.01, the logarithm is between −1 and −2.
If the number is between 0.01 and 0.001, the logarithm is between −2 and −3.
And so on.

7. If the number is less than 1, the logarithm is negative (§ 6), but is written in such a form that the *fractional part* is always *positive.*

For the number may be regarded as the product of two factors, one of which lies between 1 and 10, and the other is a negative power of 10 ; the logarithm will then take the form of a *difference* whose minuend is a positive proper fraction, and whose subtrahend is a positive integral number.

Thus, $0.48 = 4.8 \times 0.1$.
Therefore (§ 2), log $0.48 = \log 4.8 + \log 0.1 = 0.68124 - 1$.   (Page 1.)
Again, $0.0007 = 7 \times 0.0001$.
Therefore, $\log 0.0007 = \log 7 + \log 0.0001 = 0.84510 - 4$.

The logarithm $0.84510 - 4$ is often written $\overline{4}.84510$.

8. Every logarithm, therefore, consists of two parts : a positive or negative integral number, which is called the **Characteristic,** and a *positive* proper fraction, which is called the **Mantissa.**

Thus, in the logarithm 3.52179, the integral number 3 is the characteristic, and the fraction .52179 the mantissa. In the logarithm $0.78254 - 2$, the integral number $- 2$ is the characteristic, and the fraction 0.78254 is the mantissa.

9. If the logarithm is *negative*, it is customary to change the form of the difference so that the subtrahend shall be 10 or a multiple of 10. This is done by adding to both minuend and subtrahend a number which will increase the subtrahend to 10 or a multiple of 10.

Thus, the logarithm $0.78254 - 2$ is changed to $8.78254 - 10$ by adding 8 to both minuend and subtrahend. The logarithm $0.92737 - 13$ is changed to $7.92737 - 20$ by adding 7 to both minuend and subtrahend.

10. The following rules are derived from § 6 : —

If the number is *greater than* 1, make the characteristic of the logarithm *one unit less* than the *number of figures* on the left of the decimal point.

If the number is *less than* 1, make the characteristic of the logarithm *negative*, and *one unit more* than the *number of zeros* between the decimal point and the first significant figure of the given number.

If the characteristic of a given logarithm is *positive*, make the *number of figures* in the integral part of the corresponding number *one more* than the number of units in the characteristic.

If the characteristic is *negative*, make the *number of zeros* between the decimal point and the first significant figure of the corresponding number *one less* than the number of units in the characteristic.

Thus, the characteristic of log 7849.27 = 3 ;
the characteristic of log 0.037 = $-2 = 8.00000 - 10$.
If the characteristic is 4, the corresponding number has five figures in its integral part. If the characteristic is -3, that is, $7.00000 - 10$, the corresponding fraction has two zeros between the decimal point and the first significant figure.

11. The logarithms of numbers that can be derived one from another by multiplication or division by an integral power of 10 have the same mantissa.

For, multiplying or dividing a number by an integral power of 10 will increase or diminish its logarithm by the exponent of that power of 10; and since this exponent is an integer, the mantissa of the logarithm will be unaffected.

Thus, log 4.6021 $= 0.66296.$ (Page 9.)
  log 460.21 $= \log (4.6021 \times 10^2) = \log 4.6021 + \log 10^2$
      $= 0.66296 + 2 = 2.66296.$
  log 460210 $= \log (4.6021 \times 10^5) = \log 4.6021 + \log 10^5$
      $= 0.66296 + 5 = 5.66296.$
  log 0.046021 $= \log (4.6021 \div 10^2) = \log 4.6021 - \log 10^2$
      $= 0.66296 - 2 = 8.66296 - 10.$

TABLE I.

12. In this table (pp. 1–19) the vertical columns headed N contain the numbers, and the other columns the logarithms. On page 1 both the characteristic and the mantissa are printed. On pages 2–19 the mantissa only is printed.

The fractional part of a logarithm can be expressed only approximately, and in a five-place table all figures that follow the fifth are rejected. Whenever the sixth figure is 5, or more, the fifth figure is increased by 1. The figure 5̲ is written when the value of the figure in the place in which it stands, together with the succeeding figures, is more than 4½, but less than 5.

Thus, if the mantissa of a logarithm written to seven places is 5328732, it is written in this table (a five-place table) 53287. If it is 5328751, it is written 53288. If it is 5328461 or 5328499, it is written in this table 53285̲.

Again, if the mantissa is 5324981, it is written 53250̲; and if it is 4999967, it is written 5̲0000.

This distinction between 5 and $\underline{5}$, in case it is desired to curtail still further the mantissas of logarithms, removes all doubt whether a 5 in the last given place, or in the last but one followed by a zero, should be simply rejected, or whether the rejection should lead us to increase the preceding figure by one unit.

Thus, the mantissa 1392$\underline{5}$ when reduced to four places should be 1392 ; but 13925 should be 1393.

To Find the Logarithm of a Given Number.

13. If the given number consists of one or two significant figures, the logarithm is given on page 1. If zeros follow the significant figures, or if the number is a proper decimal fraction, the characteristic must be determined by § 10.

14. If the given number has three significant figures, it will be found in the column headed N (pp. 2–19), and the mantissa of its logarithm in the next column to the right, and on the same line. Thus,

Page 2.   log 145 = 2.16137,   log 14500 = 4.16137.
Page 14.   log 716 = 2.85491,   log 0.716 = 9.85491 − 10.

15. If the given number has four significant figures, the first three will be found in the column headed N, and the fourth at the top of the page in the line containing the figures **1, 2, 3,** etc. The mantissa will be found in the column headed by the fourth figure, and on the same line with the first three figures. Thus,

Page 15.   log 7682 = 3.88547,   log 76.85 = 1.88564.
Page 18.   log 93280 = 4.96979,   log 0.9468 = 9.97626 − 10.

16. If the given number has five or more significant figures, a process called **interpolation** is required.

Interpolation is based on the *assumption* that between two consecutive mantissas of the table the change in the mantissa is directly proportional to the change in the number.

Required the logarithm of 34237.

The required mantissa is (§ 11) the same as the mantissa for 3423.7 ; therefore it will be found by adding to the mantissa of 3423 seven-tenths of the difference between the mantissas for 3423 and 3424.

The mantissa for 3423 is 53441.

The difference between the mantissas for 3423 and 3424 is 12.

Hence, the mantissa for 3423.7 is 53441 + (0.7 × 12) = 53449.

Therefore, the required logarithm of 34237 is 4.53449.

Required the logarithm of 0.0015764.

The required mantissa is the same as the mantissa for 1576.4; therefore it will be found by adding to the mantissa for 1576 four-tenths of the difference between the mantissas for 1576 and 1577.

The mantissa for 1576 is 19756.

The difference between the mantissas for 1576 and 1577 is 27.

Hence, the mantissa for 1576.4 is 19756 + (0.4 × 27) = 19767.

Therefore, the required logarithm of 0.0015764 is 7.19767 − 10.

Required the logarithm of 32.6708.

The required mantissa is the same as the mantissa for 3267.08; therefore it will be found by adding to the mantissa for 3267 eight-hundredths of the difference between the mantissas for 3267 and 3268.

The mantissa for 3267 is 51415.

The difference between the mantissas for 3267 and 3268 is 13.

Hence, the mantissa for 3267.08 is 51415 + (0.08 × 13) = 51416.

Therefore, the required logarithm of 32.6708 is 1.51416.

17. When the fraction of a unit in the part to be added to the mantissa for four figures is less than 0.5 it is to be neglected; when it is 0.5 or more than 0.5 it is to be taken as one unit.

Thus, in the first example, the part to be added to the mantissa for 3423 is 8.4, and the .4 is rejected. In the second example, the part to be added to the mantissa for 1576 is 10.8, and 11 is added.

To Find the Antilogarithm; that is, the Number Corresponding to a Given Logarithm.

18. If the given mantissa can be found in the table, the first three figures of the required number will be found in the same line with the mantissa in the column headed N, and the fourth figure at the top of the column containing the mantissa.

The position of the decimal point is determined by the characteristic (§ 10).

Find the number corresponding to the logarithm 0.92002.

Page 16. The number for the mantissa 92002 is 8318.

The characteristic is 0; therefore, the required number is 8.318.

Find the number corresponding to the logarithm 6.09167.

Page 2. The number for the mantissa 09167 is 1235.

The characteristic is 6; therefore, the required number is 1235000.

Find the number corresponding to the logarithm 7.50325 − 10.

Page 6. The number for the mantissa 50325 is 3186.

The characteristic is − 3; therefore, the required number is 0.003186.

19. If the given mantissa cannot be found in the table, find in the table the two adjacent mantissas between which the given mantissa lies, and the four figures corresponding to the smaller of these two mantissas will be the first four significant figures of the required number. If more than four figures are desired, they may be found by interpolation, as in the following examples:

Find the number corresponding to the logarithm 1.48762.

Here the two adjacent mantissas of the table, between which the given mantissa 48762 lies, are found to be (page 6) 48756 and 48770. The corresponding numbers are 3073 and 3074. The smaller of these, 3073, contains the first four significant figures of the required number.

The difference between the two adjacent mantissas is 14, and the difference between the corresponding numbers is 1.

The difference between the smaller of the two adjacent mantissas, 48756, and the given mantissa, 48762, is 6. Therefore, the number to be annexed to 3073 is $\frac{6}{14}$ of $1 = 0.428$, and the fifth significant figure of the required number is 4.

Hence, the required number is 30.734.

Find the number corresponding to the logarithm $7.82326 - 10$.

The two adjacent mantissas between which 82326 lies are (page 13) 82321 and 82328. The number corresponding to the mantissa 82321 is 6656.

The difference between the two adjacent mantissas is 7, and the difference between the corresponding numbers is 1.

The difference between the smaller mantissa, 82321, and the given mantissa, 82326, is 5. Therefore, the number to be annexed to 6656 is $\frac{5}{7}$ of $1 = 0.7$, and the fifth significant figure of the required number is 7.

Hence, the required number is 0.0066567.

In using a five-place table the numbers corresponding to mantissas may be carried to five significant figures, and in the first part of the table to six figures.*

20. The logarithm of the reciprocal of a number is called the **Cologarithm** of the number.

If A denotes any number, then

$$\operatorname{colog} A = \log \frac{1}{A} = \log 1 - \log A \ (\S \ 3) = -\log A.$$

Hence, the cologarithm of a number is equal to the logarithm of the number with the minus sign prefixed, *which sign affects the entire logarithm, both characteristic and mantissa.*

* In most tables of logarithms proportional parts are given as an aid to interpolation; but, after a little practice, the operation can be performed nearly as rapidly without them. Their omission allows a page with larger-faced type and more open spacing, and consequently less trying to the eyes.

In order to avoid a negative mantissa in the cologarithm, it is customary to substitute for $-\log A$ its equivalent

$$(10 - \log A) - 10.$$

Hence, the cologarithm of a number is found by subtracting the logarithm of the number from 10, and then annexing -10 to the remainder.

The best way to perform the subtraction is to begin on the left and subtract each figure of $\log A$ from 9 until we reach the last significant figure, which must be subtracted from 10.

If $\log A$ is greater in absolute value than 10 and less than 20, then in order to avoid a negative mantissa, it is necessary to write $-\log A$ in the form

$$(20 - \log A) - 20.$$

So that, in this case, colog A is found by subtracting log A from 20, and then annexing -20 to the remainder.

Find the cologarithm of 4007.

$$\begin{array}{rl} & 10 \qquad -10 \\ \text{Page 8.} \qquad \log 4007 = & 3.60282 \\ \text{colog } 4007 = & \overline{6.39718 - 10} \end{array}$$

Find the cologarithm of 103992000000.

$$\begin{array}{rl} & 20 \qquad -20 \\ \text{Page 2.} \quad \log 103992000000 = & 11.01700 \\ \text{colog } 103992000000 = & \overline{8.98300 - 20} \end{array}$$

If the characteristic of log A is negative, then the subtrahend, -10 or -20, will vanish in finding the value of colog A.

Find the cologarithm of 0.004007.

$$\begin{array}{rl} & 10 \qquad -10 \\ \log 0.004007 = & 7.60282 - 10 \\ \text{colog } 0.004007 = & \overline{2.39718} \end{array}$$

With practice, the cologarithm of a number can be taken from the table as rapidly as the logarithm itself.

By using cologarithms the inconvenience of subtracting the logarithm of a divisor is avoided. For dividing by a number is equivalent to multiplying by its reciprocal. Hence, instead of subtracting the logarithm of a divisor its cologarithm may be added.

Find the logarithms of:

1. 6170.	4. 85.76.	7. 0.8694.	10. 67.3208.
2. 0.617.	5. 296.8.	8. 0.5908.	11. 18.5283.
3. 2867.	6. 7004.	9. 73243.	12. 0.0042003.

Find the cologarithms of:

13. 72433.	16. 869.278.	19. 0.002403.
14. 802.376.	17. 154000.	20. 0.000777.
15. 15.7643.	18. 70.0426.	21. 0.051828.

Find the antilogarithms of:

22. 2.47246.	25. 1.26784.	28. 9.79029 — 10.
23. 7.89081.	26. 3.79029.	29. 7.62328 — 10.
24. 2.91221.	· 27. 5.18752.	30. 6.15465 — 10.

COMPUTATION BY LOGARITHMS.

21. (1) Find the value of x, if $x = 72214 \times 0.08203$.

Page 14.	log 72214	= 4.85862
Page 16.	log 0.08203	= 8.91397 — 10
By § 2.	log x	= 3.77259
Page 11.	x	= 5923.63

(2) Find the value of x, if $x = 5250 \div 23487$.

Page 10.	log 5250	= 3.72016
Page 4.	colog 23487	= 5.62917 — 10
Page 4.	log x	= 9.34933 — 10 = log 0.22353
	∴ x	= 0.22353

(3) Find the value of x, if $x = \dfrac{7.56 \times 4667 \times 567}{899.1 \times 0.00337 \times 23435}$.

Page 15.	log 7.56	= 0.87852
Page 9.	log 4667	= 3.66904
Page 11.	log 567	= 2.75358
Page 17.	colog 899.1	= 7.04619 — 10
Page 6.	colog 0.00337	= 2.47237
Page 4.	colog 23435	= 5.63013 — 10
Page 5.	log x	= 2.44983 = log 281.73
	∴ x	= 281.73.

(4) Find the cube of 376.

Page 7.   log 376   = 2.57519
Multiply by 3 (§ 4),   3
Page 10.   log 376³   = $\overline{7.72557}$ = log 5315860♦
                 ∴ 376³   = 53158600.

(5) Find the square of 0.003278.

Page 6.   log 0.003278 = 7.51561 — 10
                                        2
Page 2.   log 0.003278² = $\overline{15.03122 — 20}$ = log 0.000010745
                 ∴ 0.003278² = 0.000010745.

(6) Find the square root of 8322.

Page 16.   log 8322   = 3.92023
Divide by 2 (§ 5),   2)3.92023
                 log √8322   = $\overline{1.96012}$ = log 91.226
                 ∴ √8322   = 91.226.

If the given number is a proper fraction, its logarithm will have as a subtrahend 10 or a multiple of 10. In this case, before dividing the logarithm by the index of the root, both the subtrahend and the number preceding the mantissa should be increased by such a number as will make the subtrahend, when divided by the index of the root, 10 or a multiple of 10.

(7) Find the square root of 0.000043641.

Page 8.   log 0.000043641 = 5.63989 — 1ι
                                    10   — 10
Divide by 2 (§ 5),   2)15.63989 — 20
Page 13.   log √0.000043641 = 7.81995 — 10 = log 0.0066062
                 ∴ √0.000043641 = 0.0066062.

(8) Find the sixth root of 0.076553.

Page 15.   log 0.076553   = 8.88397 — 10
                                    50   — 50
Divide by 6 (§ 5),   6)58.88397 — 60
Page 13.   log $\sqrt[6]{0.076553}$ = 9.81400 — 10 = log 0.65163
                 ∴ $\sqrt[6]{0.076553}$ = 0.65163.

EXERCISES.

Find by logarithms the value of :

1. $\dfrac{45607}{31045}$.   2. $\dfrac{5.6123}{0.01987}$.   3. $\dfrac{2.567}{0.05786}$.

4. $\dfrac{0.06547}{74.938 \times 0.05938}$.

5. $\dfrac{4.657 \times 0.03467}{3.908 \times 0.07189}$.

6. $\dfrac{0.0075389 \times 0.0079}{0.00907 \times 0\ 009784}$.

7. $\dfrac{312 \times 7.18 \times 31.82}{519 \times 8.27 \times 5.132}$.

8. $\dfrac{0.007 \times 57.83 \times 28.13}{9.317 \times 00.28 \times 476.5}$.

9. $\dfrac{5.55 \times 0.0007632 \times 0.87654}{2.79 \times 0.0009524 \times 1.46785}$.

10. $\sqrt{\dfrac{0.003457 \times 43.387 \times 99.2 \times 0.00025}{0.005824 \times 15.724 \times 1.38 \times 0.00089}}$

11. $\sqrt[3]{\dfrac{23.815 \times 29.36 \times 0.007 \times 0.62487}{0.00072 \times 9.236 \times 5.924 \times 3.0007}}$.

12. $\sqrt{\dfrac{3.1416 \times 0.031416 \times 0.0031416}{1.7285 \times 0.017285 \times 0.0017285}}$.

TABLE II.

22. This table (page 20) contains the value of the number π its most useful combinations, and their logarithms.

Find the length of an arc of 47° 32′ 57″ in a unit circle.

$$47° \ 32' \ 57'' = 171177''$$

$$\log 171177 \qquad = 5.23344$$

$$\log \frac{1}{a''} \qquad\qquad = 4.68557 - 10$$

$$\log \text{arc } 47° \ 32' \ 57'' = \overline{9.91901 - 10} = \log 0.82994$$

$$\therefore \text{ length of arc} \quad = 0.82994.$$

Find the angle if the length of its arc in a unit circle $= 0.54936$.

$$\log 0.54936 \qquad = 9.73986 - 10$$

$$\text{colog } \frac{1}{a''} = \log a'' \ = 5.31443$$

$$\log \text{ angle} \qquad\quad = \overline{5.05429} = \log 113316$$

$$\therefore \text{ angle} \qquad\qquad = 113316'' = 31° \ 28' \ 36''.$$

23. The relations between arcs and angles given in Table II. are readily deduced from the circular measure of an angle.

In **Circular Measure** an angle is defined by the equation

$$\text{angle} = \frac{\text{arc}}{\text{radius}},$$

in which the word arc denotes the length of the arc corresponding to the angle, when both arc and radius are expressed in terms of the *same linear unit*.

Since the arc and radius for a given angle in different circles vary in the same ratio, the value of the angle given by this equation is independent of the value of the radius.

The angle which is measured by a radius-arc is called a **Radian**, and is the *angular unit* in circular measure.

Since $C = 2\,\pi R$, it follows that $\frac{C}{R} = 2\,\pi$, and $\frac{\frac{1}{2}C}{R} = \pi$. Therefore,

If the arc = circumference,   the angle $= 2\,\pi$.
If the arc = semicircumference,   the angle $= \pi$.
If the arc = quadrant,   the angle $= \frac{1}{2}\,\pi$.
If the arc = radius,   the angle $= 1$.

Therefore, $\pi = 180°$, $\frac{1}{2}\pi = 90°$, $\frac{1}{3}\pi = 60°$, $\frac{1}{4}\pi = 45°$, $\frac{1}{6}\pi = 30°$, $\frac{1}{8}\pi = 22\frac{1}{2}°$, and so on.

Since 180° in common measure equals π units in circular measure,

$1°$ in common measure $= \frac{\pi}{180}$ units in circular measure;

1 unit in circular measure $= \frac{180°}{\pi}$ in common measure.

By means of these two equations, the value of an angle expressed in one measure may be changed to its value in the other measure.

Thus, the angle whose arc is equal to the radius is an angle of 1 unit in circular measure, and is equal to $\frac{180°}{\pi}$, or $57° \ 17' \ 45''$, very nearly.

TABLE III.

24. This table (pp. 21–49) contains the logarithms of the trigonometric functions of angles. In order to avoid negative characteristics, the characteristic of every logarithm is printed 10 too large. Therefore, — 10 is to be annexed to each logarithm.

On pages 28–49 the characteristic remains the same throughout each column, and is printed at the top and the bottom of the column.

But on pp. 30, 49, the characteristic changes one unit in value at the places marked with bars. Above these bars the proper characteristic is printed at the top, and below them at the bottom, of the column.

25. On pages 28–49 the log sin, log tan, log cot, and log cos, of 1° to 89°, are given to every minute. Conversely, this part of the table gives the value of the angle to the nearest minute when log sin, log tan, log cot, or log cos is known, provided log sin or log cos lies between 8.24186 and 9.99993, and log tan or log cot lies between 8.24192 and 11.75808.

If the exact value of the given logarithm of a function is not found in the table, the value nearest to it is to be taken, unless interpolation is employed as explained in § 26.

If the angle is less than 45° the number of degrees is printed at the top of the page, and the number of minutes in the column to the left of the columns containing the logarithm. If the angle is greater than 45°, the number of degrees is printed at the bottom of the page, and the number of minutes in the column to the right of the columns containing the logarithms.

If the angle is less than 45°, the names of its functions are printed at the top of the page; if greater than 45°, at the bottom of the page. Thus,

Page 38. log sin 21° 37′ = 9.56631 − 10.
Page 45. log cot 36° 53′ = 10.12473 − 10 = 0.12473.
Page 37. log cos 69° 14′ = 9.54969 − 10.
Page 49. log tan 45° 59′ = 10.01491 − 10 = 0.01491.
Page 48. If log cos = 9.87468 − 10, angle = 41° 28′.
Page 34. If log cot = 9.39353 − 10, angle = 76° 6′.

If log sin = 9.47760 − 10, the nearest log sin in the table is 9.47774 − 10 (page 36), and the angle corresponding to this value is 17° 29′.

If log tan = 0.76520 = 10.76520 − 10, the nearest log tan in the table is 10.76490 − 10 (page 32), and the angle corresponding to this value is 80° 15′.

26. If it is desired to obtain the logarithms of the functions of angles that contain seconds, or to obtain the value of the angle in degrees, minutes, and seconds, from the logarithms of its functions, interpolation must be employed. Here it must be remembered that,

The difference between two consecutive angles in the table is 60″.

Log sin and log tan increase as the angle increases; log cos and log cot diminish as the angle increases.

Find log tan 70° 46′ 8″.

Page 37. log tan 70° 46′ = 0.45731.
The difference between the mantissas of log tan 70° 46′ and log tan 70° 47′ is 41, and $\frac{8}{60}$ of 41 = 5.
As the function is increasing, the 5 must be added to the figure in the fifth place of the mantissa 45731; and
Therefore log tan 70° 46′ 8″ = 0.45736.

Find log cos 47° 35′ 4″.

Page 48. log cos 47° 35′ = 9.82899 − 10.
The difference between this mantissa and the mantissa of the next log cos is 14, and $\frac{4}{60}$ of 14 = 1.
As the function is decreasing, the 1 must be subtracted from the figure in the fifth place of the mantissa 82899; and
Therefore log cos 47° 35′ 4″ = 9.82898 − 10.

Find the angle for which log sin = 9.45359 − 10.

Page 35. The mantissa of the nearest smaller log sin in the table is 45334.
The angle corresponding to this value is 16° 30′.
The difference between 45334 and the given mantissa, 45359, is 25.
The difference between 45334 and the next following mantissa, 45377, is 43, and $\frac{25}{43}$ of 60″ = 35″.
As the function is increasing, the 35″ must be added to 16° 30′; and the required angle is 16° 30′ 35″.

Find the angle for which log cot = 0.73478.

Page 32. The mantissa of the nearest smaller log cot in the table is 73415.
The angle corresponding to this value is 10° 27′.
The difference between 73415 and the given mantissa is 63.
The difference between 73415 and the next following mantissa is 71, and $\frac{63}{71}$ of 60″ = 53″.
As the function is decreasing, the 53″ must be subtracted from 10° 27′; and the required angle is 10° 26′ 7″.

EXERCISES.

Find

1. log sin 30° 8′ 9″.
2. log sin 54° 54′ 40″.
3. log cos 43° 32′ 31″.
4. log cos 69° 25′ 11″.
5. log tan 32° 9′ 17″.
6. log tan 50° 2′ 2″.
7. log cot 44° 33′ 17″.
8. log cot 55° 9′ 32″.
9. log tan 25° 27′ 47″.
10. log cos 56° 11′ 57″.
11. log cot 62° 0′ 4″.
12. log cos 75° 26′ 58″.
13. log tan 33° 27′ 13″.
14. log cot 81° 55′ 24″.
15. log tan 89° 46′ 35″.
16. log tan 1° 25′ 56″.

Find the angle A if

17.	log sin $A =$ 9.70075.	25.	log cos $A =$ 9.40008.
18.	log sin $A =$ 9.91289.	26.	log cot $A =$ 9.78815.
19.	log cos $A =$ 9.86026.	27.	log cos $A =$ 9.34301.
20.	log cos $A =$ 9.54595.	28.	log tan $A =$ 10.52288.
21.	log tan $A =$ 9.79840.	29.	log cot $A =$ 9 65349.
22.	log tan $A =$ 10.07671.	30.	log sin $A =$ 8.39316.
23.	log cot $A =$ 10.00675.	31.	log sin $A =$ 8.06678.
24.	log cot $A =$ 9.84266.	32.	log tan $A =$ 8.11148.

27. If log sec or log csc of an angle is desired, it may be found from the table by the formulas,

$$\sec A = \frac{1}{\cos A}; \text{ hence, log sec } A = \text{colog cos } A.$$

$$\csc A = \frac{1}{\sin A}; \text{ hence, log csc } A = \text{colog sin } A.$$

Page 31. log sec 8° 28′   = colog cos 8° 28′   = 0.00476.
Page 42. log csc 59° 36′ 44″ = colog sin 59° 36′ 44″ = 0.06418.

28. If a given angle is between 0° and 1°, or between 89° and 90°; or, conversely, if a given log sin or log cos does *not* lie between the limits 8.24186 and 9.99993 in the table; or, if a given log tan or log cot does *not* lie between the limits 8.24192 and 11.75808 in the table; then pages 21–24 of Table III. must be used.

On page 21, log sin of angles between 0° and 0° 3′, or log cos of the complementary angles between 89° 57′ and 90°, are given to every second; for the angles between 0° and 0° 3′, log tan = log sin, and log cos = 0.00000; for the angles between 89° 57′ and 90°, log cot = log cos, and log sin = 0.00000.

On pages 22–24, log sin, log tan, and log cos of angles between 0° and 1°, or log cos, log cot, and log sin of the complementary angles between 89° and 90°, are given to every 10″.

Whenever log tan or log cot is not given, they may be found by the formulas,

$$\text{log tan} = \text{colog cot}. \qquad \text{log cot} = \text{colog tan}.$$

Conversely, if a given log tan or log cot is not contained in the table, then the colog must be found; this will be the log cot or log tan, as the case may be, and will be contained in the table.

On pages 25–27 the logarithms of the functions of angles between 1° and 2°, or between 88° and 90°, are given in the manner employed on pages 22–24. These pages should be used if the angle lies between these limits, and if not only degrees and minutes, but degrees, minutes, and multiples of 10″ are given or required.

When the angle is between 0° and 2°, or 88° and 90°, and a greater degree of accuracy is desired than that given by the table, interpolation may be employed; but for these angles interpolation does not always give true results, and it is better to use Table IV.

Find log tan 0° 2' 47", and log cos 89° 37' 20".

> Page 21.   log tan   0°   2' 47" = log sin 0° 2' 47" = 6.90829 − 10.
> Page 23.   log cos 89° 37' 20" = 7.81911 − 10.

Find log cot 0° 2' 15".

>   10   − 10
> Page 21.   log tan 0° 2' 15"   =   6.81591 − 10
> Therefore, log cot 0° 2' 15"   =   3.18409

Find log tan 89° 38' 30".

>   10   − 10
> Page 23.   log cot 89° 38' 30" =   7.79617 − 10
> Therefore, log tan 89° 38' 30" =   2.20383

Find the angle for which log tan = 6.92090 − 10.

> Page 21.   The nearest log tan is 6.92110 − 10.
> The corresponding angle for which is 0° 2' 52".

Find the angle for which log cos = 7.70240 − 10.

> Page 22.   The nearest log cos is 7.70261 − 10.
> The corresponding angle for which is 89° 42' 40".

Find the angle for which log cot = 2.37368.

This log cot is not contained in the table.
The colog cot = 7.62632 − 10 = log tan.
The log tan in the table nearest to this is (page 22) 7.62510 − 10, and the angle corresponding to this value of log tan is 0° 14' 30".

29. If an angle x is between 90° and 360°, it follows, from formulas established in Trigonometry, that,

between 90° and 180°,	between 180° and 270°,
$\log \sin x = \log \sin (180° − x)$,	$\log \sin x = \log \sin (x − 180°)_n$,
$\log \cos x = \log \cos (180° − x)_n$,	$\log \cos x = \log \cos (x − 180°)_n$,
$\log \tan x = \log \tan (180° − x)_n$,	$\log \tan x = \log \tan (x − 180°)$,
$\log \cot x = \log \cot (180° − x)_n$;	$\log \cot x = \log \cot (x − 180°)$;

between 270° and 360°,

$\log \sin x = \log \sin (360° − x)_n$,
$\log \cos x = \log \cos (360° − x)$,
$\log \tan x = \log \tan (360° − x)_n$,
$\log \cot x = \log \cot (360° − x)_n$.

The letter n is placed (according to custom) after the logarithms of those functions which are negative in value.

The above formulas show, without further explanation, how to find by means of Table III. the logarithms of the functions of any angle between 90° and 360°.

Thus, log sin 137° 45′ 22″ = log sin   42° 14′ 38″ = 9.82756 — 10.
        log cos 137° 45′ 22″ = log$_n$ cos 42° 14′ 38″ = 9.86940$_n$ — 10.
        log tan 137° 45′ 22″ = log$_n$ tan 42° 14′ 38″ = 9.95815$_n$ — 10.
        log cot 137° 45′ 22″ = log$_n$ cot 42° 14′ 38″ = 0.04185$_n$.
        log sin 209° 32′ 50″ = log$_n$ sin 29° 32′ 50″ = 9.69297$_n$ — 10.
        log cos 330° 27′ 10″ = log cos 29° 32′ 50″ = 9.93949 — 10.

Conversely, to a given logarithm of a trigonometric function there correspond between 0° and 360° four angles, one angle in each quadrant, and so related that if x denote the acute angle, the other three angles are $180° - x$, $180° + x$, and $360° - x$.

If besides the given logarithm it is known whether the function is positive or negative, the ambiguity is confined to *two* quadrants, therefore to *two* angles.

Thus, if the log tan = 9.47451 — 10, the angles are 16° 36′ 17″ in Quadrant I. and 196° 36′ 17″ in Quadrant III.; but if the log tan = 9.47451$_n$ — 10, the angles are 163° 23′ 43″ in Quadrant II. and 343° 23′ 43″ in Quadrant IV.

To remove all ambiguity, further conditions are required, or a knowledge of the special circumstances connected with the problem in question.

TABLE IV. ·

30. This table (page 50) must be used when great accuracy is desired in working with angles between 0° and 2°, or between 88° and 90°.

The values of S and T are such that when the angle a is expressed in seconds,

$$S = \log \sin a - \log a'',$$
$$T = \log \tan a - \log a''.$$

Hence follow the formulas given on page 50.

The values of S and T are printed with the characteristic 10 too large, and in using them — 10 must always be annexed.

Find log sin 0° 58′ 17″.

    0° 58′ 17″ = 3497″
    log 3497 = 3.54370
            S = 4.68555 — 10
    log sin 0° 58′ 17″ = 8.22925 — 10

Find log cos 88° 26′ 41.2″.

    90° — 88° 26′ 41.2″ = 1° 33′ 18.8″
                        = 5598.8″
    log 5598.8 = 3.74809
             S = 4.68552 — 10
    log cos 88° 26′ 41.2″ = 8.43361 — 10

Find log tan 0° 52' 47.5".

$$0° 52' 47.5'' = 3167.5''$$
$$\log 3167.5 = 3.50072$$
$$T = 4.68561 - 10$$
$$\log \tan 0° 52' 47.5'' = 8.18633 - 10$$

Find log tan 89° 54' 37.362".

$$90° - 89° 54' 37.362'' = 0° 5' 22.638''$$
$$= 322.638''$$
$$\log 322.638 = 2.50871$$
$$T = 4.68558 - 10$$
$$\log \cot 89° 54' 37.362'' = 7.19429 - 10$$
$$\log \tan 89° 54' 37.362'' = 2.80571$$

Find the angle, if log sin = 6.72306 — 10.

$$6.72306 - 10$$
$$S = 4.68557 - 10$$
$$\text{Subtract,} \quad 2.03749 \quad = \log 109.015$$
$$109.015'' \quad = 0° 1' 49.015''.$$

Find the angle for which log cot = 1.67604.

$$\text{colog cot} = 8.32396 - 10$$
$$T = 4.68564 - 10$$
$$\text{Subtract,} \quad 3.63832 \quad = \log 4348.3$$
$$4348.3'' \quad = 1° 12' 28.3''.$$

Find the angle for which log tan = 1.55407.

$$\text{colog tan} = 8.44593 - 10$$
$$T = 4.68569 - 10$$
$$\text{Subtract,} \quad 3.76024 \quad = \log 5757.6$$
$$5757.6'' \quad = 1° 35' 57.6'',$$
$$\text{and} \quad 90° - 1° 35' 57.6'' = 88° 24' 2.4''.$$
Therefore, the angle required is 88° 24' 2.4''.

TABLE V.

31. This table (p. 51), containing the circumferences and areas of circles, does not require explanation.

TABLE VI.

32. Table VI. (pp. 52–69) contains the natural sines, cosines, tangents, and cotangents of angles from 0° to 90°, at intervals of 1'. If greater accuracy is desired it may be obtained by interpolation.

NOTE. In preparing the preceding explanations, we have made free use of the Logarithmic Tables by F. G. Gauss. For Table VI. we are indebted to D. Carhart.

TABLE VII.

33. This table (pp. 70–75) gives the latitude and departure to three places of decimals for distances from 1 to 10, corresponding to bearings from 0° to 90° at intervals of 15'.

If the bearing does not exceed 45° it is found in the *left*-hand column, and the designations of the columns under "Distance" are taken from the *top* of the page; but if the bearing exceeds 45°, it is found in the *right*-hand column, and the designations of the columns under "Distance" are taken from the *bottom* of the page.

The method of using the table will be made plain by the following examples : —

(1) Let it be required to find the latitude and departure of the course N. 35° 15′ E. 6 chains.

On p. 75, left-hand column, look for 35° 15′; opposite this bearing, in the vertical column headed "Distance 6," are found 4.900 and 3.463 under the headings "Latitude" and "Departure" respectively. Hence, latitude or northing = 4.900 chains, and departure or easting = 3.463 chains.

(2) Let it be required to find the latitude and departure of the course S. 87° W. 2 chains.

As the bearing exceeds 45°, we look in the right-hand column of p. 70, and opposite 87° in the column marked "Distance 2" we find (taking the designations of the columns from the bottom of the page) latitude = 0.105 chains, and departure = 1.997 chains. Hence, latitude or southing = 0.105 chains, and departure or westing = 1.997 chains.

(3) Let it be required to find the latitude and departure of the course N. 15° 45′ W. 27.36 chains.

In this case we find the required numbers for each figure of the distance separately, arranging the work as in the following table. In practice, only the last columns under "Latitude" and "Departure" are written.

DISTANCE.	LATITUDE.	DEPARTURE.
20 = 2 × 10	1.925 × 10 = 19.25	0.543 × 10 = 5.43
7	6.737	1.90
0.3 = 3 ÷ 10	2.887 ÷ 10 = 0.289	0.814 ÷ 10 = 0.081
0.06 = 6 ÷ 100	5.775 ÷ 100 = 0.058	1.628 ÷ 100 = 0.016
27.36	26.334	7.427

Hence, latitude = 26.334 chains, and departure = 7.427 chains.

TABLE I.

THE

COMMON OR BRIGGS LOGARITHMS

OF THE

NATURAL NUMBERS

From 1 to 10000.

1—100

N	log	N	log	N	log	N	log	N	log
1	0. 00 000	**21**	1. 32 222	**41**	1. 61 278	**61**	1. 78 533	**81**	1. 90 849
2	0. 30 103	22	1. 34 242	42	1. 62 325	62	1. 79 239	82	1. 91 381
3	0. 47 712	23	1. 36 173	43	1. 63 347	63	1. 79 934	83	1. 91 908
4	0. 60 206	24	1. 38 021	44	1. 64 345	64	1. 80 618	84	1. 92 428
5	0. 69 897	25	1. 39 794	45	1. 65 321	65	1. 81 291	85	1. 92 942
6	0. 77 815	**26**	1. 41 497	**46**	1. 66 276	**66**	1. 81 954	**86**	1. 93 450
7	0. 84 510	27	1. 43 136	47	1. 67 210	67	1. 82 607	87	1. 93 952
8	0. 90 309	28	1. 44 716	48	1. 68 124	68	1. 83 251	88	1. 94 448
9	0. 95 424	29	1. 46 240	49	1. 69 020	69	1. 83 885	89	1. 94 939
10	1. 00 000	30	1. 47 712	50	1. 69 897	70	1. 84 510	90	1. 95 424
11	1. 04 139	**31**	1. 49 136	**51**	1. 70 757	**71**	1. 85 126	**91**	1. 95 904
12	1. 07 918	32	1. 50 515	52	1. 71 600	72	1. 85 733	92	1. 96 379
13	1. 11 394	33	1. 51 851	53	1. 72 428	73	1. 86 332	93	1. 96 848
14	1. 14 613	34	1. 53 148	54	1. 73 239	74	1. 86 923	94	1. 97 313
15	1. 17 609	35	1. 54 407	55	1. 74 036	75	1. 87 506	95	1. 97 772
16	1. 20 412	**36**	1. 55 630	**56**	1. 74 819	**76**	1. 88 081	**96**	1. 98 227
17	1. 23 045	37	1. 56 820	57	1. 75 587	77	1. 88 649	97	1. 98 677
18	1. 25 527	38	1. 57 978	58	1. 76 343	78	1. 89 209	98	1. 99 123
19	1. 27 875	39	1. 59 106	59	1. 77 085	79	1. 89 763	99	1. 99 564
20	1. 30 103	40	1. 60 206	60	1. 77 815	80	1. 90 309	100	2. 00 000
N	**log**	**N**	**log**	**N**	**log**	**N**	**log**	**N**	**log**

N	0	1	2	3	4	5	6	7	8	9
100	00 000	00 043	00 087	00 130	00 173	00 217	00 260	00 303	00 346	00 389
101	00 432	00 475	00 518	00 561	00 604	00 647	00 689	00 732	00 775	00 817
102	00 860	00 903	00 945	00 988	01 030	01 072	01 115	01 157	01 199	01 242
103	01 284	01 326	01 368	01 410	01 452	01 494	01 536	01 578	01 620	01 662
104	01 703	01 745	01 787	01 828	01 870	01 912	01 953	01 995	02 036	02 078
105	02 119	02 160	02 202	02 243	02 284	02 325	02 366	02 407	02 449	02 490
106	02 531	02 572	02 612	02 653	02 694	02 735	02 776	02 816	02 857	02 898
107	02 938	02 979	03 019	03 060	03 100	03 141	03 181	03 222	03 262	03 302
108	03 342	03 383	03 423	03 463	03 503	03 543	03 583	03 623	03 663	03 703
109	03 743	03 782	03 822	03 862	03 902	03 941	03 981	04 021	04 060	04 100
110	04 139	04 179	04 218	04 258	04 297	04 336	04 376	04 415	04 454	04 493
111	04 532	04 571	04 610	04 650	04 689	04 727	04 766	04 805	04 844	04 883
112	04 922	04 961	04 999	05 038	05 077	05 115	05 154	05 192	05 231	05 269
113	05 308	05 346	05 385	05 423	05 461	05 500	05 538	05 576	05 614	05 652
114	05 690	05 729	05 767	05 805	05 843	05 881	05 918	05 956	05 994	06 032
115	06 070	06 108	06 145	06 183	06 221	06 258	06 296	06 333	06 371	06 408
116	06 446	06 483	06 521	06 558	06 595	06 633	06 670	06 707	06 744	06 781
117	06 819	06 856	06 893	06 930	06 967	07 004	07 041	07 078	07 115	07 151
118	07 188	07 225	07 262	07 298	07 335	07 372	07 408	07 445	07 482	07 518
119	07 555	07 591	07 628	07 664	07 700	07 737	07 773	07 809	07 846	07 882
120	07 918	07 954	07 990	08 027	08 063	08 099	08 135	08 171	08 207	08 243
121	08 279	08 314	08 350	08 386	08 422	08 458	08 493	08 529	08 565	08 600
122	08 636	08 672	08 707	08 743	08 778	08 814	08 849	08 884	08 920	08 955
123	08 991	09 026	09 061	09 096	09 132	09 167	09 202	09 237	09 272	09 307
124	09 342	09 377	09 412	09 447	09 482	09 517	09 552	09 587	09 621	09 656
125	09 691	09 726	09 760	09 795	09 830	09 864	09 899	09 934	09 968	10 003
126	10 037	10 072	10 106	10 140	10 175	10 209	10 243	10 278	10 312	10 346
127	10 380	10 415	10 449	10 483	10 517	10 551	10 585	10 619	10 653	10 687
128	10 721	10 755	10 789	10 823	10 857	10 890	10 924	10 958	10 992	11 025
129	11 059	11 093	11 126	11 160	11 193	11 227	11 261	11 294	11 327	11 361
130	11 394	11 428	11 461	11 494	11 528	11 561	11 594	11 628	11 661	11 694
131	11 727	11 760	11 793	11 826	11 860	11 893	11 926	11 959	11 992	12 024
132	12 057	12 090	12 123	12 156	12 189	12 222	12 254	12 287	12 320	12 352
133	12 385	12 418	12 450	12 483	12 516	12 548	12 581	12 613	12 646	12 678
134	12 710	12 743	12 775	12 808	12 840	12 872	12 905	12 937	12 969	13 001
135	13 033	13 066	13 098	13 130	13 162	13 194	13 226	13 258	13 290	13 322
136	13 354	13 386	13 418	13 450	13 481	13 513	13 545	13 577	13 609	13 640
137	13 672	13 704	13 735	13 767	13 799	13 830	13 862	13 893	13 925	13 956
138	13 988	14 019	14 051	14 082	14 114	14 145	14 176	14 208	14 239	14 270
139	14 301	14 333	14 364	14 395	14 426	14 457	14 489	14 520	14 551	14 582
140	14 613	14 644	14 675	14 706	14 737	14 768	14 799	14 829	14 860	14 891
141	14 922	14 953	14 983	15 014	15 045	15 076	15 106	15 137	15 168	15 198
142	15 229	15 259	15 290	15 320	15 351	15 381	15 412	15 442	15 473	15 503
143	15 534	15 564	15 594	15 625	15 655	15 685	15 715	15 746	15 776	15 806
144	15 836	15 866	15 897	15 927	15 957	15 987	16 017	16 047	16 077	16 107
145	16 137	16 167	16 197	16 227	16 256	16 286	16 316	16 346	16 376	16 406
146	16 435	16 465	16 495	16 524	16 554	16 584	16 613	16 643	16 673	16 702
147	16 732	16 761	16 791	16 820	16 850	16 879	16 909	16 938	16 967	16 997
148	17 026	17 056	17 085	17 114	17 143	17 173	17 202	17 231	17 260	17 289
149	17 319	17 348	17 377	17 406	17 435	17 464	17 493	17 522	17 551	17 580
150	17 609	17 638	17 667	17 696	17 725	17 754	17 782	17 811	17 840	17 869
N	0	1	2	3	4	5	6	7	8	9

N	0	1	2	3	4	5	6	7	8	9
150	17 609	17 638	17 667	17 696	17 725	17 754	17 782	17 811	17 840	17 869
151	17 898	17 926	17 955	17 984	18 013	18 041	18 070	18 099	18 127	18 156
152	18 184	18 213	18 241	18 270	18 298	18 327	18 355	18 384	18 412	18 441
153	18 469	18 498	18 526	18 554	18 583	18 611	18 639	18 667	18 696	18 724
154	18 752	18 780	18 808	18 837	18 865	18 893	18 921	18 949	18 977	19 005
155	19 033	19 061	19 089	19 117	19 145	19 173	19 201	19 229	19 257	19 285
156	19 312	19 340	19 368	19 396	19 424	19 451	19 479	19 507	19 535	19 562
157	19 590	19 618	19 645	19 673	19 700	19 728	19 756	19 783	19 811	19 838
158	19 866	19 893	19 921	19 948	19 976	20 003	20 030	20 058	20 085	20 112
159	20 140	20 167	20 194	20 222	20 249	20 276	20 303	20 330	20 358	20 385
160	20 412	20 439	20 466	20 493	20 520	20 548	20 575	20 602	20 629	20 656
161	20 683	20 710	20 737	20 763	20 790	20 817	20 844	20 871	20 898	20 925
162	20 952	20 978	21 005	21 032	21 059	21 085	21 112	21 139	21 165	21 192
163	21 219	21 245	21 272	21 299	21 325	21 352	21 378	21 405	21 431	21 458
164	21 484	21 511	21 537	21 564	21 590	21 617	21 643	21 669	21 696	21 722
165	21 748	21 775	21 801	21 827	21 854	21 880	21 906	21 932	21 958	21 985
166	22 011	22 037	22 063	22 089	22 115	22 141	22 167	22 194	22 220	22 246
167	22 272	22 298	22 324	22 350	22 376	22 401	22 427	22 453	22 479	22 505
168	22 531	22 557	22 583	22 608	22 634	22 660	22 686	22 712	22 737	22 763
169	22 789	22 814	22 840	22 866	22 891	22 917	22 943	22 968	22 994	23 019
170	23 045	23 070	23 096	23 121	23 147	23 172	23 198	23 223	23 249	23 274
171	23 300	23 325	23 350	23 376	23 401	23 426	23 452	23 477	23 502	23 528
172	23 553	23 578	23 603	23 629	23 654	23 679	23 704	23 729	23 754	23 779
173	23 805	23 830	23 855	23 880	23 905	23 930	23 955	23 980	24 005	24 030
174	24 055	24 080	24 105	24 130	24 155	24 180	24 204	24 229	24 254	24 279
175	24 304	24 329	24 353	24 378	24 403	24 428	24 452	24 477	24 502	24 527
176	24 551	24 576	24 601	24 625	24 650	24 674	24 699	24 724	24 748	24 773
177	24 797	24 822	24 846	24 871	24 895	24 920	24 944	24 969	24 993	25 018
178	25 042	25 066	25 091	25 115	25 139	25 164	25 188	25 212	25 237	25 261
179	25 285	25 310	25 334	25 358	25 382	25 406	25 431	25 455	25 479	25 503
180	25 527	25 551	25 575	25 600	25 624	25 648	25 672	25 696	25 720	25 744
181	25 768	25 792	25 816	25 840	25 864	25 888	25 912	25 935	25 959	25 983
182	26 007	26 031	26 055	26 079	26 102	26 126	26 150	26 174	26 198	26 221
183	26 245	26 269	26 293	26 316	26 340	26 364	26 387	26 411	26 435	26 458
184	26 482	26 505	26 529	26 553	26 576	26 600	26 623	26 647	26 670	26 694
185	26 717	26 741	26 764	26 788	26 811	26 834	26 858	26 881	26 905	26 928
186	26 951	26 975	26 998	27 021	27 045	27 068	27 091	27 114	27 138	27 161
187	27 184	27 207	27 231	27 254	27 277	27 300	27 323	27 346	27 370	27 393
188	27 416	27 439	27 462	27 485	27 508	27 531	27 554	27 577	27 600	27 623
189	27 646	27 669	27 692	27 715	27 738	27 761	27 784	27 807	27 830	27 852
190	27 875	27 898	27 921	27 944	27 967	27 989	28 012	28 035	28 058	28 081
191	28 103	28 126	28 149	28 171	28 194	28 217	28 240	28 262	28 285	28 307
192	28 330	28 353	28 375	28 398	28 421	28 443	28 466	28 488	28 511	28 533
193	28 556	28 578	28 601	28 623	28 646	28 668	28 691	28 713	28 735	28 758
194	28 780	28 803	28 825	28 847	28 870	28 892	28 914	28 937	28 959	28 981
195	29 003	29 026	29 048	29 070	29 092	29 115	29 137	29 159	29 181	29 203
196	29 226	29 248	29 270	29 292	29 314	29 336	29 358	29 380	29 403	29 425
197	29 447	29 469	29 491	29 513	29 535	29 557	29 579	29 601	29 623	29 645
198	29 667	29 688	29 710	29 732	29 754	29 776	29 798	29 820	29 842	29 863
199	29 885	29 907	29 929	29 951	29 973	29 994	30 016	30 038	30 060	30 081
200	30 103	30 125	30 146	30 168	30 190	30 211	30 233	30 255	30 276	30 298
N	0	1	2	3	4	5	6	7	8	9

N	0	1	2	3	4	5	6	7	8	9
200	30 103	30 125	30 146	30 168	30 190	30 211	30 233	30 255	30 276	30 298
201	30 320	30 341	30 363	30 384	30 406	30 428	30 449	30 471	30 492	30 514
202	30 535	30 557	30 578	30 600	30 621	30 643	30 664	30 685	30 707	30 728
203	30 750	30 771	30 792	30 814	30 835	30 856	30 878	30 899	30 920	30 942
204	30 963	30 984	31 006	31 027	31 048	31 069	31 091	31 112	31 133	31 154
205	31 175	31 197	31 218	31 239	31 260	31 281	31 302	31 323	31 345	31 366
206	31 387	31 408	31 429	31 450	31 471	31 492	31 513	31 534	31 555	31 576
207	31 597	31 618	31 639	31 660	31 681	31 702	31 723	31 744	31 765	31 785
208	31 806	31 827	31 848	31 869	31 890	31 911	31 931	31 952	31 973	31 994
209	32 015	32 035	32 056	32 077	32 098	32 118	32 139	32 160	32 181	32 201
210	32 222	32 243	32 263	32 284	32 305	32 325	32 346	32 366	32 387	32 408
211	32 428	32 449	32 469	32 490	32 510	32 531	32 552	32 572	32 593	32 613
212	32 634	32 654	32 675	32 695	32 715	32 736	32 756	32 777	32 797	32 818
213	32 838	32 858	32 879	32 899	32 919	32 940	32 960	32 980	33 001	33 021
214	33 041	33 062	33 082	33 102	33 122	33 143	33 163	33 183	33 203	33 224
215	33 244	33 264	33 284	33 304	33 325	33 345	33 365	33 385	33 405	33 425
216	33 445	33 465	33 486	33 506	33 526	33 546	33 566	33 586	33 606	33 626
217	33 646	33 666	33 686	33 706	33 726	33 746	33 766	33 786	33 806	33 826
218	33 846	33 866	33 885	33 905	33 925	33 945	33 965	33 985	34 005	34 025
219	34 044	34 064	34 084	34 104	34 124	34 143	34 163	34 183	34 203	34 223
220	34 242	34 262	34 282	34 301	34 321	34 341	34 361	34 380	34 400	34 420
221	34 439	34 459	34 479	34 498	34 518	34 537	34 557	34 577	34 596	34 616
222	34 635	34 655	34 674	34 694	34 713	34 733	34 753	34 772	34 792	34 811
223	34 830	34 850	34 869	34 889	34 908	34 928	34 947	34 967	34 986	35 005
224	35 025	35 044	35 064	35 083	35 102	35 122	35 141	35 160	35 180	35 199
225	35 218	35 238	35 257	35 276	35 295	35 315	35 334	35 353	35 372	35 392
226	35 411	35 430	35 449	35 468	35 488	35 507	35 526	35 545	35 564	35 583
227	35 603	35 622	35 641	35 660	35 679	35 698	35 717	35 736	35 755	35 774
228	35 793	35 813	35 832	35 851	35 870	35 889	35 908	35 927	35 946	35 965
229	35 984	36 003	36 021	36 040	36 059	36 078	36 097	36 116	36 135	36 154
230	36 173	36 192	36 211	36 229	36 248	36 267	36 286	36 305	36 324	36 342
231	36 361	36 380	36 399	36 418	36 436	36 455	36 474	36 493	36 511	36 530
232	36 549	36 568	36 586	36 605	36 624	36 642	36 661	36 680	36 698	36 717
233	36 736	36 754	36 773	36 791	36 810	36 829	36 847	36 866	36 884	36 903
234	36 922	36 940	36 959	36 977	36 996	37 014	37 033	37 051	37 070	37 088
235	37 107	37 125	37 144	37 162	37 181	37 199	37 218	37 236	37 254	37 273
236	37 291	37 310	37 328	37 346	37 365	37 383	37 401	37 420	37 438	37 457
237	37 475	37 493	37 511	37 530	37 548	37 566	37 585	37 603	37 621	37 639
238	37 658	37 676	37 694	37 712	37 731	37 749	37 767	37 785	37 803	37 822
239	37 840	37 858	37 876	37 894	37 912	37 931	37 949	37 967	37 985	38 003
240	38 021	38 039	38 057	38 075	38 093	38 112	38 130	38 148	38 166	38 184
241	38 202	38 220	38 238	38 256	38 274	38 292	38 310	38 328	38 346	38 364
242	38 382	38 399	38 417	38 435	38 453	38 471	38 489	38 507	38 525	38 543
243	38 561	38 578	38 596	38 614	38 632	38 650	38 668	38 686	38 703	38 721
244	38 739	38 757	38 775	38 792	38 810	38 828	38 846	38 863	38 881	38 899
245	38 917	38 934	38 952	38 970	38 987	39 005	39 023	39 041	39 058	39 076
246	39 094	39 111	39 129	39 146	39 164	39 182	39 199	39 217	39 235	39 252
247	39 270	39 287	39 305	39 322	39 340	39 358	39 375	39 393	39 410	39 428
248	39 445	39 463	39 480	39 498	39 515	39 533	39 550	39 568	39 585	39 602
249	39 620	39 637	39 655	39 672	39 690	39 707	39 724	39 742	39 759	39 777
250	39 794	39 811	39 829	39 846	39 863	39 881	39 898	39 915	39 933	39 950
N	0	1	2	3	4	5	6	7	8	9

N	0	1	2	3	4	5	6	7	8	9
250	39 794	39 811	39 829	39 846	39 863	39 881	39 898	39 915	39 933	39 950
251	39 967	39 985	40 002	40 019	40 037	40 054	40 071	40 088	40 106	40 123
252	40 140	40 157	40 175	40 192	40 209	40 226	40 243	40 261	40 278	40 295
253	40 312	40 329	40 346	40 364	40 381	40 398	40 415	40 432	40 449	40 466
254	40 483	40 500	40 518	40 535	40 552	40 569	40 586	40 603	40 620	40 637
255	40 654	40 671	40 688	40 705	40 722	40 739	40 756	40 773	40 790	40 807
256	40 824	40 841	40 858	40 875	40 892	40 909	40 926	40 943	40 960	40 976
257	40 993	41 010	41 027	41 044	41 061	41 078	41 095	41 111	41 128	41 145
258	41 162	41 179	41 196	41 212	41 229	41 246	41 263	41 280	41 296	41 313
259	41 330	41 347	41 363	41 380	41 397	41 414	41 430	41 447	41 464	41 481
260	41 497	41 514	41 531	41 547	41 564	41 581	41 597	41 614	41 631	41 647
261	41 664	41 681	41 697	41 714	41 731	41 747	41 764	41 780	41 797	41 814
262	41 830	41 847	41 863	41 880	41 896	41 913	41 929	41 946	41 963	41 979
263	41 996	42 012	42 029	42 045	42 062	42 078	42 095	42 111	42 127	42 144
264	42 160	42 177	42 193	42 210	42 226	42 243	42 259	42 275	42 292	42 308
265	42 325	42 341	42 357	42 374	42 390	42 406	42 423	42 439	42 455	42 472
266	42 488	42 504	42 521	42 537	42 553	42 570	42 586	42 602	42 619	42 635
267	42 651	42 667	42 684	42 700	42 716	42 732	42 749	42 765	42 781	42 797
268	42 813	42 830	42 846	42 862	42 878	42 894	42 911	42 927	42 943	42 959
269	42 975	42 991	43 008	43 024	43 040	43 056	43 072	43 088	43 104	43 120
270	43 136	43 152	43 169	43 185	43 201	43 217	43 233	43 249	43 265	43 281
271	43 297	43 313	43 329	43 345	43 361	43 377	43 393	43 409	43 425	43 441
272	43 457	43 473	43 489	43 505	43 521	43 537	43 553	43 569	43 584	43 600
273	43 616	43 632	43 648	43 664	43 680	43 696	43 712	43 727	43 743	43 759
274	43 775	43 791	43 807	43 823	43 838	43 854	43 870	43 886	43 902	43 917
275	43 933	43 949	43 965	43 981	43 996	44 012	44 028	44 044	44 059	44 075
276	44 091	44 107	44 122	44 138	44 154	44 170	44 185	44 201	44 217	44 232
277	44 248	44 264	44 279	44 295	44 311	44 326	44 342	44 358	44 373	44 389
278	44 404	44 420	44 436	44 451	44 467	44 483	44 498	44 514	44 529	44 545
279	44 560	44 576	44 592	44 607	44 623	44 638	44 654	44 669	44 685	44 700
280	44 716	44 731	44 747	44 762	44 778	44 793	44 809	44 824	44 840	44 855
281	44 871	44 886	44 902	44 917	44 932	44 948	44 963	44 979	44 994	45 010
282	45 025	45 040	45 056	45 071	45 086	45 102	45 117	45 133	45 148	45 163
283	45 179	45 194	45 209	45 225	45 240	45 255	45 271	45 286	45 301	45 317
284	45 332	45 347	45 362	45 378	45 393	45 408	45 423	45 439	45 454	45 469
285	45 484	45 500	45 515	45 530	45 545	45 561	45 576	45 591	45 606	45 621
286	45 637	45 652	45 667	45 682	45 697	45 712	45 728	45 743	45 758	45 773
287	45 788	45 803	45 818	45 834	45 849	45 864	45 879	45 894	45 909	45 924
288	45 939	45 954	45 969	45 984	46 000	46 015	46 030	46 045	46 060	46 075
289	46 090	46 105	46 120	46 135	46 150	46 165	46 180	46 195	46 210	46 225
290	46 240	46 255	46 270	46 285	46 300	46 315	46 330	46 345	46 359	46 374
291	46 389	46 404	46 419	46 434	46 449	46 464	46 479	46 494	46 509	46 523
292	46 538	46 553	46 568	46 583	46 598	46 613	46 627	46 642	46 657	46 672
293	46 687	46 702	46 716	46 731	46 746	46 761	46 776	46 790	46 805	46 820
294	46 835	46 850	46 864	46 879	46 894	46 909	46 923	46 938	46 953	46 967
295	46 982	46 997	47 012	47 026	47 041	47 056	47 070	47 085	47 100	47 114
296	47 129	47 144	47 159	47 173	47 188	47 202	47 217	47 232	47 246	47 261
297	47 276	47 290	47 305	47 319	47 334	47 349	47 363	47 378	47 392	47 407
298	47 422	47 436	47 451	47 465	47 480	47 494	47 509	47 524	47 538	47 553
299	47 567	47 582	47 596	47 611	47 625	47 640	47 654	47 669	47 683	47 698
300	47 712	47 727	47 741	47 756	47 770	47 784	47 799	47 813	47 828	47 842
N	**0**	**1**	**2**	**3**	**4**	**5**	**6**	**7**	**8**	**9**

N	0	1	2	3	4	5	6	7	8	9
300	47 712	47 727	47 741	47 756	47 770	47 784	47 799	47 813	47 828	47 842
301	47 857	47 871	47 885	47 900	47 914	47 929	47 943	47 958	47 972	47 986
302	48 001	48 015	48 029	48 044	48 058	48 073	48 087	48 101	48 116	48 130
303	48 144	48 159	48 173	48 187	48 202	48 216	48 230	48 244	48 259	48 273
304	48 287	48 302	48 316	48 330	48 344	48 359	48 373	48 387	48 401	48 416
305	48 430	48 444	48 458	48 473	48 487	48 501	48 515	48 530	48 544	48 558
306	48 572	48 586	48 601	48 615	48 629	48 643	48 657	48 671	48 686	48 700
307	48 714	48 728	48 742	48 756	48 770	48 785	48 799	48 813	48 827	48 841
308	48 855	48 869	48 883	48 897	48 911	48 926	48 940	48 954	48 968	48 982
309	48 996	49 010	49 024	49 038	49 052	49 066	49 080	49 094	49 108	49 122
310	49 136	49 150	49 164	49 178	49 192	49 206	49 220	49 234	49 248	49 262
311	49 276	49 290	49 304	49 318	49 332	49 346	49 360	49 374	49 388	49 402
312	49 415	49 429	49 443	49 457	49 471	49 485	49 499	49 513	49 527	49 541
313	49 554	49 568	49 582	49 596	49 610	49 624	49 638	49 651	49 665	49 679
314	49 693	49 707	49 721	49 734	49 748	49 762	49 776	49 790	49 803	49 817
315	49 831	49 845	49 859	49 872	49 886	49 900	49 914	49 927	49 941	49 955
316	49 969	49 982	49 996	50 010	50 024	50 037	50 051	50 065	50 079	50 092
317	50 106	50 120	50 133	50 147	50 161	50 174	50 188	50 202	50 215	50 229
318	50 243	50 256	50 270	50 284	50 297	50 311	50 325	50 338	50 352	50 365
319	50 379	50 393	50 406	50 420	50 433	50 447	50 461	50 474	50 488	50 501
320	50 515	50 529	50 542	50 556	50 569	50 583	50 596	50 610	50 623	50 637
321	50 651	50 664	50 678	50 691	50 705	50 718	50 732	50 745	50 759	50 772
322	50 786	50 799	50 813	50 826	50 840	50 853	50 866	50 880	50 893	50 907
323	50 920	50 934	50 947	50 961	50 974	50 987	51 001	51 014	51 028	51 041
324	51 055	51 068	51 081	51 095	51 108	51 121	51 135	51 148	51 162	51 175
325	51 188	51 202	51 215	51 228	51 242	51 255	51 268	51 282	51 295	51 308
326	51 322	51 335	51 348	51 362	51 375	51 388	51 402	51 415	51 428	51 441
327	51 455	51 468	51 481	51 495	51 508	51 521	51 534	51 548	51 561	51 574
328	51 587	51 601	51 614	51 627	51 640	51 654	51 667	51 680	51 693	51 706
329	51 720	51 733	51 746	51 759	51 772	51 786	51 799	51 812	51 825	51 838
330	51 851	51 865	51 878	51 891	51 904	51 917	51 930	51 943	51 957	51 970
331	51 983	51 996	52 009	52 022	52 035	52 048	52 061	52 075	52 088	52 101
332	52 114	52 127	52 140	52 153	52 166	52 179	52 192	52 205	52 218	52 231
333	52 244	52 257	52 270	52 284	52 297	52 310	52 323	52 336	52 349	52 362
334	52 375	52 388	52 401	52 414	52 427	52 440	52 453	52 466	52 479	52 492
335	52 504	52 517	52 530	52 543	52 556	52 569	52 582	52 595	52 608	52 621
336	52 634	52 647	52 660	52 673	52 686	52 699	52 711	52 724	52 737	52 750
337	52 763	52 776	52 789	52 802	52 815	52 827	52 840	52 853	52 866	52 879
338	52 892	52 905	52 917	52 930	52 943	52 956	52 969	52 982	52 994	53 007
339	53 020	53 033	53 046	53 058	53 071	53 084	53 097	53 110	53 122	53 135
340	53 148	53 161	53 173	53 186	53 199	53 212	53 224	53 237	53 250	53 263
341	53 275	53 288	53 301	53 314	53 326	53 339	53 352	53 364	53 377	53 390
342	53 403	53 415	53 428	53 441	53 453	53 466	53 479	53 491	53 504	53 517
343	53 529	53 542	53 555	53 567	53 580	53 593	53 605	53 618	53 631	53 643
344	53 656	53 668	53 681	53 694	53 706	53 719	53 732	53 744	53 757	53 769
345	53 782	53 794	53 807	53 820	53 832	53 845	53 857	53 870	53 882	53 895
346	53 908	53 920	53 933	53 945	53 958	53 970	53 983	53 995	54 008	54 020
347	54 033	54 045	54 058	54 070	54 083	54 095	54 108	54 120	54 133	54 145
348	54 158	54 170	54 183	54 195	54 208	54 220	54 233	54 245	54 258	54 270
349	54 283	54 295	54 307	54 320	54 332	54 345	54 357	54 370	54 382	54 394
350	54 407	54 419	54 432	54 444	54 456	54 469	54 481	54 494	54 506	54 518
N	0	1	2	3	4	5	6	7	8	9

N	0	1	2	3	4	5	6	7	8	9
350	54 407	54 419	54 432	54 444	54 456	54 469	54 481	54 494	54 506	54 518
351	54 531	54 543	54 555	54 568	54 580	54 593	54 605	54 617	54 630	54 642
352	54 654	54 667	54 679	54 691	54 704	54 716	54 728	54 741	54 753	54 765
353	54 777	54 790	54 802	54 814	54 827	54 839	54 851	54 864	54 876	54 888
354	54 900	54 913	54 925	54 937	54 949	54 962	54 974	54 986	54 998	55 011
355	55 023	55 035	55 047	55 060	55 072	55 084	55 096	55 108	55 121	55 133
356	55 145	55 157	55 169	55 182	55 194	55 206	55 218	55 230	55 242	55 255
357	55 267	55 279	55 291	55 303	55 315	55 328	55 340	55 352	55 364	55 376
358	55 388	55 400	55 413	55 425	55 437	55 449	55 461	55 473	55 485	55 497
359	55 509	55 522	55 534	55 546	55 558	55 570	55 582	55 594	55 606	55 618
360	55 630	55 642	55 654	55 666	55 678	55 691	55 703	55 715	55 727	55 739
361	55 751	55 763	55 775	55 787	55 799	55 811	55 823	55 835	55 847	55 859
362	55 871	55 883	55 895	55 907	55 919	55 931	55 943	55 955	55 967	55 979
363	55 991	56 003	56 015	56 027	56 038	56 050	56 062	56 074	56 086	56 098
364	56 110	56 122	56 134	56 146	56 158	56 170	56 182	56 194	56 205	56 217
365	56 229	56 241	56 253	56 265	56 277	56 289	56 301	56 312	56 324	56 336
366	56 348	56 360	56 372	56 384	56 396	56 407	56 419	56 431	56 443	56 455
367	56 467	56 478	56 490	56 502	56 514	56 526	56 538	56 549	56 561	56 573
368	56 585	56 597	56 608	56 620	56 632	56 644	56 656	56 667	56 679	56 691
369	56 703	56 714	56 726	56 738	56 750	56 761	56 773	56 785	56 797	56 808
370	56 820	56 832	56 844	56 855	56 867	56 879	56 891	56 902	56 914	56 926
371	56 937	56 949	56 961	56 972	56 984	56 996	57 008	57 019	57 031	57 043
372	57 054	57 066	57 078	57 089	57 101	57 113	57 124	57 136	57 148	57 159
373	57 171	57 183	57 194	57 206	57 217	57 229	57 241	57 252	57 264	57 276
374	57 287	57 299	57 310	57 322	57 334	57 345	57 357	57 368	57 380	57 392
375	57 403	57 415	57 426	57 438	57 449	57 461	57 473	57 484	57 496	57 507
376	57 519	57 530	57 542	57 553	57 565	57 576	57 588	57 600	57 611	57 623
377	57 634	57 646	57 657	57 669	57 680	57 692	57 703	57 715	57 726	57 738
378	57 749	57 761	57 772	57 784	57 795	57 807	57 818	57 830	57 841	57 852
379	57 864	57 875	57 887	57 898	57 910	57 921	57 933	57 944	57 955	57 967
380	57 978	57 990	58 001	58 013	58 024	58 035	58 047	58 058	58 070	58 081
381	58 092	58 104	58 115	58 127	58 138	58 149	58 161	58 172	58 184	58 195
382	58 206	58 218	58 229	58 240	58 252	58 263	58 274	58 286	58 297	58 309
383	58 320	58 331	58 343	58 354	58 365	58 377	58 388	58 399	58 410	58 422
384	58 433	58 444	58 456	58 467	58 478	58 490	58 501	58 512	58 524	58 535
385	58 546	58 557	58 569	58 580	58 591	58 602	58 614	58 625	58 636	58 647
386	58 659	58 670	58 681	58 692	58 704	58 715	58 726	58 737	58 749	58 760
387	58 771	58 782	58 794	58 805	58 816	58 827	58 838	58 850	58 861	58 872
388	58 883	58 894	58 906	58 917	58 928	58 939	58 950	58 961	58 973	58 984
389	58 995	59 006	59 017	59 028	59 040	59 051	59 062	59 073	59 084	59 095
390	59 106	59 118	59 129	59 140	59 151	59 162	59 173	59 184	59 195	59 207
391	59 218	59 229	59 240	59 251	59 262	59 273	59 284	59 295	59 306	59 318
392	59 329	59 340	59 351	59 362	59 373	59 384	59 395	59 406	59 417	59 428
393	59 439	59 450	59 461	59 472	59 483	59 494	59 506	59 517	59 528	59 539
394	59 550	59 561	59 572	59 583	59 594	59 605	59 616	59 627	59 638	59 649
395	59 660	59 671	59 682	59 693	59 704	59 715	59 726	59 737	59 748	59 759
396	59 770	59 780	59 791	59 802	59 813	59 824	59 835	59 846	59 857	59 868
397	59 879	59 890	59 901	59 912	59 923	59 934	59 945	59 956	59 966	59 977
398	59 988	59 999	60 010	60 021	60 032	60 043	60 054	60 065	60 076	60 086
399	60 097	60 108	60 119	60 130	60 141	60 152	60 163	60 173	60 184	60 195
400	60 206	60 217	60 228	60 239	60 249	60 260	60 271	60 282	60 293	60 304
N	0	1	2	3	4	5	6	7	8	9

N	0	1	2	3	4	5	6	7	8	9
400	60 206	60 217	60 228	60 239	60 249	60 260	60 271	60 282	60 293	60 304
401	60 314	60 325	60 336	60 347	60 358	60 369	60 379	60 390	60 401	60 412
402	60 423	60 433	60 444	60 455	60 466	60 477	60 487	60 498	60 509	60 520
403	60 531	60 541	60 552	60 563	60 574	60 584	60 595	60 606	60 617	60 627
404	60 638	60 649	60 660	60 670	60 681	60 692	60 703	60 713	60 724	60 735
405	60 746	60 756	60 767	60 778	60 788	60 799	60 810	60 821	60 831	60 842
406	60 853	60 863	60 874	60 885	60 895	60 906	60 917	60 927	60 938	60 949
407	60 959	60 970	60 981	60 991	61 002	61 013	61 023	61 034	61 045	61 055
408	61 066	61 077	61 087	61 098	61 109	61 119	61 130	61 140	61 151	61 162
409	61 172	61 183	61 194	61 204	61 215	61 225	61 236	61 247	61 257	61 268
410	61 278	61 289	61 300	61 310	61 321	61 331	61 342	61 352	61 363	61 374
411	61 384	61 395	61 405	61 416	61 426	61 437	61 448	61 458	61 469	61 479
412	61 490	61 500	61 511	61 521	61 532	61 542	61 553	61 563	61 574	61 584
413	61 595	61 606	61 616	61 627	61 637	61 648	61 658	61 669	61 679	61 690
414	61 700	61 711	61 721	61 731	61 742	61 752	61 763	61 773	61 784	61 794
415	61 805	61 815	61 826	61 836	61 847	61 857	61 868	61 878	61 888	61 899
416	61 909	61 920	61 930	61 941	61 951	61 962	61 972	61 982	61 993	62 003
417	62 014	62 024	62 034	62 045	62 055	62 066	62 076	62 086	62 097	62 107
418	62 118	62 128	62 138	62 149	62 159	62 170	62 180	62 190	62 201	62 211
419	62 221	62 232	62 242	62 252	62 263	62 273	62 284	62 294	62 304	62 315
420	62 325	62 335	62 346	62 356	62 366	62 377	62 387	62 397	62 408	62 418
421	62 428	62 439	62 449	62 459	62 469	62 480	62 490	62 500	62 511	62 521
422	62 531	62 542	62 552	62 562	62 572	62 583	62 593	62 603	62 613	62 624
423	62 634	62 644	62 655	62 665	62 675	62 685	62 696	62 706	62 716	62 726
424	62 737	62 747	62 757	62 767	62 778	62 788	62 798	62 808	62 818	62 829
425	62 839	62 849	62 859	62 870	62 880	62 890	62 900	62 910	62 921	62 931
426	62 941	62 951	62 961	62 972	62 982	62 992	63 002	63 012	63 022	63 033
427	63 043	63 053	63 063	63 073	63 083	63 094	63 104	63 114	63 124	63 134
428	63 144	63 155	63 165	63 175	63 185	63 195	63 205	63 215	63 225	63 236
429	63 246	63 256	63 266	63 276	63 286	63 296	63 306	63 317	63 327	63 337
430	63 347	63 357	63 367	63 377	63 387	63 397	63 407	63 417	63 428	63 438
431	63 448	63 458	63 468	63 478	63 488	63 498	63 508	63 518	63 528	63 538
432	63 548	63 558	63 568	63 579	63 589	63 599	63 609	63 619	63 629	63 639
433	63 649	63 659	63 669	63 679	63 689	63 699	63 709	63 719	63 729	63 739
434	63 749	63 759	63 769	63 779	63 789	63 799	63 809	63 819	63 829	63 839
435	63 849	63 859	63 869	63 879	63 889	63 899	63 909	63 919	63 929	63 939
436	63 949	63 959	63 969	63 979	63 988	63 998	64 008	64 018	64 028	64 038
437	64 048	64 058	64 068	64 078	64 088	64 098	64 108	64 118	64 128	64 137
438	64 147	64 157	64 167	64 177	64 187	64 197	64 207	64 217	64 227	64 237
439	64 246	64 256	64 266	64 276	64 286	64 296	64 306	64 316	64 326	64 335
440	64 345	64 355	64 365	64 375	64 385	64 395	64 404	64 414	64 424	64 434
441	64 444	64 454	64 464	64 473	64 483	64 493	64 503	64 513	64 523	64 532
442	64 542	64 552	64 562	64 572	64 582	64 591	64 601	64 611	64 621	64 631
443	64 640	64 650	64 660	64 670	64 680	64 689	64 699	64 709	64 719	64 729
444	64 738	64 748	64 758	64 768	64 777	64 787	64 797	64 807	64 816	64 826
445	64 836	64 846	64 856	64 865	64 875	64 885	64 895	64 904	64 914	64 924
446	64 933	64 943	64 953	64 963	64 972	64 982	64 992	65 002	65 011	65 021
447	65 031	65 040	65 050	65 060	65 070	65 079	65 089	65 099	65 108	65 118
448	65 128	65 137	65 147	65 157	65 167	65 176	65 186	65 196	65 205	65 215
449	65 225	65 234	65 244	65 254	65 263	65 273	65 283	65 292	65 302	65 312
450	65 321	65 331	65 341	65 350	65 360	65 369	65 379	65 389	65 398	65 408
N	0	1	2	3	4	5	6	7	8	9

N	0	1	2	3	4	5	6	7	8	9
450	65 321	65 331	65 341	65 350	65 360	65 369	65 379	65 389	65 398	65 408
451	65 418	65 427	65 437	65 447	65 456	65 466	65 475	65 485	65 495	65 504
452	65 514	65 523	65 533	65 543	65 552	65 562	65 571	65 581	65 591	65 600
453	65 610	65 619	65 629	65 639	65 648	65 658	65 667	65 677	65 686	65 696
454	65 706	65 715	65 725	65 734	65 744	65 753	65 763	65 772	65 782	65 792
455	65 801	65 811	65 820	65 830	65 839	65 849	65 858	65 868	65 877	65 887
456	65 896	65 906	65 916	65 925	65 935	65 944	65 954	65 963	65 973	65 982
457	65 992	66 001	66 011	66 020	66 030	66 039	66 049	66 058	66 068	66 077
458	66 087	66 096	66 106	66 115	66 124	66 134	66 143	66 153	66 162	66 172
459	66 181	66 191	66 200	66 210	66 219	66 229	66 238	66 247	66 257	66 266
460	66 276	66 285	66 295	66 304	66 314	66 323	66 332	66 342	66 351	66 361
461	66 370	66 380	66 389	66 398	66 408	66 417	66 427	66 436	66 445	66 455
462	66 464	66 474	66 483	66 492	66 502	66 511	66 521	66 530	66 539	66 549
463	66 558	66 567	66 577	66 586	66 596	66 605	66 614	66 624	66 633	66 642
464	66 652	66 661	66 671	66 680	66 689	66 699	66 708	66 717	66 727	66 736
465	66 745	66 755	66 764	66 773	66 783	66 792	66 801	66 811	66 820	66 829
466	66 839	66 848	66 857	66 867	66 876	66 885	66 894	66 904	66 913	66 922
467	66 932	66 941	66 950	66 960	66 969	66 978	66 987	66 997	67 006	67 015
468	67 025	67 034	67 043	67 052	67 062	67 071	67 080	67 089	67 099	67 108
469	67 117	67 127	67 136	67 145	67 154	67 164	67 173	67 182	67 191	67 201
470	67 210	67 219	67 228	67 237	67 247	67 256	67 265	67 274	67 284	67 293
471	67 302	67 311	67 321	67 330	67 339	67 348	67 357	67 367	67 376	67 385
472	67 394	67 403	67 413	67 422	67 431	67 440	67 449	67 459	67 468	67 477
473	67 486	67 495	67 504	67 514	67 523	67 532	67 541	67 550	67 560	67 569
474	67 578	67 587	67 596	67 605	67 614	67 624	67 633	67 642	67 651	67 660
475	67 669	67 679	67 688	67 697	67 706	67 715	67 724	67 733	67 742	67 752
476	67 761	67 770	67 779	67 788	67 797	67 806	67 815	67 825	67 834	67 843
477	67 852	67 861	67 870	67 879	67 888	67 897	67 906	67 916	67 925	67 934
478	67 943	67 952	67 961	67 970	67 979	67 988	67 997	68 006	68 015	68 024
479	68 034	68 043	68 052	68 061	68 070	68 079	68 088	68 097	68 106	68 115
480	68 124	68 133	68 142	68 151	68 160	68 169	68 178	68 187	68 196	68 205
481	68 215	68 224	68 233	68 242	68 251	68 260	68 269	68 278	68 287	68 296
482	68 305	68 314	68 323	68 332	68 341	68 350	68 359	68 368	68 377	68 386
483	68 395	68 404	68 413	68 422	68 431	68 440	68 449	68 458	68 467	68 476
484	68 485	68 494	68 502	68 511	68 520	68 529	68 538	68 547	68 556	68 565
485	68 574	68 583	68 592	68 601	68 610	68 619	68 628	68 637	68 646	68 655
486	68 664	68 673	68 681	68 690	68 699	68 708	68 717	68 726	68 735	68 744
487	68 753	68 762	68 771	68 780	68 789	68 797	68 806	68 815	68 824	68 833
488	68 842	68 851	68 860	68 869	68 878	68 886	68 895	68 904	68 913	68 922
489	68 931	68 940	68 949	68 958	68 966	68 975	68 984	68 993	69 002	69 011
490	69 020	69 028	69 037	69 046	69 055	69 064	69 073	69 082	69 090	69 099
491	69 108	69 117	69 126	69 135	69 144	69 152	69 161	69 170	69 179	69 188
492	69 197	69 205	69 214	69 223	69 232	69 241	69 249	69 258	69 267	69 276
493	69 285	69 294	69 302	69 311	69 320	69 329	69 338	69 346	69 355	69 364
494	69 373	69 381	69 390	69 399	69 408	69 417	69 425	69 434	69 443	69 452
495	69 461	69 469	69 478	69 487	69 496	69 504	69 513	69 522	69 531	69 539
496	69 548	69 557	69 566	69 574	69 583	69 592	69 601	69 609	69 618	69 627
497	69 636	69 644	69 653	69 662	69 671	69 679	69 688	69 697	69 705	69 714
498	69 723	69 732	69 740	69 749	69 758	69 767	69 775	69 784	69 793	69 801
499	69 810	69 819	69 827	69 836	69 845	69 854	69 862	69 871	69 880	69 888
500	69 897	69 906	69 914	69 923	69 932	69 940	69 949	69 958	69 966	69 975
N	0	1	2	3	4	5	6	7	8	9

N	0	1	2	3	4	5	6	7	8	9
500	69 897	69 906	69 914	69 923	69 932	69 940	69 949	69 958	69 966	69 975
501	69 984	69 992	70 001	70 010	70 018	70 027	70 036	70 044	70 053	70 062
502	70 070	70 079	70 088	70 096	70 105	70 114	70 122	70 131	70 140	70 148
503	70 157	70 165	70 174	70 183	70 191	70 200	70 209	70 217	70 226	70 234
504	70 243	70 252	70 260	70 269	70 278	70 286	70 295	70 303	70 312	70 321
505	70 329	70 338	70 346	70 355	70 364	70 372	70 381	70 389	70 398	70 406
506	70 415	70 424	70 432	70 441	70 449	70 458	70 467	70 475	70 484	70 492
507	70 501	70 509	70 518	70 526	70 535	70 544	70 552	70 561	70 569	70 578
508	70 586	70 595	70 603	70 612	70 621	70 629	70 638	70 646	70 655	70 663
509	70 672	70 680	70 689	70 697	70 706	70 714	70 723	70 731	70 740	70 749
510	70 757	70 766	70 774	70 783	70 791	70 800	70 808	70 817	70 825	70 834
511	70 842	70 851	70 859	70 868	70 876	70 885	70 893	70 902	70 910	70 919
512	70 927	70 935	70 944	70 952	70 961	70 969	70 978	70 986	70 995	71 003
513	71 012	71 020	71 029	71 037	71 046	71 054	71 063	71 071	71 079	71 088
514	71 096	71 105	71 113	71 122	71 130	71 139	71 147	71 155	71 164	71 172
515	71 181	71 189	71 198	71 206	71 214	71 223	71 231	71 240	71 248	71 257
516	71 265	71 273	71 282	71 290	71 299	71 307	71 315	71 324	71 332	71 341
517	71 349	71 357	71 366	71 374	71 383	71 391	71 399	71 408	71 416	71 425
518	71 433	71 441	71 450	71 458	71 466	71 475	71 483	71 492	71 500	71 508
519	71 517	71 525	71 533	71 542	71 550	71 559	71 567	71 575	71 584	71 592
520	71 600	71 609	71 617	71 625	71 634	71 642	71 650	71 659	71 667	71 675
521	71 684	71 692	71 700	71 709	71 717	71 725	71 734	71 742	71 750	71 759
522	71 767	71 775	71 784	71 792	71 800	71 809	71 817	71 825	71 834	71 842
523	71 850	71 858	71 867	71 875	71 883	71 892	71 900	71 908	71 917	71 925
524	71 933	71 941	71 950	71 958	71 966	71 975	71 983	71 991	71 999	72 008
525	72 016	72 024	72 032	72 041	72 049	72 057	72 066	72 074	72 082	72 090
526	72 099	72 107	72 115	72 123	72 132	72 140	72 148	72 156	72 165	72 173
527	72 181	72 189	72 198	72 206	72 214	72 222	72 230	72 239	72 247	72 255
528	72 263	72 272	72 280	72 288	72 296	72 304	72 313	72 321	72 329	72 337
529	72 346	72 354	72 362	72 370	72 378	72 387	72 395	72 403	72 411	72 419
530	72 428	72 436	72 444	72 452	72 460	72 469	72 477	72 485	72 493	72 501
531	72 509	72 518	72 526	72 534	72 542	72 550	72 558	72 567	72 575	72 583
532	72 591	72 599	72 607	72 616	72 624	72 632	72 640	72 648	72 656	72 665
533	72 673	72 681	72 689	72 697	72 705	72 713	72 722	72 730	72 738	72 746
534	72 754	72 762	72 770	72 779	72 787	72 795	72 803	72 811	72 819	72 827
535	72 835	72 843	72 852	72 860	72 868	72 876	72 884	72 892	72 900	72 908
536	72 916	72 925	72 933	72 941	72 949	72 957	72 965	72 973	72 981	72 989
537	72 997	73 006	73 014	73 022	73 030	73 038	73 046	73 054	73 062	73 070
538	73 078	73 086	73 094	73 102	73 111	73 119	73 127	73 135	73 143	73 151
539	73 159	73 167	73 175	73 183	73 191	73 199	73 207	73 215	73 223	73 231
540	73 239	73 247	73 255	73 263	73 272	73 280	73 288	73 296	73 304	73 312
541	73 320	73 328	73 336	73 344	73 352	73 360	73 368	73 376	73 384	73 392
542	73 400	73 408	73 416	73 424	73 432	73 440	73 448	73 456	73 464	73 472
543	73 480	73 488	73 496	73 504	73 512	73 520	73 528	73 536	73 544	73 552
544	73 560	73 568	73 576	73 584	73 592	73 600	73 608	73 616	73 624	73 632
545	73 640	73 648	73 656	73 664	73 672	73 679	73 687	73 695	73 703	73 711
546	73 719	73 727	73 735	73 743	73 751	73 759	73 767	73 775	73 783	73 791
547	73 799	73 807	73 815	73 823	73 830	73 838	73 846	73 854	73 862	73 870
548	73 878	73 886	73 894	73 902	73 910	73 918	73 926	73 933	73 941	73 949
549	73 957	73 965	73 973	73 981	73 989	73 997	74 005	74 013	74 020	74 028
550	74 036	74 044	74 052	74 060	74 068	74 076	74 084	74 092	74 099	74 107
N	0	1	2	3	4	5	6	7	8	9

N	0	1	2	3	4	5	6	7	8	9
550	74 036	74 044	74 052	74 060	74 068	74 076	74 084	74 092	74 099	74 107
551	74 115	74 123	74 131	74 139	74 147	74 155	74 162	74 170	74 178	74 186
552	74 194	74 202	74 210	74 218	74 225	74 233	74 241	74 249	74 257	74 265
553	74 273	74 280	74 288	74 296	74 304	74 312	74 320	74 327	74 335	74 343
554	74 351	74 359	74 367	74 374	74 382	74 390	74 398	74 406	74 414	74 421
555	74 429	74 437	74 445	74 453	74 461	74 468	74 476	74 484	74 492	74 500
556	74 507	74 515	74 523	74 531	74 539	74 547	74 554	74 562	74 570	74 578
557	74 586	74 593	74 601	74 609	74 617	74 624	74 632	74 640	74 648	74 656
558	74 663	74 671	74 679	74 687	74 695	74 702	74 710	74 718	74 726	74 733
559	74 741	74 749	74 757	74 764	74 772	74 780	74 788	74 796	74 803	74 811
560	74 819	74 827	74 834	74 842	74 850	74 858	74 865	74 873	74 881	74 889
561	74 896	74 904	74 912	74 920	74 927	74 935	74 943	74 950	74 958	74 966
562	74 974	74 981	74 989	74 997	75 005	75 012	75 020	75 028	75 035	75 043
563	75 051	75 059	75 066	75 074	75 082	75 089	75 097	75 105	75 113	75 120
564	75 128	75 136	75 143	75 151	75 159	75 166	75 174	75 182	75 189	75 197
565	75 205	75 213	75 220	75 228	75 236	75 243	75 251	75 259	75 266	75 274
566	75 282	75 289	75 297	75 305	75 312	75 320	75 328	75 335	75 343	75 351
567	75 358	75 366	75 374	75 381	75 389	75 397	75 404	75 412	75 420	75 427
568	75 435	75 442	75 450	75 458	75 465	75 473	75 481	75 488	75 496	75 504
569	75 511	75 519	75 526	75 534	75 542	75 549	75 557	75 565	75 572	75 580
570	75 587	75 595	75 603	75 610	75 618	75 626	75 633	75 641	75 648	75 656
571	75 664	75 671	75 679	75 686	75 694	75 702	75 709	75 717	75 724	75 732
572	75 740	75 747	75 755	75 762	75 770	75 778	75 785	75 793	75 800	75 808
573	75 815	75 823	75 831	75 838	75 846	75 853	75 861	75 868	75 876	75 884
574	75 891	75 899	75 906	75 914	75 921	75 929	75 937	75 944	75 952	75 959
575	75 967	75 974	75 982	75 989	75 997	76 005	76 012	76 020	76 027	76 035
576	76 042	76 050	76 057	76 065	76 072	76 080	76 087	76 095	76 103	76 110
577	76 118	76 125	76 133	76 140	76 148	76 155	76 163	76 170	76 178	76 185
578	76 193	76 200	76 208	76 215	76 223	76 230	76 238	76 245	76 253	76 260
579	76 268	76 275	76 283	76 290	76 298	76 305	76 313	76 320	76 328	76 335
580	76 343	76 350	76 358	76 365	76 373	76 380	76 388	76 395	76 403	76 410
581	76 418	76 425	76 433	76 440	76 448	76 455	76 462	76 470	76 477	76 485
582	76 492	76 500	76 507	76 515	76 522	76 530	76 537	76 545	76 552	76 559
583	76 567	76 574	76 582	76 589	76 597	76 604	76 612	76 619	76 626	76 634
584	76 641	76 649	76 656	76 664	76 671	76 678	76 686	76 693	76 701	76 708
585	76 716	76 723	76 730	76 738	76 745	76 753	76 760	76 768	76 775	76 782
586	76 790	76 797	76 805	76 812	76 819	76 827	76 834	76 842	76 849	76 856
587	76 864	76 871	76 879	76 886	76 893	76 901	76 908	76 916	76 923	76 930
588	76 938	76 945	76 953	76 960	76 967	76 975	76 982	76 989	76 997	77 004
589	77 012	77 019	77 026	77 034	77 041	77 048	77 056	77 063	77 070	77 078
590	77 085	77 093	77 100	77 107	77 115	77 122	77 129	77 137	77 144	77 151
591	77 159	77 166	77 173	77 181	77 188	77 195	77 203	77 210	77 217	77 225
592	77 232	77 240	77 247	77 254	77 262	77 269	77 276	77 283	77 291	77 298
593	77 305	77 313	77 320	77 327	77 335	77 342	77 349	77 357	77 364	77 371
594	77 379	77 386	77 393	77 401	77 408	77 415	77 422	77 430	77 437	77 444
595	77 452	77 459	77 466	77 474	77 481	77 488	77 495	77 503	77 510	77 517
596	77 525	77 532	77 539	77 546	77 554	77 561	77 568	77 576	77 583	77 590
597	77 597	77 605	77 612	77 619	77 627	77 634	77 641	77 648	77 656	77 663
598	77 670	77 677	77 685	77 692	77 699	77 706	77 714	77 721	77 728	77 735
599	77 743	77 750	77 757	77 764	77 772	77 779	77 786	77 793	77 801	77 808
600	77 815	77 822	77 830	77 837	77 844	77 851	77 859	77 866	77 873	77 880
N	0	1	2	3	4	5	6	7	8	9

N	0	1	2	3	4	5	6	7	8	9
600	77 815	77 822	77 830	77 837	77 844	77 851	77 859	77 866	77 873	77 880
601	77 887	77 895	77 902	77 909	77 916	77 924	77 931	77 938	77 945	77 952
602	77 960	77 967	77 974	77 981	77 988	77 996	78 003	78 010	78 017	78 025
603	78 032	78 039	78 046	78 053	78 061	78 068	78 075	78 082	78 089	78 097
604	78 104	78 111	78 118	78 125	78 132	78 140	78 147	78 154	78 161	78 168
605	78 176	78 183	78 190	78 197	78 204	78 211	78 219	78 226	78 233	78 240
606	78 247	78 254	78 262	78 269	78 276	78 283	78 290	78 297	78 305	78 312
607	78 319	78 326	78 333	78 340	78 347	78 355	78 362	78 369	78 376	78 383
608	78 390	78 398	78 405	78 412	78 419	78 426	78 433	78 440	78 447	78 455
609	78 462	78 469	78 476	78 483	78 490	78 497	78 504	78 512	78 519	78 526
610	78 533	78 540	78 547	78 554	78 561	78 569	78 576	78 583	78 590	78 597
611	78 604	78 611	78 618	78 625	78 633	78 640	78 647	78 654	78 661	78 668
612	78 675	78 682	78 689	78 696	78 704	78 711	78 718	78 725	78 732	78 739
613	78 746	78 753	78 760	78 767	78 774	78 781	78 789	78 796	78 803	78 810
614	78 817	78 824	78 831	78 838	78 845	78 852	78 859	78 866	78 873	78 880
615	78 888	78 895	78 902	78 909	78 916	78 923	78 930	78 937	78 944	78 951
616	78 958	78 965	78 972	78 979	78 986	78 993	79 000	79 007	79 014	79 021
617	79 029	79 036	79 043	79 050	79 057	79 064	79 071	79 078	79 085	79 092
618	79 099	79 106	79 113	79 120	79 127	79 134	79 141	79 148	79 155	79 162
619	79 169	79 176	79 183	79 190	79 197	79 204	79 211	79 218	79 225	79 232
620	79 239	79 246	79 253	79 260	79 267	79 274	79 281	79 288	79 295	79 302
621	79 309	79 316	79 323	79 330	79 337	79 344	79 351	79 358	79 365	79 372
622	79 379	79 386	79 393	79 400	79 407	79 414	79 421	79 428	79 435	79 442
623	79 449	79 456	79 463	79 470	79 477	79 484	79 491	79 498	79 505	79 511
624	79 518	79 525	79 532	79 539	79 546	79 553	79 560	79 567	79 574	79 581
625	79 588	79 595	79 602	79 609	79 616	79 623	79 630	79 637	79 644	79 650
626	79 657	79 664	79 671	79 678	79 685	79 692	79 699	79 706	79 713	79 720
627	79 727	79 734	79 741	79 748	79 754	79 761	79 768	79 775	79 782	79 789
628	79 796	79 803	79 810	79 817	79 824	79 831	79 837	79 844	79 851	79 858
629	79 865	79 872	79 879	79 886	79 893	79 900	79 906	79 913	79 920	79 927
630	79 934	79 941	79 948	79 955	79 962	79 969	79 975	79 982	79 989	79 996
631	80 003	80 010	80 017	80 024	80 030	80 037	80 044	80 051	80 058	80 065
632	80 072	80 079	80 085	80 092	80 099	80 106	80 113	80 120	80 127	80 134
633	80 140	80 147	80 154	80 161	80 168	80 175	80 182	80 188	80 195	80 202
634	80 209	80 216	80 223	80 229	80 236	80 243	80 250	80 257	80 264	80 271
635	80 277	80 284	80 291	80 298	80 305	80 312	80 318	80 325	80 332	80 339
636	80 346	80 353	80 359	80 366	80 373	80 380	80 387	80 393	80 400	80 407
637	80 414	80 421	80 428	80 434	80 441	80 448	80 455	80 462	80 468	80 475
638	80 482	80 489	80 496	80 502	80 509	80 516	80 523	80 530	80 536	80 543
639	80 550	80 557	80 564	80 570	80 577	80 584	80 591	80 598	80 604	80 611
640	80 618	80 625	80 632	80 638	80 645	80 652	80 659	80 665	80 672	80 679
641	80 686	80 693	80 699	80 706	80 713	80 720	80 726	80 733	80 740	80 747
642	80 754	80 760	80 767	80 774	80 781	80 787	80 794	80 801	80 808	80 814
643	80 821	80 828	80 835	80 841	80 848	80 855	80 862	80 868	80 875	80 882
644	80 889	80 895	80 902	80 909	80 916	80 922	80 929	80 936	80 943	80 949
645	80 956	80 963	80 969	80 976	80 983	80 990	80 996	81 003	81 010	81 017
646	81 023	81 030	81 037	81 043	81 050	81 057	81 064	81 070	81 077	81 084
647	81 090	81 097	81 104	81 111	81 117	81 124	81 131	81 137	81 144	81 151
648	81 158	81 164	81 171	81 178	81 184	81 191	81 198	81 204	81 211	81 218
649	81 224	81 231	81 238	81 245	81 251	81 258	81 265	81 271	81 278	81 285
650	81 291	81 298	81 305	81 311	81 318	81 325	81 331	81 338	81 345	81 351
N	0	1	2	3	4	5	6	7	8	9

N	0	1	2	3	4	5	6	7	8	9
650	81 291	81 298	81 305	81 311	81 318	81 325	81 331	81 338	81 345	81 351
651	81 358	81 365	81 371	81 378	81 385	81 391	81 398	81 405	81 411	81 418
652	81 425	81 431	81 438	81 445	81 451	81 458	81 465	81 471	81 478	81 485
653	81 491	81 498	81 505	81 511	81 518	81 525	81 531	81 538	81 544	81 551
654	81 558	81 564	81 571	81 578	81 584	81 591	81 598	81 604	81 611	81 617
655	81 624	81 631	81 637	81 644	81 651	81 657	81 664	81 671	81 677	81 684
656	81 690	81 697	81 704	81 710	81 717	81 723	81 730	81 737	81 743	81 750
657	81 757	81 763	81 770	81 776	81 783	81 790	81 796	81 803	81 809	81 816
658	81 823	81 829	81 836	81 842	81 849	81 856	81 862	81 869	81 875	81 882
659	81 889	81 895	81 902	81 908	81 915	81 921	81 928	81 935	81 941	81 948
660	81 954	81 961	81 968	81 974	81 981	81 987	81 994	82 000	82 007	82 014
661	82 020	82 027	82 033	82 040	82 046	82 053	82 060	82 066	82 073	82 079
662	82 086	82 092	82 099	82 105	82 112	82 119	82 125	82 132	82 138	82 145
663	82 151	82 158	82 164	82 171	82 178	82 184	82 191	82 197	82 204	82 210
664	82 217	82 223	82 230	82 236	82 243	82 249	82 256	82 263	82 269	82 276
665	82 282	82 289	82 295	82 302	82 308	82 315	82 321	82 328	82 334	82 341
666	82 347	82 354	82 360	82 367	82 373	82 380	82 387	82 393	82 400	82 406
667	82 413	82 419	82 426	82 432	82 439	82 445	82 452	82 458	82 465	82 471
668	82 478	82 484	82 491	82 497	82 504	82 510	82 517	82 523	82 530	82 536
669	82 543	82 549	82 556	82 562	82 569	82 575	82 582	82 588	82 595	82 601
670	82 607	82 614	82 620	82 627	82 633	82 640	82 646	82 653	82 659	82 666
671	82 672	82 679	82 685	82 692	82 698	82 705	82 711	82 718	82 724	82 730
672	82 737	82 743	82 750	82 756	82 763	82 769	82 776	82 782	82 789	82 795
673	82 802	82 808	82 814	82 821	82 827	82 834	82 840	82 847	82 853	82 860
674	82 866	82 872	82 879	82 885	82 892	82 898	82 905	82 911	82 918	82 924
675	82 930	82 937	82 943	82 950	82 956	82 963	82 969	82 975	82 982	82 988
676	82 995	83 001	83 008	83 014	83 020	83 027	83 033	83 040	83 046	83 052
677	83 059	83 065	83 072	83 078	83 085	83 091	83 097	83 104	83 110	83 117
678	83 123	83 129	83 136	83 142	83 149	83 155	83 161	83 168	83 174	83 181
679	83 187	83 193	83 200	83 206	83 213	83 219	83 225	83 232	83 238	83 245
680	83 251	83 257	83 264	83 270	83 276	83 283	83 289	83 296	83 302	83 308
681	83 315	83 321	83 327	83 334	83 340	83 347	83 353	83 359	83 366	83 372
682	83 378	83 385	83 391	83 398	83 404	83 410	83 417	83 423	83 429	83 436
683	83 442	83 448	83 455	83 461	83 467	83 474	83 480	83 487	83 493	83 499
684	83 506	83 512	83 518	83 525	83 531	83 537	83 544	83 550	83 556	83 563
685	83 569	83 575	83 582	83 588	83 594	83 601	83 607	83 613	83 620	83 626
686	83 632	83 639	83 645	83 651	83 658	83 664	83 670	83 677	83 683	83 689
687	83 696	83 702	83 708	83 715	83 721	83 727	83 734	83 740	83 746	83 753
688	83 759	83 765	83 771	83 778	83 784	83 790	83 797	83 803	83 809	83 816
689	83 822	83 828	83 835	83 841	83 847	83 853	83 860	83 866	83 872	83 879
690	83 885	83 891	83 897	83 904	83 910	83 916	83 923	83 929	83 935	83 942
691	83 948	83 954	83 960	83 967	83 973	83 979	83 985	83 992	83 998	84 004
692	84 011	84 017	84 023	84 029	84 036	84 042	84 048	84 055	84 061	84 067
693	84 073	84 080	84 086	84 092	84 098	84 105	84 111	84 117	84 123	84 130
694	84 136	84 142	84 148	84 155	84 161	84 167	84 173	84 180	84 186	84 192
695	84 198	84 205	84 211	84 217	84 223	84 230	84 236	84 242	84 248	84 255
696	84 261	84 267	84 273	84 280	84 286	84 292	84 298	84 305	84 311	84 317
697	84 323	84 330	84 336	84 342	84 348	84 354	84 361	84 367	84 373	84 379
698	84 386	84 392	84 398	84 404	84 410	84 417	84 423	84 429	84 435	84 442
699	84 448	84 454	84 460	84 466	84 473	84 479	84 485	84 491	84 497	84 504
700	84 510	84 516	84 522	84 528	84 535	84 541	84 547	84 553	84 559	84 566
N	0	1	2	3	4	5	6	7	8	9

N	0	1	2	3	4	5	6	7	8	9
700	84 510	84 516	84 522	84 528	84 535	84 541	84 547	84 553	84 559	84 566
701	84 572	84 578	84 584	84 590	84 597	84 603	84 609	84 615	84 621	84 628
702	84 634	84 640	84 646	84 652	84 658	84 665	84 671	84 677	84 683	84 689
703	84 696	84 702	84 708	84 714	84 720	84 726	84 733	84 739	84 745	84 751
704	84 757	84 763	84 770	84 776	84 782	84 788	84 794	84 800	84 807	84 813
705	84 819	84 825	84 831	84 837	84 844	84 850	84 856	84 862	84 868	84 874
706	84 880	84 887	84 893	84 899	84 905	84 911	84 917	84 924	84 930	84 936
707	84 942	84 948	84 954	84 960	84 967	84 973	84 979	84 985	84 991	84 997
708	85 003	85 009	85 016	85 022	85 028	85 034	85 040	85 046	85 052	85 058
709	85 065	85 071	85 077	85 083	85 089	85 095	85 101	85 107	85 114	85 120
710	85 126	85 132	85 138	85 144	85 150	85 156	85 163	85 169	85 175	85 181
711	85 187	85 193	85 199	85 205	85 211	85 217	85 224	85 230	85 236	85 242
712	85 248	85 254	85 260	85 266	85 272	85 278	85 285	85 291	85 297	85 303
713	85 309	85 315	85 321	85 327	85 333	85 339	85 345	85 352	85 358	85 364
714	85 370	85 376	85 382	85 388	85 394	85 400	85 406	85 412	85 418	85 425
715	85 431	85 437	85 443	85 449	85 455	85 461	85 467	85 473	85 479	85 485
716	85 491	85 497	85 503	85 509	85 516	85 522	85 528	85 534	85 540	85 546
717	85 552	85 558	85 564	85 570	85 576	85 582	85 588	85 594	85 600	85 606
718	85 612	85 618	85 625	85 631	85 637	85 643	85 649	85 655	85 661	85 667
719	85 673	85 679	85 685	85 691	85 697	85 703	85 709	85 715	85 721	85 727
720	85 733	85 739	85 745	85 751	85 757	85 763	85 769	85 775	85 781	85 788
721	85 794	85 800	85 806	85 812	85 818	85 824	85 830	85 836	85 842	85 848
722	85 854	85 860	85 866	85 872	85 878	85 884	85 890	85 896	85 902	85 908
723	85 914	85 920	85 926	85 932	85 938	85 944	85 950	85 956	85 962	85 968
724	85 974	85 980	85 986	85 992	85 998	86 004	86 010	86 016	86 022	86 028
725	86 034	86 040	86 046	86 052	86 058	86 064	86 070	86 076	86 082	86 088
726	86 094	86 100	86 106	86 112	86 118	86 124	86 130	86 136	86 141	86 147
727	86 153	86 159	86 165	86 171	86 177	86 183	86 189	86 195	86 201	86 207
728	86 213	86 219	86 225	86 231	86 237	86 243	86 249	86 255	86 261	86 267
729	86 273	86 279	86 285	86 291	86 297	86 303	86 308	86 314	86 320	86 326
730	86 332	86 338	86 344	86 350	86 356	86 362	86 368	86 374	86 380	86 386
731	86 392	86 398	86 404	86 410	86 415	86 421	86 427	86 433	86 439	86 445
732	86 451	86 457	86 463	86 469	86 475	86 481	86 487	86 493	86 499	86 504
733	86 510	86 516	86 522	86 528	86 534	86 540	86 546	86 552	86 558	86 564
734	86 570	86 576	86 581	86 587	86 593	86 599	86 605	86 611	86 617	86 623
735	86 629	86 635	86 641	86 646	86 652	86 658	86 664	86 670	86 676	86 682
736	86 688	86 694	86 700	86 705	86 711	86 717	86 723	86 729	86 735	86 741
737	86 747	86 753	86 759	86 764	86 770	86 776	86 782	86 788	86 794	86 800
738	86 806	86 812	86 817	86 823	86 829	86 835	86 841	86 847	86 853	86 859
739	86 864	86 870	86 876	86 882	86 888	86 894	86 900	86 906	86 911	86 917
740	86 923	86 929	86 935	86 941	86 947	86 953	86 958	86 964	86 970	86 976
741	86 982	86 988	86 994	86 999	87 005	87 011	87 017	87 023	87 029	87 035
742	87 040	87 046	87 052	87 058	87 064	87 070	87 075	87 081	87 087	87 093
743	87 099	87 105	87 111	87 116	87 122	87 128	87 134	87 140	87 146	87 151
744	87 157	87 163	87 169	87 175	87 181	87 186	87 192	87 198	87 204	87 210
745	87 216	87 221	87 227	87 233	87 239	87 245	87 251	87 256	87 262	87 268
746	87 274	87 280	87 286	87 291	87 297	87 303	87 309	87 315	87 320	87 326
747	87 332	87 338	87 344	87 349	87 355	87 361	87 367	87 373	87 379	87 384
748	87 390	87 396	87 402	87 408	87 413	87 419	87 425	87 431	87 437	87 442
749	87 448	87 454	87 460	87 466	87 471	87 477	87 483	87 489	87 495	87 500
750	87 506	87 512	87 518	87 523	87 529	87 535	87 541	87 547	87 552	87 558
N	0	1	2	3	4	5	6	7	8	9

N	0	1	2	3	4	5	6	7	8	9
750	87 506	87 512	87 518	87 523	87 529	87 535	87 541	87 547	87 552	87 558
751	87 564	87 570	87 576	87 581	87 587	87 593	87 599	87 604	87 610	87 616
752	87 622	87 628	87 633	87 639	87 645	87 651	87 656	87 662	87 668	87 674
753	87 679	87 685	87 691	87 697	87 703	87 708	87 714	87 720	87 726	87 731
754	87 737	87 743	87 749	87 754	87 760	87 766	87 772	87 777	87 783	87 789
755	87 795	87 800	87 806	87 812	87 818	87 823	87 829	87 835	87 841	87 846
756	87 852	87 858	87 864	87 869	87 875	87 881	87 887	87 892	87 898	87 904
757	87 910	87 915	87 921	87 927	87 933	87 938	87 944	87 950	87 955	87 961
758	87 967	87 973	87 978	87 984	87 990	87 996	88 001	88 007	88 013	88 018
759	88 024	88 030	88 036	88 041	88 047	88 053	88 058	88 064	88 070	88 076
760	88 081	88 087	88 093	88 098	88 104	88 110	88 116	88 121	88 127	88 133
761	88 138	88 144	88 150	88 156	88 161	88 167	88 173	88 178	88 184	88 190
762	88 195	88 201	88 207	88 213	88 218	88 224	88 230	88 235	88 241	88 247
763	88 252	88 258	88 264	88 270	88 275	88 281	88 287	88 292	88 298	88 304
764	88 309	88 315	88 321	88 326	88 332	88 338	88 343	88 349	88 355	88 360
765	88 366	88 372	88 377	88 383	88 389	88 395	88 400	88 406	88 412	88 417
766	88 423	88 429	88 434	88 440	88 446	88 451	88 457	88 463	88 468	88 474
767	88 480	88 485	88 491	88 497	88 502	88 508	88 513	88 519	88 525	88 530
768	88 536	88 542	88 547	88 553	88 559	88 564	88 570	88 576	88 581	88 587
769	88 593	88 598	88 604	88 610	88 615	88 621	88 627	88 632	88 638	88 643
770	88 649	88 655	88 660	88 666	88 672	88 677	88 683	88 689	88 694	88 700
771	88 705	88 711	88 717	88 722	88 728	88 734	88 739	88 745	88 750	88 756
772	88 762	88 767	88 773	88 779	88 784	88 790	88 795	88 801	88 807	88 812
773	88 818	88 824	88 829	88 835	88 840	88 846	88 852	88 857	88 863	88 868
774	88 874	88 880	88 885	88 891	88 897	88 902	88 908	88 913	88 919	88 925
775	88 930	88 936	88 941	88 947	88 953	88 958	88 964	88 969	88 975	88 981
776	88 986	88 992	88 997	89 003	89 009	89 014	89 020	89 025	89 031	89 037
777	89 042	89 048	89 053	89 059	89 064	89 070	89 076	89 081	89 087	89 092
778	89 098	89 104	89 109	89 115	89 120	89 126	89 131	89 137	89 143	89 148
779	89 154	89 159	89 165	89 170	89 176	89 182	89 187	89 193	89 198	89 204
780	89 209	89 215	89 221	89 226	89 232	89 237	89 243	89 248	89 254	89 260
781	89 265	89 271	89 276	89 282	89 287	89 293	89 298	89 304	89 310	89 315
782	89 321	89 326	89 332	89 337	89 343	89 348	89 354	89 360	89 365	89 371
783	89 376	89 382	89 387	89 393	89 398	89 404	89 409	89 415	89 421	89 426
784	89 432	89 437	89 443	89 448	89 454	89 459	89 465	89 470	89 476	89 481
785	89 487	89 492	89 498	89 504	89 509	89 515	89 520	89 526	89 531	89 537
786	89 542	89 548	89 553	89 559	89 564	89 570	89 575	89 581	89 586	89 592
787	89 597	89 603	89 609	89 614	89 620	89 625	89 631	89 636	89 642	89 647
788	89 653	89 658	89 664	89 669	89 675	89 680	89 686	89 691	89 697	89 702
789	89 708	89 713	89 719	89 724	89 730	89 735	89 741	89 746	89 752	89 757
790	89 763	89 768	89 774	89 779	89 785	89 790	89 796	89 801	89 807	89 812
791	89 818	89 823	89 829	89 834	89 840	89 845	89 851	89 856	89 862	89 867
792	89 873	89 878	89 883	89 889	89 894	89 900	89 905	89 911	89 916	89 922
793	89 927	89 933	89 938	89 944	89 949	89 955	89 960	89 966	89 971	89 977
794	89 982	89 988	89 993	89 998	90 004	90 009	90 015	90 020	90 026	90 031
795	90 037	90 042	90 048	90 053	90 059	90 064	90 069	90 075	90 080	90 086
796	90 091	90 097	90 102	90 108	90 113	90 119	90 124	90 129	90 135	90 140
797	90 146	90 151	90 157	90 162	90 168	90 173	90 179	90 184	90 189	90 195
798	90 200	90 206	90 211	90 217	90 222	90 227	90 233	90 238	90 244	90 249
799	90 255	90 260	90 266	90 271	90 276	90 282	90 287	90 293	90 298	90 304
800	90 309	90 314	90 320	90 325	90 331	90 336	90 342	90 347	90 352	90 358
N	0	1	2	3	4	5	6	7	8	9

N	0	1	2	3	4	5	6	7	8	9
800	90 309	90 314	90 320	90 325	90 331	90 336	90 342	90 347	90 352	90 358
801	90 363	90 369	90 374	90 380	90 385	90 390	90 396	90 401	90 407	90 412
802	90 417	90 423	90 428	90 434	90 439	90 445	90 450	90 455	90 461	90 466
803	90 472	90 477	90 482	90 488	90 493	90 499	90 504	90 509	90 515	90 520
804	90 526	90 531	90 536	90 542	90 547	90 553	90 558	90 563	90 569	90 574
805	90 580	90 585	90 590	90 596	90 601	90 607	90 612	90 617	90 623	90 628
806	90 634	90 639	90 644	90 650	90 655	90 660	90 666	90 671	90 677	90 682
807	90 687	90 693	90 698	90 703	90 709	90 714	90 720	90 725	90 730	90 736
808	90 741	90 747	90 752	90 757	90 763	90 768	90 773	90 779	90 784	90 789
809	90 795	90 800	90 806	90 811	90 816	90 822	90 827	90 832	90 838	90 843
810	90 849	90 854	90 859	90 865	90 870	90 875	90 881	90 886	90 891	90 897
811	90 902	90 907	90 913	90 918	90 924	90 929	90 934	90 940	90 945	90 950
812	90 956	90 961	90 966	90 972	90 977	90 982	90 988	90 993	90 998	91 004
813	91 009	91 014	91 020	91 025	91 030	91 036	91 041	91 046	91 052	91 057
814	91 062	91 068	91 073	91 078	91 084	91 089	91 094	91 100	91 105	91 110
815	91 116	91 121	91 126	91 132	91 137	91 142	91 148	91 153	91 158	91 164
816	91 169	91 174	91 180	91 185	91 190	91 196	91 201	91 206	91 212	91 217
817	91 222	91 228	91 233	91 238	91 243	91 249	91 254	91 259	91 265	91 270
818	91 275	91 281	91 286	91 291	91 297	91 302	91 307	91 312	91 318	91 323
819	91 328	91 334	91 339	91 344	91 350	91 355	91 360	91 365	91 371	91 376
820	91 381	91 387	91 392	91 397	91 403	91 408	91 413	91 418	91 424	91 429
821	91 434	91 440	91 445	91 450	91 455	91 461	91 466	91 471	91 477	91 482
822	91 487	91 492	91 498	91 503	91 508	91 514	91 519	91 524	91 529	91 535
823	91 540	91 545	91 551	91 556	91 561	91 566	91 572	91 577	91 582	91 587
824	91 593	91 598	91 603	91 609	91 614	91 619	91 624	91 630	91 635	91 640
825	91 645	91 651	91 656	91 661	91 666	91 672	91 677	91 682	91 687	91 693
826	91 698	91 703	91 709	91 714	91 719	91 724	91 730	91 735	91 740	91 745
827	91 751	91 756	91 761	91 766	91 771	91 777	91 782	91 787	91 793	91 798
828	91 803	91 808	91 814	91 819	91 824	91 829	91 834	91 840	91 845	91 850
829	91 855	91 861	91 866	91 871	91 876	91 882	91 887	91 892	91 897	91 903
830	91 908	91 913	91 918	91 924	91 929	91 934	91 939	91 944	91 950	91 955
831	91 960	91 965	91 971	91 976	91 981	91 986	91 991	91 997	92 002	92 007
832	92 012	92 018	92 023	92 028	92 033	92 038	92 044	92 049	92 054	92 059
833	92 065	92 070	92 075	92 080	92 085	92 091	92 096	92 101	92 106	92 111
834	92 117	92 122	92 127	92 132	92 137	92 143	92 148	92 153	92 158	92 163
835	92 169	92 174	92 179	92 184	92 189	92 195	92 200	92 205	92 210	92 215
836	92 221	92 226	92 231	92 236	92 241	92 247	92 252	92 257	92 262	92 267
837	92 273	92 278	92 283	92 288	92 293	92 298	92 304	92 309	92 314	92 319
838	92 324	92 330	92 335	92 340	92 345	92 350	92 355	92 361	92 366	92 371
839	92 376	92 381	92 387	92 392	92 397	92 402	92 407	92 412	92 418	92 423
840	92 428	92 433	92 438	92 443	92 449	92 454	92 459	92 464	92 469	92 474
841	92 480	92 485	92 490	92 495	92 500	92 505	92 511	92 516	92 521	92 526
842	92 531	92 536	92 542	92 547	92 552	92 557	92 562	92 567	92 572	92 578
843	92 583	92 588	92 593	92 598	92 603	92 609	92 614	92 619	92 624	92 629
844	92 634	92 639	92 645	92 650	92 655	92 660	92 665	92 670	92 675	92 681
845	92 686	92 691	92 696	92 701	92 706	92 711	92 716	92 722	92 727	92 732
846	92 737	92 742	92 747	92 752	92 758	92 763	92 768	92 773	92 778	92 783
847	92 788	92 793	92 799	92 804	92 809	92 814	92 819	92 824	92 829	92 834
848	92 840	92 845	92 850	92 855	92 860	92 865	92 870	92 875	92 881	92 886
849	92 891	92 896	92 901	92 906	92 911	92 916	92 921	92 927	92 932	92 937
850	92 942	92 947	92 952	92 957	92 962	92 967	92 973	92 978	92 983	92 988
N	0	1	2	3	4	5	6	7	8	9

N	0	1	2	3	4	5	6	7	8	9
850	92 942	92 947	92 952	92 957	92 962	92 967	92 973	92 978	92 983	92 988
851	92 993	92 998	93 003	93 008	93 013	93 018	93 024	93 029	93 034	93 039
852	93 044	93 049	93 054	93 059	93 064	93 069	93 075	93 080	93 085	93 090
853	93 095	93 100	93 105	93 110	93 115	93 120	93 125	93 131	93 136	93 141
854	93 146	93 151	93 156	93 161	93 166	93 171	93 176	93 181	93 186	93 192
855	93 197	93 202	93 207	93 212	93 217	93 222	93 227	93 232	93 237	93 242
856	93 247	93 252	93 258	93 263	93 268	93 273	93 278	93 283	93 288	93 293
857	93 298	93 303	93 308	93 313	93 318	93 323	93 328	93 334	93 339	93 344
858	93 349	93 354	93 359	93 364	93 369	93 374	93 379	93 384	93 389	93 394
859	93 399	93 404	93 409	93 414	93 420	93 425	93 430	93 435	93 440	93 445
860	93 450	93 455	93 460	93 465	93 470	93 475	93 480	93 485	93 490	93 495
861	93 500	93 505	93 510	93 515	93 520	93 526	93 531	93 536	93 541	93 546
862	93 551	93 556	93 561	93 566	93 571	93 576	93 581	93 586	93 591	93 596
863	93 601	93 606	93 611	93 616	93 621	93 626	93 631	93 636	93 641	93 646
864	93 651	93 656	93 661	93 666	93 671	93 676	93 682	93 687	93 692	93 697
865	93 702	93 707	93 712	93 717	93 722	93 727	93 732	93 737	93 742	93 747
866	93 752	93 757	93 762	93 767	93 772	93 777	93 782	93 787	93 792	93 797
867	93 802	93 807	93 812	93 817	93 822	93 827	93 832	93 837	93 842	93 847
868	93 852	93 857	93 862	93 867	93 872	93 877	93 882	93 887	93 892	93 897
869	93 902	93 907	93 912	93 917	93 922	93 927	93 932	93 937	93 942	93 947
870	93 952	93 957	93 962	93 967	93 972	93 977	93 982	93 987	93 992	93 997
871	94 002	94 007	94 012	94 017	94 022	94 032	94 037	94 042	94 047	
872	94 052	94 057	94 062	94 067	94 072	94 077	94 082	94 086	94 091	94 096
873	94 101	94 106	94 111	94 116	94 121	94 126	94 131	94 136	94 141	94 146
874	94 151	94 156	94 161	94 166	94 171	94 176	94 181	94 186	94 191	94 196
875	94 201	94 206	94 211	94 216	94 221	94 226	94 231	94 236	94 240	94 245
876	94 250	94 255	94 260	94 265	94 270	94 275	94 280	94 285	94 290	94 295
877	94 300	94 305	94 310	94 315	94 320	94 325	94 330	94 335	94 340	94 345
878	94 349	94 354	94 359	94 364	94 369	94 374	94 379	94 384	94 389	94 394
879	94 399	94 404	94 409	94 414	94 419	94 424	94 429	94 433	94 438	94 443
880	94 448	94 453	94 458	94 463	94 468	94 473	94 478	94 483	94 488	94 493
881	94 498	94 503	94 507	94 512	94 517	94 522	94 527	94 532	94 537	94 542
882	94 547	94 552	94 557	94 562	94 567	94 571	94 576	94 581	94 586	94 591
883	94 596	94 601	94 606	94 611	94 616	94 621	94 626	94 630	94 635	94 640
884	94 645	94 650	94 655	94 660	94 665	94 670	94 675	94 680	94 685	94 689
885	94 694	94 699	94 704	94 709	94 714	94 719	94 724	94 729	94 734	94 738
886	94 743	94 748	94 753	94 758	94 763	94 768	94 773	94 778	94 783	94 787
887	94 792	94 797	94 802	94 807	94 812	94 817	94 822	94 827	94 832	94 836
888	94 841	94 846	94 851	94 856	94 861	94 866	94 871	94 876	94 880	94 885
889	94 890	94 895	94 900	94 905	94 910	94 915	94 919	94 924	94 929	94 934
890	94 939	94 944	94 949	94 954	94 959	94 963	94 968	94 973	94 978	94 983
891	94 988	94 993	94 998	95 002	95 007	95 012	95 017	95 022	95 027	95 032
892	95 036	95 041	95 046	95 051	95 056	95 061	95 066	95 071	95 075	95 080
893	95 085	95 090	95 095	95 100	95 105	95 109	95 114	95 119	95 124	95 129
894	95 134	95 139	95 143	95 148	95 153	95 158	95 163	95 168	95 173	95 177
895	95 182	95 187	95 192	95 197	95 202	95 207	95 211	95 216	95 221	95 226
896	95 231	95 236	95 240	95 245	95 250	95 255	95 260	95 265	95 270	95 274
897	95 279	95 284	95 289	95 294	95 299	95 303	95 308	95 313	95 318	95 323
898	95 328	95 332	95 337	95 342	95 347	95 352	95 357	95 361	95 366	95 371
899	95 376	95 381	95 386	95 390	95 395	95 400	95 405	95 410	95 415	95 419
900	95 424	95 429	95 434	95 439	95 444	95 448	95 453	95 458	95 463	95 468
N	0	1	2	3	4	5	6	7	8	9

N	0	1	2	3	4	5	6	7	8	9
900	95 424	95 429	95 434	95 439	95 444	95 448	95 453	95 458	95 463	95 468
901	95 472	95 477	95 482	95 487	95 492	95 497	95 501	95 506	95 511	95 516
902	95 521	95 525	95 530	95 535	95 540	95 545	95 550	95 554	95 559	95 564
903	95 569	95 574	95 578	95 583	95 588	95 593	95 598	95 602	95 607	95 612
904	95 617	95 622	95 626	95 631	95 636	95 641	95 646	95 650	95 655	95 660
905	95 665	95 670	95 674	95 679	95 684	95 689	95 694	95 698	95 703	95 708
906	95 713	95 718	95 722	95 727	95 732	95 737	95 742	95 746	95 751	95 756
907	95 761	95 766	95 770	95 775	95 780	95 785	95 789	95 794	95 799	95 804
908	95 809	95 813	95 818	95 823	95 828	95 832	95 837	95 842	95 847	95 852
909	95 856	95 861	95 866	95 871	95 875	95 880	95 885	95 890	95 895	95 899
910	95 904	95 909	95 914	95 918	95 923	95 928	95 933	95 938	95 942	95 947
911	95 952	95 957	95 961	95 966	95 971	95 976	95 980	95 985	95 990	95 995
912	95 999	96 004	96 009	96 014	96 019	96 023	96 028	96 033	96 038	96 042
913	96 047	96 052	96 057	96 061	96 066	96 071	96 076	96 080	96 085	96 090
914	96 095	96 099	96 104	96 109	96 114	96 118	96 123	96 128	96 133	96 137
915	96 142	96 147	96 152	96 156	96 161	96 166	96 171	96 175	96 180	96 185
916	96 190	96 194	96 199	96 204	96 209	96 213	96 218	96 223	96 227	96 232
917	96 237	96 242	96 246	96 251	96 256	96 261	96 265	96 270	96 275	96 280
918	96 284	96 289	96 294	96 298	96 303	96 308	96 313	96 317	96 322	96 327
919	96 332	96 336	96 341	96 346	96 350	96 355	96 360	96 365	96 369	96 374
920	96 379	96 384	96 388	96 393	96 398	96 402	96 407	96 412	96 417	96 421
921	96 426	96 431	96 435	96 440	96 445	96 450	96 454	96 459	96 464	96 468
922	96 473	96 478	96 483	96 487	96 492	96 497	96 501	96 506	96 511	96 515
923	96 520	96 525	96 530	96 534	96 539	96 544	96 548	96 553	96 558	96 562
924	96 567	96 572	96 577	96 581	96 586	96 591	96 595	96 600	96 605	96 609
925	96 614	96 619	96 624	96 628	96 633	96 638	96 642	96 647	96 652	96 656
926	96 661	96 666	96 670	96 675	96 680	96 685	96 689	96 694	96 699	96 703
927	96 708	96 713	96 717	96 722	96 727	96 731	96 736	96 741	96 745	96 750
928	96 755	96 759	96 764	96 769	96 774	96 778	96 783	96 788	96 792	96 797
929	96 802	96 806	96 811	96 816	96 820	96 825	96 830	96 834	96 839	96 844
930	96 848	96 853	96 858	96 862	96 867	96 872	96 876	96 881	96 886	96 890
931	96 895	96 900	96 904	96 909	96 914	96 918	96 923	96 928	96 932	96 937
932	96 942	96 946	96 951	96 956	96 960	96 965	96 970	96 974	96 979	96 984
933	96 988	96 993	96 997	97 002	97 007	97 011	97 016	97 021	97 025	97 030
934	97 035	97 039	97 044	97 049	97 053	97 058	97 063	97 067	97 072	97 077
935	97 081	97 086	97 090	97 095	97 100	97 104	97 109	97 114	97 118	97 123
936	97 128	97 132	97 137	97 142	97 146	97 151	97 155	97 160	97 165	97 169
937	97 174	97 179	97 183	97 188	97 192	97 197	97 202	97 206	97 211	97 216
938	97 220	97 225	97 230	97 234	97 239	97 243	97 248	97 253	97 257	97 262
939	97 267	97 271	97 276	97 280	97 285	97 290	97 294	97 299	97 304	97 308
940	97 313	97 317	97 322	97 327	97 331	97 336	97 340	97 345	97 350	97 354
941	97 359	97 364	97 368	97 373	97 377	97 382	97 387	97 391	97 396	97 400
942	97 405	97 410	97 414	97 419	97 424	97 428	97 433	97 437	97 442	97 447
943	97 451	97 456	97 460	97 465	97 470	97 474	97 479	97 483	97 488	97 493
944	97 497	97 502	97 506	97 511	97 516	97 520	97 525	97 529	97 534	97 539
945	97 543	97 548	97 552	97 557	97 562	97 566	97 571	97 575	97 580	97 585
946	97 589	97 594	97 598	97 603	97 607	97 612	97 617	97 621	97 626	97 630
947	97 635	97 640	97 644	97 649	97 653	97 658	97 663	97 667	97 672	97 676
948	97 681	97 685	97 690	97 695	97 699	97 704	97 708	97 713	97 717	97 722
949	97 727	97 731	97 736	97 740	97 745	97 749	97 754	97 759	97 763	97 768
950	97 772	97 777	97 782	97 786	97 791	97 795	97 800	97 804	97 809	97 813
N	0	1	2	3	4	5	6	7	8	9

N	0	1	2	3	4	5	6	7	8	9
950	97 772	97 777	97 782	97 786	97 791	97 795	97 800	97 804	97 809	97 813
951	97 818	97 823	97 827	97 832	97 836	97 841	97 845	97 850	97 855	97 859
952	97 864	97 868	97 873	97 877	97 882	97 886	97 891	97 896	97 900	97 905
953	97 909	97 914	97 918	97 923	97 928	97 932	97 937	97 941	97 946	97 950
954	97 955	97 959	97 964	97 968	97 973	97 978	97 982	97 987	97 991	97 996
955	98 000	98 005	98 009	98 014	98 019	98 023	98 028	98 032	98 037	98 041
956	98 046	98 050	98 055	98 059	98 064	98 068	98 073	98 078	98 082	98 087
957	98 091	98 096	98 100	98 105	98 109	98 114	98 118	98 123	98 127	98 132
958	98 137	98 141	98 146	98 150	98 155	98 159	98 164	98 168	98 173	98 177
959	98 182	98 186	98 191	98 195	98 200	98 204	98 209	98 214	98 218	98 223
960	98 227	98 232	98 236	98 241	98 245	98 250	98 254	98 259	98 263	98 268
961	98 272	98 277	98 281	98 286	98 290	98 295	98 299	98 304	98 308	98 313
962	98 318	98 322	98 327	98 331	98 336	98 340	98 345	98 349	98 354	98 358
963	98 363	98 367	98 372	98 376	98 381	98 385	98 390	98 394	98 399	98 403
964	98 408	98 412	98 417	98 421	98 426	98 430	98 435	98 439	98 444	98 448
965	98 453	98 457	98 462	98 466	98 471	98 475	98 480	98 484	98 489	98 493
966	98 498	98 502	98 507	98 511	98 516	98 520	98 525	98 529	98 534	98 538
967	98 543	98 547	98 552	98 556	98 561	98 565	98 570	98 574	98 579	98 583
968	98 588	98 592	98 597	98 601	98 605	98 610	98 614	98 619	98 623	98 628
969	98 632	98 637	98 641	98 646	98 650	98 655	98 659	98 664	98 668	98 673
970	98 677	98 682	98 686	98 691	98 695	98 700	98 704	98 709	98 713	98 717
971	98 722	98 726	98 731	98 735	98 740	98 744	98 749	98 753	98 758	98 762
972	98 767	98 771	98 776	98 780	98 784	98 789	98 793	98 798	98 802	98 807
973	98 811	98 816	98 820	98 825	98 829	98 834	98 838	98 843	98 847	98 851
974	98 856	98 860	98 865	98 869	98 874	98 878	98 883	98 887	98 892	98 896
975	98 900	98 905	98 909	98 914	98 918	98 923	98 927	98 932	98 936	98 941
976	98 945	98 949	98 954	98 958	98 963	98 967	98 972	98 976	98 981	98 985
977	98 989	98 994	98 998	99 003	99 007	99 012	99 016	99 021	99 025	99 029
978	99 034	99 038	99 043	99 047	99 052	99 056	99 061	99 065	99 069	99 074
979	99 078	99 083	99 087	99 092	99 096	99 100	99 105	99 109	99 114	99 118
980	99 123	99 127	99 131	99 136	99 140	99 145	99 149	99 154	99 158	99 162
981	99 167	99 171	99 176	99 180	99 185	99 189	99 193	99 198	99 202	99 207
982	99 211	99 216	99 220	99 224	99 229	99 233	99 238	99 242	99 247	99 251
983	99 255	99 260	99 264	99 269	99 273	99 277	99 282	99 286	99 291	99 295
984	99 300	99 304	99 308	99 313	99 317	99 322	99 326	99 330	99 335	99 339
985	99 344	99 348	99 352	99 357	99 361	99 366	99 370	99 374	99 379	99 383
986	99 388	99 392	99 396	99 401	99 405	99 410	99 414	99 419	99 423	99 427
987	99 432	99 436	99 441	99 445	99 449	99 454	99 458	99 463	99 467	99 471
988	99 476	99 480	99 484	99 489	99 493	99 498	99 502	99 506	99 511	99 515
989	99 520	99 524	99 528	99 533	99 537	99 542	99 546	99 550	99 555	99 559
990	99 564	99 568	99 572	99 577	99 581	99 585	99 590	99 594	99 599	99 603
991	99 607	99 612	99 616	99 621	99 625	99 629	99 634	99 638	99 642	99 647
992	99 651	99 656	99 660	99 664	99 669	99 673	99 677	99 682	99 686	99 691
993	99 695	99 699	99 704	99 708	99 712	99 717	99 721	99 726	99 730	99 734
994	99 739	99 743	99 747	99 752	99 756	99 760	99 765	99 769	99 774	99 778
995	99 782	99 787	99 791	99 795	99 800	99 804	99 808	99 813	99 817	99 822
996	99 826	99 830	99 835	99 839	99 843	99 848	99 852	99 856	99 861	99 865
997	99 870	99 874	99 878	99 883	99 887	99 891	99 896	99 900	99 904	99 909
998	99 913	99 917	99 922	99 926	99 930	99 935	99 939	99 944	99 948	99 952
999	99 957	99 961	99 965	99 970	99 974	99 978	99 983	99 987	99 991	99 996
1000	00 000	00 004	00 009	00 013	00 017	00 022	00 026	00 030	00 035	00 039
N	0	1	2	3	4	5	6	7	8	9

		log
Circumference of the Circle in degrees............ =	360	2. 55 630 250
Circumference of the Circle in minutes........... =	21 600	4. 33 445 375
Circumference of the Circle in seconds............ =	1 296 000	6. 11 260 500
If the radius $r = 1$, half the Circumference of the Circle is		
$\pi = 3.\,14\,159\,265\,358\,979\,323\,846\,264\,338\,328$...............		0. 49 714 987

Also:

		log			log
$2\pi =$	6. 28 318 531	0. 79 817 987	$\pi^2 =$	9. 86 960 440	0. 99 429 975
$4\pi =$	12. 56 637 061	1. 09 920 986	$\dfrac{1}{\pi^2} =$	0. 10 132 118	9. 00 570 025 − 10
$\dfrac{\pi}{2} =$	1. 57 079 633	0. 19 611 988	$\sqrt{\pi} =$	1. 77 245 385	0. 24 857 494
$\dfrac{\pi}{3} =$	1. 04 719 755	0. 02 002 862	$\dfrac{1}{\sqrt{\pi}} =$	0. 56 418 958	9. 75 142 506 − 10
$\dfrac{4\pi}{3} =$	4. 18 879 020	0. 62 208 861	$\sqrt{\dfrac{3}{\pi}} =$	0. 97 720 502	9. 98 998 569 − 10
$\dfrac{\pi}{4} =$	0. 78 539 816	9. 89 508 988 − 10	$\sqrt{\dfrac{4}{\pi}} =$	1. 12 837 917	0. 05 245 506
$\dfrac{\pi}{6} =$	0. 52 359 878	9. 71 899 862 − 10	$\sqrt[3]{\pi} =$	1. 46 459 189	0. 16 571 662
$\dfrac{1}{\pi} =$	0. 31 830 989	9. 50 285 013 − 10	$\dfrac{1}{\sqrt[3]{\pi}} =$	0. 68 278 406	9. 83 428 338 − 10
$\dfrac{1}{2\pi} =$	0. 15 915 494	9. 20 182 013 − 10	$\sqrt[3]{\pi^2} =$	2. 14 502 940	0. 33 143 325
$\dfrac{3}{\pi} =$	0. 95 492 966	9. 97 997 138 − 10	$\sqrt[3]{\dfrac{3}{4\pi}} =$	0. 62 035 049	9. 79 263 713 − 10
$\dfrac{4}{\pi} =$	1. 27 323 954	0. 10 491 012			
$\dfrac{3}{4\pi} =$	0. 23 873 241	9. 37 791 139 − 10	$\sqrt[3]{\dfrac{\pi}{6}} =$	0. 80 599 598	9. 90 633 287 − 10

Arc a, whose length is equal to the radius r, is :

			log
in degrees ...... a° ..... $= \dfrac{180}{\pi}$ .... $= 57.\,29\,577\,951^\circ.$			1. 75 812 263
in minutes ...... a' ..... $= \dfrac{10\,800}{\pi}$ ...... $= 3\,437.\,74\,677'\,.$			3. 53 627 388
in seconds ...... a'' ..... $= \dfrac{648\,000}{\pi}$ ..... $= 206\,264.\,806''\,.$			5. 31 442 513

Arc $2\,a$, whose length is equal to twice the radius, $2\,r$, is :

in degrees ...... $2\,a^\circ$ ..... $= \dfrac{360}{\pi}$ .... $= 114.\,59\,155\,903^\circ$		2. 05 915 263
in minutes...... $2\,a'$ ..... $= \dfrac{21\,600}{\pi}$ ..... $= 6\,875.\,49\,354'\,.$		3. 83 730 388
in seconds ...... $2\,a''$ .... $= \dfrac{1\,296\,000}{\pi}$ ... $= 412\,529.\,612''\,.$		5. 61 545 513

If the radius $r = 1$, the length of the arc is :

for 1 degree ..... $\dfrac{1}{a^\circ}$ ..... $= \dfrac{\pi}{180}$ .... $= 0.\,01\,745\,329...$			8. 24 187 737 − 10
for 1 minute..... $\dfrac{1}{a'}$ ...... $= \dfrac{\pi}{10\,800}$ ...... $= 0.\,00\,029\,089...$			6. 46 372 612 − 10
for 1 second ..... $\dfrac{1}{a''}$ ..... $= \dfrac{\pi}{648\,000}$ ..... $= 0.\,00\,000\,485...$			4. 68 557 487 − 10
for $\tfrac{1}{2}$ degree ..... $\dfrac{1}{2\,a^\circ}$ .... $= \dfrac{\pi}{360}$ .... $= 0.\,00\,872\,665...$			7. 94 084 737 − 10
for $\tfrac{1}{2}$ minute..... $\dfrac{1}{2\,a'}$ ..... $= \dfrac{\pi}{21\,600}$ .... $= 0.\,00\,014\,544...$			6. 16 269 612 − 10
for $\tfrac{1}{2}$ second ..... $\dfrac{1}{2\,a''}$ ... $= \dfrac{\pi}{1\,296\,000}$ .... $= 0.\,00\,000\,242...$			4. 38 454 487 − 10
Sin $1''$ in the unit circle .... $= 0.\,00\,000\,485...$			4. 68 557 487 − 10

TABLE III.

THE LOGARITHMS

OF THE

TRIGONOMETRIC FUNCTIONS:

From 0° to 0° 3', or 89° 57' to 90°, for every second;
From 0° to 2°, or 88° to 90°, for every ten seconds;
From 1° to 89°, for every minute.

NOTE. To all the logarithms −10 is to be appended.

| **log sin** | | | | **0°** | | log tan = log sin
log cos = 10.00 000 | | |

''	0'	1'	2'	''	''	0'	1'	2'	''
0	—	6.16 373	6.76 476	**60**	**30**	6.16 270	6.63 982	6.86 167	**30**
1	4.68 557	6.47 090	6.76 836	59	31	6.17 694	6.64 462	6.86 455	29
2	4.98 660	6.47 797	6.77 193	58	32	6.19 072	6.64 936	6.86 742	28
3	5.16 270	6.48 492	6.77 548	57	33	6.20 409	6.65 406	6.87 027	27
4	5.28 763	6.49 175	6.77 900	56	34	6.21 705	6.65 870	6.87 310	26
5	5.38 454	6.49 849	6.78 248	**55**	**35**	6.22 964	6.66 330	6.87 591	**25**
6	5.46 373	6.50 512	6.78 595	54	36	6.24 188	6.66 785	6.87 870	24
7	5.53 067	6.51 165	6.78 938	53	37	6.25 378	6.67 235	6.88 147	23
8	5.58 866	6.51 808	6.79 278	52	38	6.26 536	6.67 680	6.88 423	22
9	5.63 982	6.52 442	6.79 616	51	39	6.27 664	6.68 121	6.88 697	21
10	5.68 557	6.53 067	6.79 952	**50**	**40**	6.28 763	6.68 557	6.88 969	**20**
11	5.72 697	6.53 683	6.80 285	49	41	6.29 836	6.68 990	6.89 240	19
12	5.76 476	6.54 291	6.80 615	48	42	6.30 882	6.69 418	6.89 509	18
13	5.79 952	6.54 890	6.80 943	47	43	6.31 904	6.69 841	6.89 776	17
14	5.83 170	6.55 481	6.81 268	46	44	6.32 903	6.70 261	6.90 042	16
15	5.86 167	6.56 064	6.81 591	**45**	**45**	6.33 879	6.70 676	6.90 306	**15**
16	5.88 969	6.56 639	6.81 911	44	46	6.34 833	6.71 088	6.90 568	14
17	5.91 602	6.57 207	6.82 230	43	47	6.35 767	6.71 496	6.90 829	13
18	5.94 085	6.57 767	6.82 545	42	48	6.36 682	6.71 900	6.91 088	12
19	5.96 433	6.58 320	6.82 859	41	49	6.37 577	6.72 300	6.91 346	11
20	5.98 660	6.58 866	6.83 170	**40**	**50**	6.38 454	6.72 697	6.91 602	**10**
21	6.00 779	6.59 406	6.83 479	39	51	6.39 315	6.73 090	6.91 857	9
22	6.02 800	6.59 939	6.83 786	38	52	6.40 158	6.73 479	6.92 110	8
23	6.04 730	6.60 465	6.84 091	37	53	6.40 985	6.73 865	6.92 362	7
24	6.06 579	6.60 985	6.84 394	36	54	6.41 797	6.74 248	6.92 612	6
25	6.08 351	6.61 499	6.84 694	**35**	**55**	6.42 594	6.74 627	6.92 861	**5**
26	6.10 055	6.62 007	6.84 993	34	56	6.43 376	6.75 003	6.93 109	4
27	6.11 694	6.62 509	6.85 289	33	57	6.44 145	6.75 376	6.93 355	3
28	6.13 273	6.63 006	6.85 584	32	58	6.44 900	6.75 746	6.93 599	2
29	6.14 797	6.63 496	6.85 876	31	59	6.45 643	6.76 112	6.93 843	1
30	6.16 270	6.63 982	6.86 167	**30**	**60**	6.46 373	6.76 476	6.94 085	**0**
''	59'	58'	57'	''	''	59'	58'	57'	''

log cot = log cos
log sin = 10.00 000   **89°**   **log cos**

' ''	log sin	log tan	log cos	'' '	' ''	log sin	log tan	log cos	'' '
0 0	—	—	10.00000	0 60	10 0	7.46 373	7.46 373	10.00000	0 50
10	5.68 557	5.68 557	10.00000	50	10	7.47 090	7.47 091	10.00000	50
20	5.98 660	5.98 660	10.00000	40	20	7.47 797	7.47 797	10.00000	40
30	6.16 270	6.16 270	10.00000	30	30	7.48 491	7.48 492	10.00000	30
40	6.28 763	6.28 763	10.00000	20	40	7.49 175	7.49 176	10.00000	20
50	6.38 454	6.38 454	10.00000	10	50	7.49 849	7.49 849	10.00000	10
1 0	6.46 373	6.46 373	10.00000	0 59	11 0	7.50 512	7.50 512	10.00000	0 49
10	6.53 067	6.53 067	10.00000	50	10	7.51 165	7.51 165	10.00000	50
20	6.58 866	6.58 866	10.00000	40	20	7.51 808	7.51 809	10.00000	40
30	6.63 982	6.63 982	10.00000	30	30	7.52 442	7.52 443	10.00000	30
40	6.68 557	6.68 557	10.00000	20	40	7.53 067	7.53 067	10.00000	20
50	6.72 697	6.72 697	10.00000	10	50	7.53 683	7.53 683	10.00000	10
2 0	6.76 476	6.76 476	10 00000	0 58	12 0	7.54 291	7.54 291	10 00000	0 48
10	6.79 952	6.79 952	10.00000	50	10	7.54 890	7.54 890	10.00000	50
20	6 83 170	6.83 170	10.00000	40	20	7.55 481	7.55 481·	10 00000	40
30	6.86 167	6.86 167	10.00000	30	30	7.56 064	7.56 064	10.00000	30
40	6.88 969	6.88 969	10.00000	20	40	7.56 639	7.56 639	10.00000	20
50	6.91 602	6.91 602	10.00000	10	50	7.57 206	7.57 207	10.00000	10
3 0	6.94 085	6.94 085	10 00000	0 57	13 0	7.57 767	7.57 767	10.00000	0 47
10	6.96 433	6.96 433	10.00000	50	10	7.58 320	7.58 320	10.00000	50
20	6.98 660	6.98 661	10 00000	40	20	7.58 866	7.58 867	10 00000	40
30	7.00 779	7.00 779	10 00000	30	30	7.59 406	7.59 406	10.00000	30
40	7.02 800	7.02 800	10 00000	20	40	7.59 939	7.59 939	10.00000	20
50	7.04 730	7.04 730	10 00000	10	50	7.60 465	7.60 466	10.00000	10
4 0	7.06 579	7.06 579	10.00000	0 56	14 0	7.60 985	7.60 986	10.00000	0 46
10	7.08 351	7.08 352	10.00000	50	10	7.61 499	7.61 500	10.00000	50
20	7.10 055	7.10 055	10.00000	40	20	7.62 007	7.62 008	10.00000	40
30	7.11 694	7.11 694	10.00000	30	30	7.62 509	7.62 510	10.00000	30
40	7.13 273	7.13 273	10.00000	20	40	7.63 006	7.63 006	10.00000	20
50	7.14 797	7.14 797	10.00000	10	50	7.63 496	7.63 497	10.00000	10
5 0	7.16 270	7.16 270	10.00000	0 55	15 0	7.63 982	7.63 982	10.00000	0 45
10	7.17 694	7.17 694	10.00000	50	10	7.64 461	7.64 462	10.00000	50
20	7.19 072	7.19 073	10.00000	40	20	7.64 936	7.64 937	10.00000	40
30	7.20 409	7.20 409	10 00000	30	30	7.65 406	7.65 406	10.00000	30
40	7.21 705	7.21 705	10 00000	20	40	7.65 870	7.65 871	10.00000	20
50	7.22 964	7.22 964	10.00000	10	50	7.66 330	7.66 330	10.00000	10
6 0	7.24 188	7.24 188	10.00000	0 54	16 0	7.66 784	7.66 785	10.00000	0 44
10	7.25 378	7.25 378	10.00000	50	10	7.67 235	7.67 235	10.00000	50
20	7.26 536	7.26 536	10.00000	40	20	7.67 680	7.67 680	10.00000	40
30	7.27 664	7.27 664	10.00000	30	30	7.68 121	7.68 121	10.00000	30
40	7.28 763	7.28 764	10.00000	20	40	7.68 557	7.68 558	9.99999	20
50	7.29 836	7.29 836	10.00000	10	50	7.68 989	7.68 990	9.99999	10
7 0	7.30 882	7.30 882	10.00000	0 53	17 0	7.69 417	7.69 418	9.99 999	0 43
10	7.31 904	7.31 904	10.00000	50	10	7.69 841	7.69 842	9.99 999	50
20	7.32 903	7.32 903	10.00000	40	20	7.70 261	7.70 261	9.99 999	40
30	7.33 879	7.33 879	10.00000	30	30	7.70 676	7.70 677	9.99 999	30
40	7.34 833	7.34 833	10.00000	20	40	7.71 088	7.71 088	9.99 999	20
50	7.35 767	7.35 767	10.00000	10	50	7.71 496	7.71 496	9.99 999	10
8 0	7.36 682	7.36 682	10.00000	0 52	18 0	7.71 900	7.71 900	9.99 999	0 42
10	7.37 577	7.37 577	10.00000	50	10	7.72 300	7.72 301	9.99 999	50
20	7.38 454	7.38 455	10.00000	40	20	7.72 697	7.72 697	9.99 999	40
30	7.39 314	7.39 315	10.00000	30	30	7.73 090	7.73 090	9.99 999	30
40	7.40 158	7.40 158	10.00000	20	40	7.73 479	7.73 480	9.99 999	20
50	7.40 985	7.40 985	10.00000	10	50	7.73 865	7.73 866	9.99 999	10
9 0	7.41 797	7.41 797	10.00000	0 51	19 0	7.74 248	7.74 248	9.99 999	0 41
10	7.42 594	7.42 594	10.00000	50	10	7.74 627	7.74 628	9.99 999	50
20	7.43 376	7.43 376	10.00000	40	20	7.75 003	7.75 004	9.99 999	40
30	7.44 145	7.44 145	10.00000	30	30	7.75 376	7.75 377	9.99 999	30
40	7.44 900	7.44 900	10.00000	20	40	7.75 745	7.75 746	9.99 999	20
50	7.45 643	7.45 643	10.00000	10	50	7.76 112	7.76 113	9.99 999	10
100	7.46 373	7.46 373	10.00000	0 50	20 0	7.76 475	7.76 476	9.99 999	0 40
' ''	log cos	log cot	log sin	'' '	' ''	log cos	log cot	log sin	'' '

′ ″	log sin	log tan	log cos	″ ′	′ ″	log sin	log tan	log cos	″ ′
20 0	7. 76 475	7. 76 476	9. 99 999	0 **40**	**30** 0	7. 94 084	7. 94 086	9. 99 998	0 **30**
10	7. 76 836	7. 76 837	9. 99 999	50	10	7. 94 325	7. 94 326	9. 99 998	50
20	7. 77 193	7. 77 194	9. 99 999	40	20	7. 94 564	7. 94 566	9. 99 998	40
30	7. 77 548	7. 77 549	9. 99 999	30	30	7. 94 802	7. 94 804	9. 99 998	30
40	7. 77 899	7. 77 900	9. 99 999	20	40	7. 95 039	7. 95 040	9. 99 998	20
50	7. 78 248	7. 78 249	9. 99 999	10	50	7. 95 274	7. 95 276	9. 99 998	10
21 0	7. 78 594	7. 78 595	9. 99 999	0 **39**	**31** 0	7. 95 508	7. 95 510	9. 99 998	0 **29**
10	7. 78 938	7. 78 938	9. 99 999	50	10	7. 95 741	7. 95 743	9. 99 998	50
20	7. 79 278	7. 79 279	9. 99 999	40	20	7. 95 973	7. 95 974	9. 99 998	40
30	7. 79 616	7. 79 617	9. 99 999	30	30	7. 96 203	7. 96 205	9. 99 998	30
40	7. 79 952	7. 79 952	9. 99 999	20	40	7. 96 432	7. 96 434	9. 99 998	20
50	7. 80 284	7. 80 285	9. 99 999	10	50	7. 96 660	7. 96 662	9. 99 998	10
22 0	7. 80 615	7. 80 615	9. 99 999	0 **38**	**32** 0	7. 96 887	7. 96 889	9. 99 998	0 **28**
10	7. 80 942	7. 80 943	9. 99 999	50	10	7. 97 113	7. 97 114	9. 99 998	50
20	7. 81 268	7. 81 269	9. 99 999	40	20	7. 97 337	7. 97 339	9. 99 998	40
30	7. 81 591	7. 81 591	9. 99 999	30	30	7. 97 560	7. 97 562	9. 99 998	30
40	7. 81 911	7. 81 912	9. 99 999	20	40	7. 97 782	7. 97 784	9. 99 998	20
50	7. 82 229	7. 82 230	9. 99 999	10	50	7. 98 003	7. 98 005	9. 99 998	10
23 0	7. 82 545	7. 82 546	9. 99 999	0 **37**	**33** 0	7. 98 223	7. 98 225	9. 99 998	0 **27**
10	7. 82 859	7. 82 860	9. 99 999	50	10	7. 98 442	7. 98 444	9. 99 998	50
20	7. 83 170	7. 83 171	9. 99 999	40	20	7. 98 660	7. 98 662	9. 99 998	40
30	7. 83 479	7. 83 480	9. 99 999	30	30	7. 98 876	7. 98 878	9. 99 998	30
40	7. 83 786	7. 83 787	9. 99 999	20	40	7. 99 092	7. 99 094	9. 99 998	20
50	7. 84 091	7. 84 092	9. 99 999	10	50	7. 99 306	7. 99 308	9. 99 998	10
24 0	7. 84 393	7. 84 394	9. 99 999	0 **36**	**34** 0	7. 99 520	7. 99 522	9. 99 998	0 **26**
10	7. 84 694	7. 84 695	9. 99 999	50	10	7. 99 732	7. 99 734	9. 99 998	50
20	7. 84 992	7. 84 994	9. 99 999	40	20	7. 99 943	7. 99 946	9. 99 998	40
30	7. 85 289	7. 85 290	9. 99 999	30	30	8. 00 154	8. 00 156	9. 99 998	30
40	7. 85 583	7. 85 584	9. 99 999	20	40	8. 00 363	8. 00 365	9. 99 998	20
50	7. 85 876	7. 85 877	9. 99 999	10	50	8. 00 571	8. 00 574	9. 99 998	10
25 0	7. 86 166	7. 86 167	9. 99 999	0 **35**	**35** 0	8. 00 779	8. 00 781	9. 99 998	0 **25**
10	7. 86 455	7. 86 456	9. 99 999	50	10	8. 00 985	8. 00 987	9. 99 998	50
20	7. 86 741	7. 86 743	9. 99 999	40	20	8. 01 190	8. 01 193	9. 99 998	40
30	7. 87 026	7. 87 027	9. 99 999	30	30	8. 01 395	8. 01 397	9. 99 998	30
40	7. 87 309	7. 87 310	9. 99 999	20	40	8. 01 598	8. 01 600	9. 99 998	20
50	7. 87 590	7. 87 591	9. 99 999	10	50	8. 01 801	8. 01 803	9. 99 998	10
26 0	7. 87 870	7. 87 871	9. 99 999	0 **34**	**36** 0	8. 02 002	8. 02 004	9. 99 998	0 **24**
10	7. 88 147	7. 88 148	9. 99 999	50	10	8. 02 203	8. 02 205	9. 99 998	50
20	7. 88 423	7. 88 424	9. 99 999	40	20	8. 02 402	8. 02 405	9. 99 998	40
30	7. 88 697	7. 88 698	9. 99 999	30	30	8. 02 601	8. 02 604	9. 99 998	30
40	7. 88 969	7. 88 970	9. 99 999	20	40	8. 02 799	8. 02 801	9. 99 998	20
50	7. 89 240	7. 89 241	9. 99 999	10	50	8. 02 996	8. 02 998	9. 99 998	10
27 0	7. 89 509	7. 89 510	9. 99 999	0 **33**	**37** 0	8. 03 192	8. 03 194	9. 99 997	0 **23**
10	7. 89 776	7. 89 777	9. 99 999	50	10	8. 03 387	8. 03 390	9. 99 997	50
20	7. 90 041	7. 90 043	9. 99 999	40	20	8. 03 581	8. 03 584	9. 99 997	40
30	7. 90 305	7. 90 307	9. 99 999	30	30	8. 03 775	8. 03 777	9. 99 997	30
40	7. 90 568	7. 90 569	9. 99 999	20	40	8. 03 967	8. 03 970	9. 99 997	20
50	7. 90 829	7. 90 830	9. 99 999	10	50	8. 04 159	8. 04 162	9. 99 997	10
28 0	7. 91 088	7. 91 089	9. 99 999	0 **32**	**38** 0	8. 04 350	8. 04 353	9. 99 997	0 **22**
10	7. 91 346	7. 91 347	9. 99 999	50	10	8. 04 540	8. 04 543	9. 99 997	50
20	7. 91 602	7. 91 603	9. 99 999	40	20	8. 04 729	8. 04 732	9. 99 997	40
30	7. 91 857	7. 91 858	9. 99 999	30	30	8. 04 918	8. 04 921	9. 99 997	30
40	7. 92 110	7. 92 111	9. 99 998	20	40	8. 05 105	8. 05 108	9. 99 997	20
50	7. 92 362	7. 92 363	9. 99 998	10	50	8. 05 292	8. 05 295	9. 99 997	10
29 0	7. 92 612	7. 92 613	9. 99 998	0 **31**	**39** 0	8. 05 478	8. 05 481	9. 99 997	0 **21**
10	7. 92 861	7. 92 862	9. 99 998	50	10	8. 05 663	8. 05 666	9. 99 997	50
20	7. 93 108	7. 93 110	9. 99 998	40	20	8. 05 848	8. 05 851	9. 99 997	40
30	7. 93 354	7. 93 356	9. 99 998	30	30	8. 06 031	8. 06 034	9. 99 997	30
40	7. 93 599	7. 93 601	9. 99 998	20	40	8. 06 214	8. 06 217	9. 99 997	20
50	7. 93 842	7. 93 844	9. 99 998	10	50	8. 06 396	8. 06 399	9. 99 997	10
30 0	7. 94 084	7. 94 086	9. 99 998	0 **30**	**40** 0	8. 06 578	8. 06 581	9. 99 997	0 **20**
′ ″	log cos	log cot	log sin	″ ′	′ ″	log cos	log cot	log sin	″ ′

´ ´´	log sin	log tan	log cos	´´ ´	´ ´´	log sin	log tan	log cos	´´ ´
40 0	8.06578	8.06581	9.99997	0 **20**	**50** 0	8.16268	8.16273	9.99995	0 **10**
10	8.06758	8.06761	9.99997	50	10	8.16413	8.16417	9.99995	50
20	8.06938	8.06941	9.99997	40	20	8.16557	8.16561	9.99995	40
30	8.07117	8.07120	9.99997	30	30	8.16700	8.16705	9.99995	30
40	8.07295	8.07299	9.99997	20	40	8.16843	8.16848	9.99995	20
50	8.07473	8.07476	9.99997	10	50	8.16986	8.16991	9.99995	10
41 0	8.07650	8.07653	9.99997	0 **19**	**51** 0	8.17128	8.17133	9.99995	0 **9**
10	8.07826	8.07829	9.99997	50	10	8.17270	8.17275	9.99995	50
20	8.08002	8.08005	9.99997	40	20	8.17411	8.17416	9.99995	40
30	8.08176	8.08180	9.99997	30	30	8.17552	8.17557	9.99995	30
40	8.08350	8.08354	9.99997	20	40	8.17692	8.17697	9.99995	20
50	8.08524	8.08527	9.99997	10	50	8.17832	8.17837	9.99995	10
42 0	8.08696	8.08700	9.99997	0 **18**	**52** 0	8.17971	8.17976	9.99995	0 **8**
10	8.08868	8.08872	9.99997	50	10	8.18110	8.18115	9.99995	50
20	8.09040	8.09043	9.99997	40	20	8.18249	8.18254	9.99995	40
30	8.09210	8.09214	9.99997	30	30	8.18387	8.18392	9.99995	30
40	8.09380	8.09384	9.99997	20	40	8.18524	8.18530	9.99995	20
50	8.09550	8.09553	9.99997	10	50	8.18662	8.18667	9.99995	10
43 0	8.09718	8.09722	9.99997	0 **17**	**53** 0	8.18798	8.18804	9.99995	0 **7**
10	8.09886	8.09890	9.99997	50	10	8.18935	8.18940	9.99995	50
20	8.10054	8.10057	9.99997	40	20	8.19071	8.19076	9.99995	40
30	8.10220	8.10224	9.99997	30	30	8.19206	8.19212	9.99995	30
40	8.10386	8.10390	9.99997	20	40	8.19341	8.19347	9.99995	20
50	8.10552	8.10555	9.99996	10	50	8.19476	8.19481	9.99995	10
44 0	8.10717	8.10720	9.99996	0 **16**	**54** 0	8.19610	8.19616	9.99995	0 **6**
10	8.10881	8.10884	9.99996	50	10	8.19744	8.19749	9.99995	50
20	8.11044	8.11048	9.99996	40	20	8.19877	8.19883	9.99995	40
30	8.11207	8.11211	9.99996	30	30	8.20010	8.20016	9.99995	30
40	8.11370	8.11373	9.99996	20	40	8.20143	8.20149	9.99995	20
50	8.11531	8.11535	9.99996	10	50	8.20275	8.20281	9.99994	10
45 0	8.11693	8.11696	9.99996	0 **15**	**55** 0	8.20407	8.20413	9.99994	0 **5**
10	8.11853	8.11857	9.99996	50	10	8.20538	8.20544	9.99994	50
20	8.12013	8.12017	9.99996	40	20	8.20669	8.20675	9.99994	40
30	8.12172	8.12176	9.99996	30	30	8.20800	8.20806	9.99994	30
40	8.12331	8.12335	9.99996	20	40	8.20930	8.20936	9.99994	20
50	8.12489	8.12493	9.99996	10	50	8.21060	8.21066	9.99994	10
46 0	8.12647	8.12651	9.99996	0 **14**	**56** 0	8.21189	8.21195	9.99994	0 **4**
10	8.12804	8.12808	9.99996	50	10	8.21319	8.21324	9.99994	50
20	8.12961	8.12965	9.99996	40	20	8.21447	8.21453	9.99994	40
30	8.13117	8.13121	9.99996	30	30	8.21576	8.21581	9.99994	30
40	8.13272	8.13276	9.99996	20	40	8.21703	8.21709	9.99994	20
50	8.13427	8.13431	9.99996	10	50	8.21831	8.21837	9.99994	10
47 0	8.13581	8.13585	9.99996	0 **13**	**57** 0	8.21958	8.21964	9.99994	0 **3**
10	8.13735	8.13739	9.99996	50	10	8.22085	8.22091	9.99994	50
20	8.13888	8.13892	9.99996	40	20	8.22211	8.22217	9.99994	40
30	8.14041	8.14045	9.99996	30	30	8.22337	8.22343	9.99994	30
40	8.14193	8.14197	9.99996	20	40	8.22463	8.22469	9.99994	20
50	8.14344	8.14348	9.99996	10	50	8.22588	8.22595	9.99994	10
48 0	8.14495	8.14500	9.99996	0 **12**	**58** 0	8.22713	8.22720	9.99994	0 **2**
10	8.14646	8.14650	9.99996	50	10	8.22838	8.22844	9.99994	50
20	8.14796	8.14800	9.99996	40	20	8.22962	8.22968	9.99994	40
30	8.14945	8.14950	9.99996	30	30	8.23086	8.23092	9.99994	30
40	8.15094	8.15099	9.99996	20	40	8.23210	8.23216	9.99994	20
50	8.15243	8.15247	9.99996	10	50	8.23333	8.23339	9.99994	10
49 0	8.15391	8.15395	9.99996	0 **11**	**59** 0	8.23456	8.23462	9.99994	0 **1**
10	8.15538	8.15543	9.99996	50	10	8.23578	8.23585	9.99994	50
20	8.15685	8.15690	9.99996	40	20	8.23700	8.23707	9.99994	40
30	8.15832	8.15836	9.99996	30	30	8.23822	8.23829	9.99993	30
40	8.15978	8.15982	9.99996	20	40	8.23944	8.23950	9.99993	20
50	8.16123	8.16128	9.99995	10	50	8.24065	8.24071	9.99993	10
50 0	8.16268	8.16273	9.99995	0 **10**	**60** 0	8.24186	8.24192	9.99993	0 **0**
´ ´´	log cos	log cot	log sin	´´ ´	´ ´´	log cos	log cot	log sin	´´ ´

' ''	log sin	log tan	log cos	'' !	' ''	log sin	log tan	log cos	'' '
0 0	8. 24 186	8. 24 192	9. 99 993	0 **60**	10 0	8. 30 879	8. 30 888	9. 99 991	0 **50**
10	8. 24 306	8. 24 313	9. 99 993	50	10	8. 30 983	8. 30 992	9. 99 991	50
20	8. 24 426	8. 24 433	9. 99 993	40	20	8. 31 086	8. 31 095	9. 99 991	40
30	8. 24 546	8. 24 553	9. 99 993	30	30	8. 31 188	8. 31 198	9. 99 991	30
40	8. 24 665	8. 24 672	9. 99 993	20	40	8. 31 291	8. 31 300	9. 99 991	20
50	8. 24 785	8. 24 791	9. 99 993	10	50	8. 31 393	8. 31 403	9. 99 991	10
1 0	8. 24 903	8. 24 910	9. 99 993	0 **59**	11 0	8. 31 495	8. 31 505	9. 99 991	0 **49**
10	8. 25 022	8. 25 029	9. 99 993	50	10	8. 31 597	8. 31 606	9. 99 991	50
20	8. 25 140	8. 25 147	9. 99 993	40	20	8. 31 699	8. 31 708	9. 99 991	40
30	8. 25 258	8. 25 265	9. 99 993	30	30	8. 31 800	8. 31 809	9. 99 991	30
40	8. 25 375	8. 25 382	9. 99 993	20	40	8. 31 901	8. 31 911	9. 99 991	20
50	8. 25 493	8. 25 500	9. 99 993	10	50	8. 32 002	8. 32 012	9. 99 991	10
2 0	8. 25 609	8. 25 616	9. 99 993	0 **58**	12 0	8. 32 103	8. 32 112	9. 99 990	0 **48**
10	8. 25 726	8. 25 733	9. 99 993	50	10	8. 32 203	8. 32 213	9. 99 990	50
20	8. 25 842	8. 25 849	9. 99 993	40	20	8 32 303	8. 32 313	9. 99 990	40
30	8. 25 958	8. 25 965	9. 99 993	30	30	8. 32 403	8. 32 413	9. 99 990	30
40	8. 26 074	8. 26 081	9. 99 993	20	40	8. 32 503	8. 32 513	9. 99 990	20
50	8. 26 189	8. 26 196	9. 99 993	10	50	8. 32 602	8. 32 612	9. 99 990	10
3 0	8. 26 304	8. 26 312	9. 99 993	0 **57**	13 0	8. 32 702	8. 32 711	9. 99 990	0 **47**
10	8. 26 419	8. 26 426	9. 99 993	50	10	8. 32 801	8. 32 811	9. 99 990	50
20	8. 26 533	8. 26 541	9. 99 993	40	20	8. 32 899	8. 32 909	9. 99 990	40
30	8. 26 648	8. 26 655	9. 99 993	30	30	8. 32 998	8. 33 008	9. 99 990	30
40	8. 26 761	8. 26 769	9. 99 993	20	40	8. 33 096	8. 33 106	9. 99 990	20
50	8. 26 875	8. 26 882	9. 99 993	10	50	8. 33 195	8, 33 205	9. 99 990	10
4 0	8. 26 988	8. 26 996	9. 99 992	0 **56**	14 0	8. 33 292	8. 33 302	9. 99 990	0 **46**
10	8. 27 101	8. 27 109	9. 99 992	50	10	8. 33 390	8. 33 400	9. 99 990	50
20	8. 27 214	8. 27 221	9. 99 992	40	20	8. 33 488	8. 33 498	9. 99 990	40
30	8. 27 326	8. 27 334	9. 99 992	30	30	8. 33 585	8. 33 595	9. 99 990	30
40	8. 27 438	8. 27 446	9. 99 992	20	40	8. 33 682	8. 33 692	9. 99 990	20
50	8. 27 550	8. 27 558	9. 99 992	10	50	8. 33 779	8. 33 789	9. 99 990	10
5 0	8. 27 661	8. 27 669	9. 99 992	0 **55**	15 0	8. 33 875	8. 33 886	9. 99 990	0 **45**
10	8. 27 773	8. 27 780	9. 99 992	50	10	8. 33 972	8. 33 982	9. 99 990	50
20	8. 27 883	8. 27 891	9. 99 992	40	20	8. 34 068	8. 34 078	9. 99 990	40
30	8. 27 994	8. 28 002	9. 99 992	30	30	8. 34 164	8. 34 174	9. 99 990	30
40	8. 28 104	8. 28 112	9. 99 992	20	40	8. 34 260	8. 34 270	9. 99 989	20
50	8. 28 215	8. 28 223	9. 99 992	10	50	8. 34 355	8. 34 366	9. 99 989	10
6 0	8. 28 324	8. 28 332	9. 99 992	0 **54**	16 0	8. 34 450	8. 34 461	9. 99 989	0 **44**
10	8. 28 434	8. 28 442	9. 99 992	50	10	8. 34 546	8. 34 556	9. 99 989	50
20	8. 28 543	8. 28 551	9. 99 992	40	20	8. 34 640	8. 34 651	9. 99 989	40
30	8. 28 652	8. 28 660	9. 99 992	30	30	8. 34 735	8. 34 746	9. 99 989	30
40	8. 28 761	8. 28 769	9. 99 992	20	40	8. 34 830	8. 34 840	9. 99 989	20
50	8. 28 869	8. 28 877	9. 99 992	10	50	8. 34 924	8. 34 935	9. 99 989	10
7 0	8. 28 977	8. 28 986	9. 99 992	0 **53**	17 0	8. 35 018	8. 35 029	9. 99 989	0 **43**
10	8. 29 085	8. 29 094	9. 99 992	50	10	8. 35 112	8. 35 123	9. 99 989	50
20	8. 29 193	8. 29 201	9. 99 992	40	20	8. 35 206	8. 35 217	9. 99 989	40
30	8. 29 300	8. 29 309	9. 99 992	30	30	8. 35 299	8. 35 310	9. 99 989	30
40	8. 29 407	8. 29 416	9. 99 992	20	40	8. 35 392	8. 35 403	9. 99 989	20
50	8. 29 514	8. 29 523	9. 99 992	10	50	8. 35 485	8. 35 497	9. 99 989	10
8 0	8. 29 621	8. 29 629	9. 99 992	0 **52**	18 0	8. 35 578	8. 35 590	9. 99 989	0 **42**
10	8. 29 727	8. 29 736	9. 99 991	50	10	8. 35 671	8. 35 682	9. 99 989	50
20	8. 29 833	8. 29 842	9. 99 991	40	20	8. 35 764	8. 35 775	9. 99 989	40
30	8. 29 939	8. 29 947	9. 99 991	30	30	8. 35 856	8. 35 867	9. 99 989	30
40	8. 30 044	8. 30 053	9. 99 991	20	40	8. 35 948	8. 35 959	9. 99 989	20
50	8. 30 150	8. 30 158	9. 99 991	10	50	8. 36 040	8. 36 051	9. 99 989	10
9 0	8. 30 255	8. 30 263	9. 99 991	0 **51**	19 0	8. 36 131	8. 36 143	9. 99 989	0 **41**
10	8. 30 359	8. 30 368	9. 99 991	50	10	8. 36 223	8. 36 235	9. 99 988	50
20	8. 30 464	8. 30 473	9. 99 991	40	20	8. 36 314	8. 36 326	9. 99 988	40
30	8. 30 568	8. 30 577	9. 99 991	30	30	8. 36 405	8. 36 417	9. 99 988	30
40	8. 30 672	8. 30 681	9. 99 991	20	40	8. 36 496	8. 36 508	9. 99 988	20
50	8. 30 776	8. 30 785	9. 99 991	10	50	8. 36 587	8. 36 599	9. 99 988	10
10 0	8. 30 879	8. 30 888	9. 99 991	0 **50**	20 0	8. 36 678	8. 36 689	9. 99 988	0 **40**
' ''	log cos	log cot	log sin	'' '	' ''	log cos	log cot	log sin	'' '

′ ″	log sin	log tan	log cos	″ ′	′ ″	log sin	log tan	log cos	″ ′
20 0	8. 36 678	8. 36 689	9. 99 988	0 **40**	**30** 0	8. 41 792	8. 41 807	9. 99 985	0 **30**
10	8. 36 768	8. 36 780	9. 99 988	50	10	8. 41 872	8. 41 887	9. 99 985	50
20	8. 36 858	8. 36 870	9. 99 988	40	20	8. 41 952	8. 41 967	9. 99 985	40
30	8. 36 948	8. 36 960	9. 99 988	30	30	8. 42 032	8. 42 048	9. 99 985	30
40	8. 37 038	8. 37 050	9. 99 988	20	40	8. 42 112	8. 42 127	9. 99 985	20
50	8. 37 128	8. 37 140	9. 99 988	10	50	8. 42 192	8. 42 207	9. 99 985	10
21 0	8. 37 217	8. 37 229	9. 99 988	0 **39**	**31** 0	8. 42 272	8. 42 287	9. 99 985	0 **29**
10	8. 37 306	8. 37 318	9. 99 988	50	10	8. 42 351	8. 42 366	9. 99 985	50
20	8. 37 395	8. 37 408	9. 99 988	40	20	8. 42 430	8. 42 446	9. 99 985	40
30	8. 37 484	8. 37 497	9. 99 988	30	30	8. 42 510	8. 42 525	9. 99 985	30
40	8. 37 573	8. 37 585	9. 99 988	20	40	8. 42 589	8. 42 604	9. 99 985	20
50	8. 37 662	8. 37 674	9. 99 988	10	50	8. 42 667	8. 42 683	9. 99 985	10
22 0	8. 37 750	8. 37 762	9. 99 988	0 **38**	**32** 0	8. 42 746	8. 42 762	9. 99 984	0 **28**
10	8. 37 838	8. 37 850	9. 99 988	50	10	8. 42 825	8. 42 840	9. 99 984	50
20	8. 37 926	8. 37 938	9. 99 988	40	20	8. 42 903	8. 42 919	9. 99 984	40
30	8. 38 014	8. 38 026	9. 99 987	30	30	8. 42 982	8. 42 997	9. 99 984	30
40	8. 38 101	8. 38 114	9. 99 987	20	40	8. 43 060	8. 43 075	9. 99 984	20
50	8. 38 189	8. 38 202	9. 99 987	10	50	8. 43 138	8. 43 154	9. 99 984	10
23 0	8. 38 276	8. 38 289	9. 99 987	0 **37**	**33** 0	8. 43 216	8. 43 232	9. 99 984	0 **27**
10	8. 38 363	8. 38 376	9. 99 987	50	10	8. 43 293	8. 43 309	9. 99 984	50
20	8. 38 450	8. 38 463	9. 99 987	40	20	8. 43 371	8. 43 387	9. 99 984	40
30	8. 38 537	8. 38 550	9. 99 987	30	30	8. 43 448	8. 43 464	9. 99 984	30
40	8. 38 624	8. 38 636	9. 99 987	20	40	8. 43 526	8. 43 542	9. 99 984	20
50	8. 38 710	8. 38 723	9. 99 987	10	50	8. 43 603	8. 43 619	9. 99 984	10
24 0	8. 38 796	8. 38 809	9. 99 987	0 **36**	**34** 0	8. 43 680	8. 43 696	9. 99 984	0 **26**
10	8. 38 882	8. 38 895	9. 99 987	50	10	8. 43 757	8. 43 773	9. 99 984	50
20	8. 38 968	8. 38 981	9. 99 987	40	20	8. 43 834	8. 43 850	9. 99 984	40
30	8. 39 054	8. 39 067	9. 99 987	30	30	8. 43 910	8. 43 927	9. 99 984	30
40	8. 39 139	8. 39 153	9. 99 987	20	40	8. 43 987	8. 44 003	9. 99 984	20
50	8. 39 225	8. 39 238	9. 99 987	10	50	8. 44 063	8. 44 080	9. 99 983	10
25 0	8. 39 310	8. 39 323	9. 99 987	0 **35**	**35** 0	8. 44 139	8. 44 156	9. 99 983	0 **25**
10	8. 39 395	8. 39 408	9. 99 987	50	10	8. 44 216	8. 44 232	9. 99 983	50
20	8. 39 480	8. 39 493	9. 99 987	40	20	8. 44 292	8. 44 308	9. 99 983	40
30	8. 39 565	8. 39 578	9. 99 987	30	30	8. 44 367	8. 44 384	9. 99 983	30
40	8. 39 649	8. 39 663	9. 99 987	20	40	8. 44 443	8. 44 460	9. 99 983	20
50	8. 39 734	8. 39 747	9. 99 986	10	50	8. 44 519	8. 44 536	9. 99 983	10
26 0	8. 39 818	8. 39 832	9. 99 986	0 **34**	**36** 0	8. 44 594	8. 44 611	9. 99 983	0 **24**
10	8. 39 902	8. 39 916	9. 99 986	50	10	8. 44 669	8. 44 686	9. 99 983	50
20	8. 39 986	8. 40 000	9. 99 986	40	20	8. 44 745	8. 44 762	9. 99 983	40
30	8. 40 070	8. 40 083	9. 99 986	30	30	8. 44 820	8. 44 837	9. 99 983	30
40	8. 40 153	8. 40 167	9. 99 986	20	40	8. 44 895	8. 44 912	9. 99 983	20
50	8. 40 237	8. 40 251	9. 99 986	10	50	8. 44 969	8. 44 987	9. 99 983	10
27 0	8. 40 320	8. 40 334	9. 99 986	0 **33**	**37** 0	8. 45 044	8. 45 061	9. 99 983	0 **23**
10	8. 40 403	8. 40 417	9. 99 986	50	10	8. 45 119	8. 45 136	9. 99 983	50
20	8. 40 486	8. 40 500	9. 99 986	40	20	8. 45 193	8. 45 210	9. 99 983	40
30	8. 40 569	8. 40 583	9. 99 986	30	30	8. 45 267	8. 45 285	9. 99 982	30
40	8. 40 651	8. 40 665	9. 99 986	20	40	8. 45 341	8. 45 359	9. 99 982	20
50	8. 40 734	8. 40 748	9. 99 986	10	50	8. 45 415	8. 45 433	9. 99 982	10
28 0	8. 40 816	8. 40 830	9. 99 986	0 **32**	**38** 0	8. 45 489	8. 45 507	9. 99 982	0 **22**
10	8. 40 898	8. 40 913	9. 99 986	50	10	8. 45 563	8. 45 581	9. 99 982	50
20	8. 40 980	8. 40 995	9. 99 986	40	20	8. 45 637	8. 45 655	9. 99 982	40
30	8. 41 062	8. 41 077	9. 99 986	30	30	8. 45 710	8. 45 728	9. 99 982	30
40	8. 41 144	8. 41 158	9. 99 986	20	40	8. 45 784	8. 45 802	9. 99 982	20
50	8. 41 225	8. 41 240	9. 99 986	10	50	8. 45 857	8. 45 875	9. 99 982	10
29 0	8. 41 307	8. 41 321	9. 99 985	0 **31**	**39** 0	8. 45 930	8. 45 948	9. 99 982	0 **21**
10	8. 41 388	8. 41 403	9. 99 985	50	10	8. 46 003	8. 46 021	9. 99 982	50
20	8. 41 469	8. 41 484	9. 99 985	40	20	8. 46 076	8. 46 094	9. 99 982	40
30	8. 41 550	8. 41 565	9. 99 985	30	30	8. 46 149	8. 46 167	9. 99 982	30
40	8. 41 631	8. 41 646	9. 99 985	20	40	8. 46 222	8. 46 240	9. 99 982	20
50	8. 41 711	8. 41 726	9. 99 985	10	50	8. 46 294	8. 46 312	9. 99 982	10
30 0	8. 41 792	8. 41 807	9. 99 985	0 **30**	**40** 0	8. 46 366	8. 46 385	9. 99 982	0 **20**
′ ″	log cos	log cot	log sin	″ ′	′ ″	log cos	log cot	log sin	″ ′

88°

′ ″	log sin	log tan	log cos	″ ′	′ ″	log sin	log tan	log cos	″ ′
40 0	8. 46 366	8. 46 385	9. 99 982	0 **20**	**50** 0	8. 50 504	8. 50 527	9. 99 978	0 **10**
10	8. 46 439	8. 46 457	9. 99 982	50	10	8. 50 570	8. 50 593	9. 99 978	50
20	8. 46 511	8. 46 529	9. 99 982	40	20	8. 50 636	8. 50 658	9. 99 978	40
30	8. 46 583	8. 46 602	9. 99 981	30	30	8. 50 701	8. 50 724	9. 99 978	30
40	8. 46 655	8. 46 674	9. 99 981	20	40	8. 50 767	8. 50 789	9. 99 977	20
50	8. 46 727	8. 46 745	9. 99 981	10	50	8. 50 832	8. 50 855	9. 99 977	10
41 0	8. 46 799	8. 46 817	9. 99 981	0 **19**	**51** 0	8. 50 897	8. 50 920	9. 99 977	0 **9**
10	8. 46 870	8. 46 889	9. 99 981	50	10	8. 50 963	8. 50 985	9. 99 977	50
20	8. 46 942	8. 46 960	9. 99 981	40	20	8. 51 028	8. 51 050	9. 99 977	40
30	8. 47 013	8. 47 032	9. 99 981	30	30	8. 51 092	8. 51 115	9. 99 977	30
40	8. 47 084	8. 47 103	9. 99 981	20	40	8. 51 157	8. 51 180	9. 99 977	20
50	8. 47 155	8. 47 174	9. 99 981	10	50	8. 51 222	8. 51 245	9. 99 977	10
42 0	8. 47 226	8. 47 245	9. 99 981	0 **18**	**52** 0	8. 51 287	8. 51 310	9. 99 977	0 **8**
10	8. 47 297	8. 47 316	9. 99 981	50	10	8. 51 351	8. 51 374	9. 99 977	50
20	8. 47 368	8. 47 387	9. 99 981	40	20	8. 51 416	8. 51 439	9. 99 977	40
30	8. 47 439	8. 47 458	9. 99 981	30	30	8. 51 480	8. 51 503	9. 99 977	30
40	8. 47 509	8. 47 528	9. 99 981	20	40	8. 51 544	8. 51 568	9. 99 977	20
50	8. 47 580	8. 47 599	9. 99 981	10	50	8. 51 609	8. 51 632	9. 99 977	10
43 0	8. 47 650	8. 47 669	9. 99 981	0 **17**	**53** 0	8. 51 673	8. 51 696	9. 99 977	0 **7**
10	8. 47 720	8. 47 740	9. 99 980	50	10	8. 51 737	8. 51 760	9. 99 976	50
20	8. 47 790	8. 47 810	9. 99 980	40	20	8. 51 801	8. 51 824	9. 99 976	40
30	8. 47 860	8. 47 880	9. 99 980	30	30	8. 51 864	8. 51 888	9. 99 976	30
40	8. 47 930	8 47 950	9. 99 980	20	40	8. 51 928	8. 51 952	9. 99 976	20
50	8. 48 000	8. 48 020	9. 99 980	10	50	8. 51 992	8. 52 015	9. 99 976	10
44 0	8. 48 069	8. 48 090	9. 99 980	0 **16**	**54** 0	8. 52 055	8. 52 079	9. 99 976	0 **6**
10	8. 48 139	8. 48 159	9. 99 980	50	10	8. 52 119	8. 52 143	9. 99 976	50
20	8. 48 208	8. 48 228	9. 99 980	40	20	8. 52 182	8. 52 206	9. 99 976	40
30	8. 48 278	8. 48 298	9. 99 980	30	30	8. 52 245	8. 52 269	9. 99 976	30
40	8. 48 347	8. 48 367	9. 99 980	20	40	8. 52 308	8. 52 332	9. 99 976	20
50	8. 48 416	8. 48 436	9. 99 980	10	50	8. 52 371	8. 52 396	9. 99 976	10
45 0	8. 48 485	8. 48 505	9. 99 980	0 **15**	**55** 0	8. 52 434	8. 52 459	9. 99 976	0 **5**
10	8. 48 554	8. 48 574	9. 99 980	50	10	8. 52 497	8. 52 522	9. 99 976	50
20	8. 48 622	8. 48 643	9. 99 980	40	20	8. 52 560	8. 52 584	9. 99 976	40
30	8. 48 691	8. 48 711	9. 99 980	30	30	8. 52 623	8. 52 647	9. 99 975	30
40	8. 48 760	8. 48 780	9. 99 979	20	40	8. 52 685	8. 52 710	9. 99 975	20
50	8. 48 828	8. 48 849	9. 99 979	10	50	8. 52 748	8. 52 772	9. 99 975	10
46 0	8. 48 896	8. 48 917	9. 99 979	0 **14**	**56** 0	8. 52 810	8. 52 835	9. 99 975	0 **4**
10	8. 48 965	8. 48 985	9. 99 979	50	10	8. 52 872	8. 52 897	9. 99 975	50
20	8. 49 033	8. 49 053	9. 99 979	40	20	8. 52 935	8. 52 960	9. 99 975	40
30	8. 49 101	8. 49 121	9. 99 979	30	30	8. 52 997	8. 53 022	9. 99 975	30
40	8. 49 169	8. 49 189	9. 99 979	20	40	8. 53 059	8. 53 084	9. 99 975	20
50	8. 49 236	8. 49 257	9. 99 979	10	50	8. 53 121	8. 53 146	9. 99 975	10
47 0	8. 49 304	8. 49 325	9. 99 979	0 **13**	**57** 0	8. 53 183	8. 53 208	9. 99 975	0 **3**
10	8. 49 372	8. 49 393	9. 99 979	50	10	8. 53 245	8. 53 270	9. 99 975	50
20	8. 49 439	8. 49 460	9. 99 979	40	20	8. 53 306	8. 53 332	9. 99 975	40
30	8. 49 506	8. 49 528	9. 99 979	30	30	8. 53 368	8. 53 393	9. 99 975	30
40	8. 49 574	8. 49 595	9. 99 979	20	40	8. 53 429	8. 53 455	9. 99 975	20
50	8. 49 641	8. 49 662	9. 99 979	10	50	8. 53 491	8. 53 516	9. 99 974	10
48 0	8. 49 708	8. 49 729	9. 99 979	0 **12**	**58** 0	8. 53 552	8. 53 578	9. 99 974	0 **2**
10	8. 49 775	8. 49 796	9. 99 979	50	10	8. 53 614	8. 53 639	9. 99 974	50
20	8. 49 842	8. 49 863	9. 99 978	40	20	8. 53 675	8. 53 700	9. 99 974	40
30	8. 49 908	8. 49 930	9. 99 978	30	30	8. 53 736	8. 53 762	9. 99 974	30
40	8. 49 975	8. 49 997	9. 99 978	20	40	8. 53 797	8. 53 823	9. 99 974	20
50	8. 50 042	8. 50 063	9. 99 978	10	50	8. 53 858	8. 53 884	9. 99 974	10
49 0	8. 50 108	8. 50 130	9. 99 978	0 **11**	**59** 0	8. 53 919	8. 53 945	9. 99 974	0 **1**
10	8. 50 174	8. 50 196	9. 99 978	50	10	8. 53 979	8. 54 005	9. 99 974	50
20	8. 50 241	8. 50 263	9. 99 978	40	20	8. 54 040	8. 54 066	9. 99 974	40
30	8. 50 307	8. 50 329	9. 99 978	30	30	8. 54 101	8. 54 127	9. 99 974	30
40	8. 50 373	8. 50 395	9. 99 978	20	40	8. 54 161	8. 54 187	9. 99 974	20
50	8. 50 439	8. 50 461	9. 99 978	10	50	8. 54 222	8. 54 248	9. 99 974	10
50 0	8. 50 504	8. 50 527	9. 99 978	0 **10**	**60** 0	8. 54 282	8. 54 308	9. 99 974	0 **0**
′ ″	log cos	log cot	log sin	″ ′	′ ″	log cos	log cot	log sin	″ ′

′	log sin 8	log tan 8	log cot 11	log cos 9	′
0	24 186	24 192	75 808	99 993	60
1	24 903	24 910	75 090	99 993	59
2	25 609	25 616	74 384	99 993	58
3	26 304	26 312	73 688	99 993	57
4	26 988	26 996	73 004	99 992	56
5	27 661	27 669	72 331	99 992	55
6	28 324	28 332	71 668	99 992	54
7	28 977	28 986	71 014	99 992	53
8	29 621	29 629	70 371	99 992	52
9	30 255	30 263	69 737	99 991	51
10	30 879	30 888	69 112	99 991	50
11	31 495	31 505	68 495	99 991	49
12	32 103	32 112	67 888	99 990	48
13	32 702	32 711	67 289	99 990	47
14	33 292	33 302	66 698	99 990	46
15	33 875	33 886	66 114	99 990	45
16	34 450	34 461	65 539	99 989	44
17	35 018	35 029	64 971	99 989	43
18	35 578	35 590	64 410	99 989	42
19	36 131	36 143	63 857	99 989	41
20	36 678	36 689	63 311	99 988	40
21	37 217	37 229	62 771	99 988	39
22	37 750	37 762	62 238	99 988	38
23	38 276	38 289	61 711	99 987	37
24	38 796	38 809	61 191	99 987	36
25	39 310	39 323	60 677	99 987	35
26	39 818	39 832	60 168	99 986	34
27	40 320	40 334	59 666	99 986	33
28	40 816	40 830	59 170	99 986	32
29	41 307	41 321	58 679	99 985	31
30	41 792	41 807	58 193	99 985	30
31	42 272	42 287	57 713	99 985	29
32	42 746	42 762	57 238	99 984	28
33	43 216	43 232	56 768	99 984	27
34	43 680	43 696	56 304	99 984	26
35	44 139	44 156	55 844	99 983	25
36	44 594	44 611	55 389	99 983	24
37	45 044	45 061	54 939	99 983	23
38	45 489	45 507	54 493	99 982	22
39	45 930	45 948	54 052	99 982	21
40	46 366	46 385	53 615	99 981	20
41	46 799	46 817	53 183	99 981	19
42	47 226	47 245	52 755	99 981	18
43	47 650	47 669	52 331	99 981	17
44	48 069	48 089	51 911	99 980	16
45	48 485	48 505	51 495	99 980	15
46	48 896	48 917	51 083	99 979	14
47	49 304	49 325	50 675	99 979	13
48	49 708	49 729	50 271	99 979	12
49	50 108	50 130	49 870	99 978	11
50	50 504	50 527	49 473	99 978	10
51	50 897	50 920	49 080	99 977	9
52	51 287	51 310	48 690	99 977	8
53	51 673	51 696	48 304	99 977	7
54	52 055	52 079	47 921	99 976	6
55	52 434	52 459	47 541	99 976	5
56	52 810	52 835	47 165	99 975	4
57	53 183	53 208	46 792	99 975	3
58	53 552	53 578	46 422	99 974	2
59	53 919	53 945	46 055	99 974	1
60	54 282	54 308	45 692	99 974	0
′	log cos 8	log cot 8	log tan 11	log sin 9	′

88°

′	log sin 8	log tan 8	log cot 11	log cos 9	′
0	54 282	54 308	45 692	99 974	60
1	54 642	54 669	45 331	99 973	59
2	54 999	55 027	44 973	99 973	58
3	55 354	55 382	44 618	99 972	57
4	55 705	55 734	44 266	99 972	56
5	56 054	56 083	43 917	99 971	55
6	56 400	56 429	43 571	99 971	54
7	56 743	56 773	43 227	99 970	53
8	57 084	57 114	42 886	99 970	52
9	57 421	57 452	42 548	99 969	51
10	57 757	57 788	42 212	99 969	50
11	58 089	58 121	41 879	99 968	49
12	58 419	58 451	41 549	99 968	48
13	58 747	58 779	41 221	99 967	47
14	59 072	59 105	40 895	99 967	46
15	59 395	59 428	40 572	99 967	45
16	59 715	59 749	40 251	99 966	44
17	60 033	60 068	39 932	99 966	43
18	60 349	60 384	39 616	99 965	42
19	60 662	60 698	39 302	99 964	41
20	60 973	61 009	38 991	99 964	40
21	61 282	61 319	38 681	99 963	39
22	61 589	61 626	38 374	99 963	38
23	61 894	61 931	38 069	99 962	37
24	62 196	62 234	37 766	99 962	36
25	62 497	62 535	37 465	99 961	35
26	62 795	62 834	37 166	99 961	34
27	63 091	63 131	36 869	99 960	33
28	63 385	63 426	36 574	99 960	32
29	63 678	63 718	36 282	99 959	31
30	63 968	64 009	35 991	99 959	30
31	64 256	64 298	35 702	99 958	29
32	64 543	64 585	35 415	99 958	28
33	64 827	64 870	35 130	99 957	27
34	65 110	65 154	34 846	99 956	26
35	65 391	65 435	34 565	99 956	25
36	65 670	65 715	34 285	99 955	24
37	65 947	65 993	34 007	99 955	23
38	66 223	66 269	33 731	99 954	22
39	66 497	66 543	33 457	99 954	21
40	66 769	66 816	33 184	99 953	20
41	67 039	67 087	32 913	99 952	19
42	67 308	67 356	32 644	99 952	18
43	67 575	67 624	32 376	99 951	17
44	67 841	67 890	32 110	99 951	16
45	68 104	68 154	31 846	99 950	15
46	68 367	68 417	31 583	99 949	14
47	68 627	68 678	31 322	99 949	13
48	68 886	68 938	31 062	99 948	12
49	69 144	69 196	30 804	99 948	11
50	69 400	69 453	30 547	99 947	10
51	69 654	69 708	30 292	99 946	9
52	69 907	69 962	30 038	99 946	8
53	70 159	70 214	29 786	99 945	7
54	70 409	70 465	29 535	99 944	6
55	70 658	70 714	29 286	99 944	5
56	70 905	70 962	29 038	99 943	4
57	71 151	71 208	28 792	99 942	3
58	71 395	71 453	28 547	99 942	2
59	71 638	71 697	28 303	99 941	1
60	71 880	71 940	28 060	99 940	0
′	log cos 8	log cot 8	log tan 11	log sin 9	′

87°

′	log sin	log tan	log cot	log cos	′
	8	8	11	9	
0	71 880	71 940	28 060	99 940	60
1	72 120	72 181	27 819	99 940	59
2	72 359	72 420	27 580	99 939	58
3	72 597	72 659	27 341	99 938	57
4	72 834	72 896	27 104	99 938	56
5	73 069	73 132	26 868	99 937	55
6	73 303	73 366	26 634	99 936	54
7	73 535	73 600	26 400	99 936	53
8	73 767	73 832	26 168	99 935	52
9	73 997	74 063	25 937	99 934	51
10	74 226	74 292	25 708	99 934	50
11	74 454	74 521	25 479	99 933	49
12	74 680	74 748	25 252	99 932	48
13	74 906	74 974	25 026	99 932	47
14	75 130	75 199	24 801	99 931	46
15	75 353	75 423	24 577	99 930	45
16	75 575	75 645	24 355	99 929	44
17	75 795	75 867	24 133	99 929	43
18	76 015	76 087	23 913	99 928	42
19	76 234	76 306	23 694	99 927	41
20	76 451	76 525	23 475	99 926	40
21	76 667	76 742	23 258	99 926	39
22	76 883	76 958	23 042	99 925	38
23	77 097	77 173	22 827	99 924	37
24	77 310	77 387	22 613	99 923	36
25	77 522	77 600	22 400	99 923	35
26	77 733	77 811	22 189	99 922	34
27	77 943	78 022	21 978	99 921	33
28	78 152	78 232	21 768	99 920	32
29	78 360	78 441	21 559	99 920	31
30	78 568	78 649	21 351	99 919	30
31	78 774	78 855	21 145	99 918	29
32	78 979	79 061	20 939	99 917	28
33	79 183	79 266	20 734	99 917	27
34	79 386	79 470	20 530	99 916	26
35	79 588	79 673	20 327	99 915	25
36	79 789	79 875	20 125	99 914	24
37	79 990	80 076	19 924	99 913	23
38	80 189	80 277	19 723	99 913	22
39	80 388	80 476	19 524	99 912	21
40	80 585	80 674	19 326	99 911	20
41	80 782	80 872	19 128	99 910	19
42	80 978	81 068	18 932	99 909	18
43	81 173	81 264	18 736	99 909	17
44	81 367	81 459	18 541	99 908	16
45	81 560	81 653	18 347	99 907	15
46	81 752	81 846	18 154	99 906	14
47	81 944	82 038	17 962	99 905	13
48	82 134	82 230	17 770	99 904	12
49	82 324	82 420	17 580	99 904	11
50	82 513	82 610	17 390	99 903	10
51	82 701	82 799	17 201	99 902	9
52	82 888	82 987	17 013	99 901	8
53	83 075	83 175	16 825	99 900	7
54	83 261	83 361	16 639	99 899	6
55	83 446	83 547	16 453	99 898	5
56	83 630	83 732	16 268	99 898	4
57	83 813	83 916	16 084	99 897	3
58	83 996	84 100	15 900	99 896	2
59	84 177	84 282	15 718	99 895	1
60	84 358	84 464	15 536	99 894	0
	8	8	11	9	
′	log cos	log cot	log tan	log sin	′

′	log sin	log tan	log cot	log cos	′
	8	8	11	9	
0	84 358	84 464	15 536	99 894	60
1	84 539	84 646	15 354	99 893	59
2	84 718	84 826	15 174	99 892	58
3	84 897	85 006	14 994	99 891	57
4	85 075	85 185	14 815	99 891	56
5	85 252	85 363	14 637	99 890	55
6	85 429	85 540	14 460	99 889	54
7	85 605	85 717	14 283	99 888	53
8	85 780	85 893	14 107	99 887	52
9	85 955	86 069	13 931	99 886	51
10	86 128	86 243	13 757	99 885	50
11	86 301	86 417	13 583	99 884	49
12	86 474	86 591	13 409	99 883	48
13	86 645	86 763	13 237	99 882	47
14	86 816	86 935	13 065	99 881	46
15	86 987	87 106	12 894	99 880	45
16	87 156	87 277	12 723	99 879	44
17	87 325	87 447	12 553	99 879	43
18	87 494	87 616	12 384	99 878	42
19	87 661	87 785	12 215	99 877	41
20	87 829	87 953	12 047	99 876	40
21	87 995	88 120	11 880	99 875	39
22	88 161	88 287	11 713	99 874	38
23	88 326	88 453	11 547	99 873	37
24	88 490	88 618	11 382	99 872	36
25	88 654	88 783	11 217	99 871	35
26	88 817	88 948	11 052	99 870	34
27	88 980	89 111	10 889	99 869	33
28	89 142	89 274	10 726	99 868	32
29	89 304	89 437	10 563	99 867	31
30	89 464	89 598	10 402	99 866	30
31	89 625	89 760	10 240	99 865	29
32	89 784	89 920	10 080	99 864	28
33	89 943	90 080	09 920	99 863	27
34	90 102	90 240	09 760	99 862	26
35	90 260	90 399	09 601	99 861	25
36	90 417	90 557	09 443	99 860	24
37	90 574	90 715	09 285	99 859	23
38	90 730	90 872	09 128	99 858	22
39	90 885	91 029	08 971	99 857	21
40	91 040	91 185	08 815	99 856	20
41	91 195	91 340	08 660	99 855	19
42	91 349	91 495	08 505	99 854	18
43	91 502	91 650	08 350	99 853	17
44	91 655	91 803	08 197	99 852	16
45	91 807	91 957	08 043	99 851	15
46	91 959	92 110	07 890	99 850	14
47	92 110	92 262	07 738	99 848	13
48	92 261	92 414	07 586	99 847	12
49	92 411	92 565	07 435	99 846	11
50	92 561	92 716	07 284	99 845	10
51	92 710	92 866	07 134	99 844	9
52	92 859	93 016	06 984	99 843	8
53	93 007	93 165	06 835	99 842	7
54	93 154	93 313	06 687	99 841	6
55	93 301	93 462	06 538	99 840	5
56	93 448	93 609	06 391	99 839	4
57	93 594	93 756	06 244	99 838	3
58	93 740	93 903	06 097	99 837	2
59	93 885	94 049	05 951	99 836	1
60	94 030	94 195	05 805	99 834	0
	8	8	11	9	
′	log cos	log cot	log tan	log sin	′

84°

′	log sin 8	log tan 8	log cot 11	log cos 9	′
0	94 030	94 195	05 805	99 834	60
1	94 174	94 340	05 660	99 833	59
2	94 317	94 485	05 515	99 832	58
3	94 461	94 630	05 370	99 831	57
4	94 603	94 773	05 227	99 830	56
5	94 746	94 917	05 083	99 829	55
6	94 887	95 060	04 940	99 828	54
7	95 029	95 202	04 798	99 827	53
8	95 170	95 344	04 656	99 825	52
9	95 310	95 486	04 514	99 824	51
10	95 450	95 627	04 373	99 823	50
11	95 589	95 767	04 233	99 822	49
12	95 728	95 908	04 092	99 821	48
13	95 867	96 047	03 953	99 820	47
14	96 005	96 187	03 813	99 819	46
15	96 143	96 325	03 675	99 817	45
16	96 280	96 464	03 536	99 816	44
17	96 417	96 602	03 398	99 815	43
18	96 553	96 739	03 261	99 814	42
19	96 689	96 877	03 123	99 813	41
20	96 825	97 013	02 987	99 812	40
21	96 960	97 150	02 850	99 810	39
22	97 095	97 285	02 715	99 809	38
23	97 229	97 421	02 579	99 808	37
24	97 363	97 556	02 444	99 807	36
25	97 496	97 691	02 309	99 806	35
26	97 629	97 825	02 175	99 804	34
27	97 762	97 959	02 041	99 803	33
28	97 894	98 092	01 908	99 802	32
29	98 026	98 225	01 775	99 801	31
30	98 157	98 358	01 642	99 800	30
31	98 288	98 490	01 510	99 798	29
32	98 419	98 622	01 378	99 797	28
33	98 549	98 753	01 247	99 796	27
34	98 679	98 884	01 116	99 795	26
35	98 808	99 015	00 985	99 793	25
36	98 937	99 145	00 855	99 792	24
37	99 066	99 275	00 725	99 791	23
38	99 194	99 405	00 595	99 790	22
39	99 322	99 534	00 466	99 788	21
40	99 450	99 662	00 338	99 787	20
41	99 577	99 791	00 209	99 786	19
42	99 704	99 919	00 081	99 785	18
43	99 830	00 046	99 954	99 783	17
44	99 956	00 174	99 826	99 782	16
45	00 082	00 301	99 699	99 781	15
46	00 207	00 427	99 573	99 780	14
47	00 332	00 553	99 447	99 778	13
48	00 456	00 679	99 321	99 777	12
49	00 581	00 805	99 195	99 776	11
50	00 704	00 930	99 070	99 775	10
51	00 828	01 055	98 945	99 773	9
52	00 951	01 179	98 821	99 772	8
53	01 074	01 303	98 697	99 771	7
54	01 196	01 427	98 573	99 769	6
55	01 318	01 550	98 450	99 768	5
56	01 440	01 673	98 327	99 767	4
57	01 561	01 796	98 204	99 765	3
58	01 682	01 918	98 082	99 764	2
59	01 803	02 040	97 960	99 763	1
60	01 923	02 162	97 838	99 761	0
′	log cos 9	log cot 9	log tan 10	log sin 9	′

83°

′	log sin 9	log tan 9	log cot 10	log cos 9	′
0	01 923	02 162	97 838	99 761	60
1	02 043	02 283	97 717	99 760	59
2	02 163	02 404	97 596	99 759	58
3	02 283	02 525	97 475	99 757	57
4	02 402	02 645	97 355	99 756	56
5	02 520	02 766	97 234	99 755	55
6	02 639	02 885	97 115	99 753	54
7	02 757	03 005	96 995	99 752	53
8	02 874	03 124	96 876	99 751	52
9	02 992	03 242	96 758	99 749	51
10	03 109	03 361	96 639	99 748	50
11	03 226	03 479	96 521	99 747	49
12	03 342	03 597	96 403	99 745	48
13	03 458	03 714	96 286	99 744	47
14	03 574	03 832	96 168	99 742	46
15	03 690	03 948	96 052	99 741	45
16	03 805	04 065	95 935	99 740	44
17	03 920	04 181	95 819	99 738	43
18	04 034	04 297	95 703	99 737	42
19	04 149	04 413	95 587	99 736	41
20	04 262	04 528	95 472	99 734	40
21	04 376	04 643	95 357	99 733	39
22	04 490	04 758	95 242	99 731	38
23	04 603	04 873	95 127	99 730	37
24	04 715	04 987	95 013	99 728	36
25	04 828	05 101	94 899	99 727	35
26	04 940	05 214	94 786	99 726	34
27	05 052	05 328	94 672	99 724	33
28	05 164	05 441	94 559	99 723	32
29	05 275	05 553	94 447	99 721	31
30	05 386	05 666	94 334	99 720	30
31	05 497	05 778	94 222	99 718	29
32	05 607	05 890	94 110	99 717	28
33	05 717	06 002	93 998	99 716	27
34	05 827	06 113	93 887	99 714	26
35	05 937	06 224	93 776	99 713	25
36	06 046	06 335	93 665	99 711	24
37	06 155	06 445	93 555	99 710	23
38	06 264	06 556	93 444	99 708	22
39	06 372	06 666	93 334	99 707	21
40	06 481	06 775	93 225	99 705	20
41	06 589	06 885	93 115	99 704	19
42	06 696	06 994	93 006	99 702	18
43	06 804	07 103	92 897	99 701	17
44	06 911	07 211	92 789	99 699	16
45	07 018	07 320	92 680	99 698	15
46	07 124	07 428	92 572	99 696	14
47	07 231	07 536	92 464	99 695	13
48	07 337	07 643	92 357	99 693	12
49	07 442	07 751	92 249	99 692	11
50	07 548	07 858	92 142	99 690	10
51	07 653	07 964	92 036	99 689	9
52	07 758	08 071	91 929	99 687	8
53	07 863	08 177	91 823	99 686	7
54	07 968	08 283	91 717	99 684	6
55	08 072	08 389	91 611	99 683	5
56	08 176	08 495	91 505	99 681	4
57	08 280	08 600	91 400	99 680	3
58	08 383	08 705	91 295	99 678	2
59	08 486	08 810	91 190	99 677	1
60	08 589	08 914	91 086	99 675	0
′	log cos 9	log cot 9	log tan 10	log sin 9	′

84°   **83°**

′	log sin 9	log tan 9	log cot 10	log cos 9	′
0	08 589	08 914	91 086	99 675	60
1	08 692	09 019	90 981	99 674	59
2	08 795	09 123	90 877	99 672	58
3	08 897	09 227	90 773	99 670	57
4	08 999	09 330	90 670	99 669	56
5	09 101	09 434	90 566	99 667	55
6	09 202	09 537	90 463	99 666	54
7	09 304	09 640	90 360	99 664	53
8	09 405	09 742	90 258	99 663	52
9	09 506	09 845	90 155	99 661	51
10	09 606	09 947	90 053	99 659	50
11	09 707	10 019	89 951	99 658	49
12	09 807	10 150	89 850	99 656	48
13	09 907	10 252	89 748	99 655	47
14	10 006	10 353	89 647	99 653	46
15	10 106	10 454	89 546	99 651	45
16	10 205	10 555	89 445	99 650	44
17	10 304	10 656	89 344	99 648	43
18	10 402	10 756	89 244	99 647	42
19	10 501	10 856	89 144	99 645	41
20	10 599	10 956	89 044	99 643	40
21	10 697	11 056	88 944	99 642	39
22	10 795	11 155	88 845	99 640	38
23	10 893	11 254	88 746	99 638	37
24	10 990	11 353	88 647	99 637	36
25	11 087	11 452	88 548	99 635	35
26	11 184	11 551	88 449	99 633	34
27	11 281	11 649	88 351	99 632	33
28	11 377	11 747	88 253	99 630	32
29	11 474	11 845	88 155	99 629	31
30	11 570	11 943	88 057	99 627	30
31	11 666	12 040	87 960	99 625	29
32	11 761	12 138	87 862	99 624	28
33	11 857	12 235	87 765	99 622	27
34	11 952	12 332	87 668	99 620	26
35	12 047	12 428	87 572	99 618	25
36	12 142	12 525	87 475	99 617	24
37	12 236	12 621	87 379	99 615	23
38	12 331	12 717	87 283	99 613	22
39	12 425	12 813	87 187	99 612	21
40	12 519	12 909	87 091	99 610	20
41	12 612	13 004	86 996	99 608	19
42	12 706	13 099	86 901	99 607	18
43	12 799	13 194	86 806	99 605	17
44	12 892	13 289	86 711	99 603	16
45	12 985	13 384	86 616	99 601	15
46	13 078	13 478	86 522	99 600	14
47	13 171	13 573	86 427	99 598	13
48	13 263	13 667	86 333	99 596	12
49	13 355	13 761	86 239	99 595	11
50	13 447	13 854	86 146	99 593	10
51	13 539	13 948	86 052	99 591	9
52	13 630	14 041	85 959	99 589	8
53	13 722	14 134	85 866	99 588	7
54	13 813	14 227	85 773	99 586	6
55	13 904	14 320	85 680	99 584	5
56	13 994	14 412	85 588	99 582	4
57	14 085	14 504	85 496	99 581	3
58	14 175	14 597	85 403	99 579	2
59	14 266	14 688	85 312	99 577	1
60	14 356	14 780	85 220	99 575	0
′	log cos 9	log cot 9	log tan 10	log sin 9	′

′	log sin 9	log tan 9	log cot 10	log cos 9	′
0	14 356	14 780	85 220	99 575	60
1	14 445	14 872	85 128	99 574	59
2	14 535	14 963	85 037	99 572	58
3	14 624	15 054	84 946	99 570	57
4	14 714	15 145	84 855	99 568	56
5	14 803	15 236	84 764	99 566	55
6	14 891	15 327	84 673	99 565	54
7	14 980	15 417	84 583	99 563	53
8	15 069	15 508	84 492	99 561	52
9	15 157	15 598	84 402	99 559	51
10	15 245	15 688	84 312	99 557	50
11	15 333	15 777	84 223	99 556	49
12	15 421	15 867	84 133	99 554	48
13	15 508	15 956	84 044	99 552	47
14	15 596	16 046	83 954	99 550	46
15	15 683	16 135	83 865	99 548	45
16	15 770	16 224	83 776	99 546	44
17	15 857	16 312	83 688	99 545	43
18	15 944	16 401	83 599	99 543	42
19	16 030	16 489	83 511	99 541	41
20	16 116	16 577	83 423	99 539	40
21	16 203	16 665	83 335	99 537	39
22	16 289	16 753	83 247	99 535	38
23	16 374	16 841	83 159	99 533	37
24	16 460	16 928	83 072	99 532	36
25	16 545	17 016	82 984	99 530	35
26	16 631	17 103	82 897	99 528	34
27	16 716	17 190	82 810	99 526	33
28	16 801	17 277	82 723	99 524	32
29	16 886	17 363	82 637	99 522	31
30	16 970	17 450	82 550	99 520	30
31	17 055	17 536	82 464	99 518	29
32	17 139	17 622	82 378	99 517	28
33	17 223	17 708	82 292	99 515	27
34	17 307	17 794	82 206	99 513	26
35	17 391	17 880	82 120	99 511	25
36	17 474	17 965	82 035	99 509	24
37	17 558	18 051	81 949	99 507	23
38	17 641	18 136	81 864	99 505	22
39	17 724	18 221	81 779	99 503	21
40	17 807	18 306	81 694	99 501	20
41	17 890	18 391	81 609	99 499	19
42	17 973	18 475	81 525	99 497	18
43	18 055	18 560	81 440	99 495	17
44	18 137	18 644	81 356	99 494	16
45	18 220	18 728	81 272	99 492	15
46	18 302	18 812	81 188	99 490	14
47	18 383	18 896	81 104	99 488	13
48	18 465	18 979	81 021	99 486	12
49	18 547	19 063	80 937	99 484	11
50	18 628	19 146	80 854	99 482	10
51	18 709	19 229	80 771	99 480	9
52	18 790	19 312	80 688	99 478	8
53	18 871	19 395	80 605	99 476	7
54	18 952	19 478	80 522	99 474	6
55	19 033	19 561	80 439	99 472	5
56	19 113	19 643	80 357	99 470	4
57	19 193	19 725	80 275	99 468	3
58	19 273	19 807	80 193	99 466	2
59	19 353	19 889	80 111	99 464	1
60	19 433	19 971	80 029	99 462	0
′	log cos 9	log cot 9	log tan 10	log sin 9	′

'	log sin 9	log tan 9	log cot 10	log cos 9	'
0	19 433	19 971	80 029	99 462	60
1	19 513	20 053	79 947	99 460	59
2	19 592	20 134	79 866	99 458	58
3	19 672	20 216	79 784	99 456	57
4	19 751	20 297	79 703	99 454	56
5	19 830	20 378	79 622	99 452	55
6	19 909	20 459	79 541	99 450	54
7	19 988	20 540	79 460	99 448	53
8	20 067	20 621	79 379	99 446	52
9	20 145	20 701	79 299	99 444	51
10	20 223	20 782	79 218	99 442	50
11	20 302	20 862	79 138	99 440	49
12	20 380	20 942	79 058	99 438	48
13	20 458	21 022	78 978	99 436	47
14	20 535	21 102	78 898	99 434	46
15	20 613	21 182	78 818	99 432	45
16	20 691	21 261	78 739	99 429	44
17	20 768	21 341	78 659	99 427	43
18	20 845	21 420	78 580	99 425	42
19	20 922	21 499	78 501	99 423	41
20	20 999	21 578	78 422	99 421	40
21	21 076	21 657	78 343	99 419	39
22	21 153	21 736	78 264	99 417	38
23	21 229	21 814	78 186	99 415	37
24	21 306	21 893	78 107	99 413	36
25	21 382	21 971	78 029	99 411	35
26	21 458	22 049	77 951	99 409	34
27	21 534	22 127	77 873	99 407	33
28	21 610	22 205	77 795	99 404	32
29	21 685	22 283	77 717	99 402	31
30	21 761	22 361	77 639	99 400	30
31	21 836	22 438	77 562	99 398	29
32	21 912	22 516	77 484	99 396	28
33	21 987	22 593	77 407	99 394	27
34	22 062	22 670	77 330	99 392	26
35	22 137	22 747	77 253	99 390	25
36	22 211	22 824	77 176	99 388	24
37	22 286	22 901	77 099	99 385	23
38	22 361	22 977	77 023	99 383	22
39	22 435	23 054	76 946	99 381	21
40	22 509	23 130	76 870	99 379	20
41	22 583	23 206	76 794	99 377	19
42	22 657	23 283	76 717	99 375	18
43	22 731	23 359	76 641	99 372	17
44	22 805	23 435	76 565	99 370	16
45	22 878	23 510	76 490	99 368	15
46	22 952	23 586	76 414	99 366	14
47	23 025	23 661	76 339	99 364	13
48	23 098	23 737	76 263	99 362	12
49	23 171	23 812	76 188	99 359	11
50	23 244	23 887	76 113	99 357	10
51	23 317	23 962	76 038	99 355	9
52	23 390	24 037	75 963	99 353	8
53	23 462	24 112	75 888	99 351	7
54	23 535	24 186	75 814	99 348	6
55	23 607	24 261	75 739	99 346	5
56	23 679	24 335	75 665	99 344	4
57	23 752	24 410	75 590	99 342	3
58	23 823	24 484	75 516	99 340	2
59	23 895	24 558	75 442	99 337	1
60	23 967	24 632	75 368	99 335	0
'	log cos 9	log cot 9	log tan 10	log sin 9	'

'	log sin 9	log tan 9	log cot 10	log cos 9	'
0	23 967	24 632	75 368	99 335	60
1	24 039	24 706	75 294	99 333	59
2	24 110	24 779	75 221	99 331	58
3	24 181	24 853	75 147	99 328	57
4	24 253	24 926	75 074	99 326	56
5	24 324	25 000	75 000	99 324	55
6	24 395	25 073	74 927	99 322	54
7	24 466	25 146	74 854	99 319	53
8	24 536	25 219	74 781	99 317	52
9	24 607	25 292	74 708	99 315	51
10	24 677	25 365	74 635	99 313	50
11	24 748	25 437	74 563	99 310	49
12	24 818	25 510	74 490	99 308	48
13	24 888	25 582	74 418	99 306	47
14	24 958	25 655	74 345	99 304	46
15	25 028	25 727	74 273	99 301	45
16	25 098	25 799	74 201	99 299	44
17	25 168	25 871	74 129	99 297	43
18	25 237	25 943	74 057	99 294	42
19	25 307	26 015	73 985	99 292	41
20	25 376	26 086	73 914	99 290	40
21	25 445	26 158	73 842	99 288	39
22	25 514	26 229	73 771	99 285	38
23	25 583	26 301	73 699	99 283	37
24	25 652	26 372	73 628	99 281	36
25	25 721	26 443	73 557	99 278	35
26	25 790	26 514	73 486	99 276	34
27	25 858	26 585	73 415	99 274	33
28	25 927	26 655	73 345	99 271	32
29	25 995	26 726	73 274	99 269	31
30	26 063	26 797	73 203	99 267	30
31	26 131	26 867	73 133	99 264	29
32	26 199	26 937	73 063	99 262	28
33	26 267	27 008	72 992	99 260	27
34	26 335	27 078	72 922	99 257	26
35	26 403	27 148	72 852	99 255	25
36	26 470	27 218	72 782	99 252	24
37	26 538	27 288	72 712	99 250	23
38	26 605	27 357	72 643	99 248	22
39	26 672	27 427	72 573	99 245	21
40	26 739	27 496	72 504	99 243	20
41	26 806	27 566	72 434	99 241	19
42	26 873	27 635	72 365	99 238	18
43	26 940	27 704	72 296	99 236	17
44	27 007	27 773	72 227	99 233	16
45	27 073	27 842	72 158	99 231	15
46	27 140	27 911	72 089	99 229	14
47	27 206	27 980	72 020	99 226	13
48	27 273	28 049	71 951	99 224	12
49	27 339	28 117	71 883	99 221	11
50	27 405	28 186	71 814	99 219	10
51	27 471	28 254	71 746	99 217	9
52	27 537	28 323	71 677	99 214	8
53	27 602	28 391	71 609	99 212	7
54	27 668	28 459	71 541	99 209	6
55	27 734	28 527	71 473	99 206	5
56	27 799	28 595	71 405	99 204	4
57	27 864	28 662	71 338	99 202	3
58	27 930	28 730	71 270	99 200	2
59	27 995	28 798	71 202	99 197	1
60	28 060	28 865	71 135	99 195	0
'	log cos 9	log cot 9	log tan 10	log sin 9	'

′	log sin	log tan	log cot	log cos	′
	9	9	10	9	
0	28 060	28 865	71 135	99 195	60
1	28 125	28 933	71 067	99 192	59
2	28 190	29 000	71 000	99 190	58
3	28 254	29 067	70 933	99 187	57
4	28 319	29 134	70 866	99 185	56
5	28 384	29 201	70 799	99 182	55
6	28 448	29 268	70 732	99 180	54
7	28 512	29 335	70 665	99 177	53
8	28 577	29 402	70 598	99 175	52
9	28 641	29 468	70 532	99 172	51
10	28 705	29 535	70 465	99 170	50
11	28 769	29 601	70 399	99 167	49
12	28 833	29 668	70 332	99 165	48
13	28 896	29 734	70 266	99 162	47
14	28 960	29 800	70 200	99 160	46
15	29 024	29 866	70 134	99 157	45
16	29 087	29 932	70 068	99 155	44
17	29 150	29 998	70 002	99 152	43
18	29 214	30 064	69 936	99 150	42
19	29 277	30 130	69 870	99 147	41
20	29 340	30 195	69 805	99 145	40
21	29 403	30 261	69 739	99 142	39
22	29 466	30 326	69 674	99 140	38
23	29 529	30 391	69 609	99 137	37
24	29 591	30 457	69 543	99 135	36
25	29 654	30 522	69 478	99 132	35
26	29 716	30 587	69 413	99 130	34
27	29 779	30 652	69 348	99 127	33
28	29 841	30 717	69 283	99 124	32
29	29 903	30 782	69 218	99 122	31
30	29 966	30 846	69 154	99 119	30
31	30 028	30 911	69 089	99 117	29
32	30 090	30 975	69 025	99 114	28
33	30 151	31 040	68 960	99 112	27
34	30 213	31 104	68 896	99 109	26
35	30 275	31 168	68 832	99 106	25
36	30 336	31 233	68 767	99 104	24
37	30 398	31 297	68 703	99 101	23
38	30 459	31 361	68 639	99 099	22
39	30 521	31 425	68 575	99 096	21
40	30 582	31 489	68 511	99 093	20
41	30 643	31 552	68 448	99 091	19
42	30 704	31 616	68 384	99 088	18
43	30 765	31 679	68 321	99 086	17
44	30 826	31 743	68 257	99 083	16
45	30 887	31 806	68 194	99 080	15
46	30 947	31 870	68 130	99 078	14
47	31 008	31 933	68 067	99 075	13
48	31 068	31 996	68 004	99 072	12
49	31 129	32 059	67 941	99 070	11
50	31 189	32 122	67 878	99 067	10
51	31 250	32 185	67 815	99 064	9
52	31 310	32 248	67 752	99 062	8
53	31 370	32 311	67 689	99 059	7
54	31 430	32 373	67 627	99 056	6
55	31 490	32 436	67 564	99 054	5
56	31 549	32 498	67 502	99 051	4
57	31 609	32 561	67 439	99 048	3
58	31 669	32 623	67 377	99 046	2
59	31 728	32 685	67 315	99 043	1
60	31 788	32 747	67 253	99 040	0
	9	9	10	9	
′	log cos	log cot	log tan	log sin	′

78°

′	log sin	log tan	log cot	log cos	′
	9	9	10	9	
0	31 788	32 747	67 253	99 040	60
1	31 847	32 810	67 190	99 038	59
2	31 907	32 872	67 128	99 035	58
3	31 966	32 933	67 067	99 032	57
4	32 025	32 995	67 005	99 030	56
5	32 084	33 057	66 943	99 027	55
6	32 143	33 119	66 881	99 024	54
7	32 202	33 180	66 820	99 022	53
8	32 261	33 242	66 758	99 019	52
9	32 319	33 303	66 697	99 016	51
10	32 378	33 365	66 635	99 013	50
11	32 437	33 426	66 574	99 011	49
12	32 495	33 487	66 513	99 008	48
13	32 553	33 548	66 452	99 005	47
14	32 612	33 609	66 391	99 002	46
15	32 670	33 670	66 330	99 000	45
16	32 728	33 731	66 269	98 997	44
17	32 786	33 792	66 208	98 994	43
18	32 844	33 853	66 147	98 991	42
19	32 902	33 913	66 087	98 989	41
20	32 960	33 974	66 026	98 986	40
21	33 018	34 034	65 966	98 983	39
22	33 075	34 095	65 905	98 980	38
23	33 133	34 155	65 845	98 978	37
24	33 190	34 215	65 785	98 975	36
25	33 248	34 276	65 724	98 972	35
26	33 305	34 336	65 664	98 969	34
27	33 362	34 396	65 604	98 967	33
28	33 420	34 456	65 544	98 964	32
29	33 477	34 516	65 484	98 961	31
30	33 534	34 576	65 424	98 958	30
31	33 591	34 635	65 365	98 955	29
32	33 647	34 695	65 305	98 953	28
33	33 704	34 755	65 245	98 950	27
34	33 761	34 814	65 186	98 947	26
35	33 818	34 874	65 126	98 944	25
36	33 874	34 933	65 067	98 941	24
37	33 931	34 992	65 008	98 938	23
38	33 987	35 051	64 949	98 936	22
39	34 043	35 111	64 889	98 933	21
40	34 100	35 170	64 830	98 930	20
41	34 156	35 229	64 771	98 927	19
42	34 212	35 288	64 712	98 924	18
43	34 268	35 347	64 653	98 921	17
44	34 324	35 405	64 595	98 919	16
45	34 380	35 464	64 536	98 916	15
46	34 436	35 523	64 477	98 913	14
47	34 491	35 581	64 419	98 910	13
48	34 547	35 640	64 360	98 907	12
49	34 602	35 698	64 302	98 904	11
50	34 658	35 757	64 243	98 901	10
51	34 713	35 815	64 185	98 898	9
52	34 769	35 873	64 127	98 896	8
53	34 824	35 931	64 069	98 893	7
54	34 879	35 989	64 011	98 890	6
55	34 934	36 047	63 953	98 887	5
56	34 989	36 105	63 895	98 884	4
57	35 044	36 163	63 837	98 881	3
58	35 099	36 221	63 779	98 878	2
59	35 154	36 279	63 721	98 875	1
60	35 209	36 336	63 664	98 872	0
	9	9	10	9	
′	log cos	log cot	log tan	log sin	′

77°

′	log sin	log tan	log cot	log cos	′		′	log sin	log tan	log cot	log cos	′
	9	9	10	9				9	9	10	9	
0	35 209	36 336	63 664	98 872	60		0	38 368	39 677	60 323	98 690	60
1	35 263	36 394	63 606	98 869	59		1	38 418	39 731	60 269	98 687	59
2	35 318	36 452	63 548	98 867	58		2	38 469	39 785	60 215	98 684	58
3	35 373	36 509	63 491	98 864	57		3	38 519	39 838	60 162	98 681	57
4	35 427	36 566	63 434	98 861	56		4	38 570	39 892	60 108	98 678	56
5	35 481	36 624	63 376	98 858	55		5	38 620	39 945	60 055	98 675	55
6	35 536	36 681	63 319	98 855	54		6	38 670	39 999	60 001	98 671	54
7	35 590	36 738	63 262	98 852	53		7	38 721	40 052	59 948	98 668	53
8	35 644	36 795	63 205	98 849	52		8	38 771	40 106	59 894	98 665	52
9	35 698	36 852	63 148	98 846	51		9	38 821	40 159	59 841	98 662	51
10	35 752	36 909	63 091	98 843	50		10	38 871	40 212	59 788	98 659	50
11	35 806	36 966	63 034	98 840	49		11	38 921	40 266	59 734	98 656	49
12	35 860	37 023	62 977	98 837	48		12	38 971	40 319	59 681	98 652	48
13	35 914	37 080	62 920	98 834	47		13	39 021	40 372	59 628	98 649	47
14	35 968	37 137	62 863	98 831	46		14	39 071	40 425	59 575	98 646	46
15	36 022	37 193	62 807	98 828	45		15	39 121	40 478	59 522	98 643	45
16	36 075	37 250	62 750	98 825	44		16	39 170	40 531	59 469	98 640	44
17	36 129	37 306	62 694	98 822	43		17	39 220	40 584	59 416	98 636	43
18	36 182	37 363	62 637	98 819	42		18	39 270	40 636	59 364	98 633	42
19	36 236	37 419	62 581	98 816	41		19	39 319	40 689	59 311	98 630	41
20	36 289	37 476	62 524	98 813	40		20	39 369	40 742	59 258	98 627	40
21	36 342	37 532	62 468	98 810	39		21	39 418	40 795	59 205	98 623	39
22	36 395	37 588	62 412	98 807	38		22	39 467	40 847	59 153	98 620	38
23	36 449	37 644	62 356	98 804	37		23	39 517	40 900	59 100	98 617	37
24	36 502	37 700	62 300	98 801	36		24	39 566	40 952	59 048	98 614	36
25	36 555	37 756	62 244	98 798	35		25	39 615	41 005	58 995	98 610	35
26	36 608	37 812	62 188	98 795	34		26	39 664	41 057	58 943	98 607	34
27	36 660	37 868	62 132	98 792	33		27	39 713	41 109	58 891	98 604	33
28	36 713	37 924	62 076	98 789	32		28	39 762	41 161	58 839	98 601	32
29	36 766	37 980	62 020	98 786	31		29	39 811	41 214	58 786	98 597	31
30	36 819	38 035	61 965	98 783	30		30	39 860	41 266	58 734	98 594	30
31	36 871	38 091	61 909	98 780	29		31	39 909	41 318	58 682	98 591	29
32	36 924	38 147	61 853	98 777	28		32	39 958	41 370	58 630	98 588	28
33	36 976	38 202	61 798	98 774	27		33	40 006	41 422	58 578	98 584	27
34	37 028	38 257	61 743	98 771	26		34	40 055	41 474	58 526	98 581	26
35	37 081	38 313	61 687	98 768	25		35	40 103	41 526	58 474	98 578	25
36	37 133	38 368	61 632	98 765	24		36	40 152	41 578	58 422	98 574	24
37	37 185	38 423	61 577	98 762	23		37	40 200	41 629	58 371	98 571	23
38	37 237	38 479	61 521	98 759	22		38	40 249	41 681	58 319	98 568	22
39	37 289	38 534	61 466	98 756	21		39	40 297	41 733	58 267	98 565	21
40	37 341	38 589	61 411	98 753	20		40	40 346	41 784	58 216	98 561	20
41	37 393	38 644	61 356	98 750	19		41	40 394	41 836	58 164	98 558	19
42	37 445	38 699	61 301	98 746	18		42	40 442	41 887	58 113	98 555	18
43	37 497	38 754	61 246	98 743	17		43	40 490	41 939	58 061	98 551	17
44	37 549	38 808	61 192	98 740	16		44	40 538	41 990	58 010	98 548	16
45	37 600	38 863	61 137	98 737	15		45	40 586	42 041	57 959	98 545	15
46	37 652	38 918	61 082	98 734	14		46	40 634	42 093	57 907	98 541	14
47	37 703	38 972	61 028	98 731	13		47	40 682	42 144	57 856	98 538	13
48	37 755	39 027	60 973	98 728	12		48	40 730	42 195	57 805	98 535	12
49	37 806	39 082	60 918	98 725	11		49	40 778	42 246	57 754	98 531	11
50	37 858	39 136	60 864	98 722	10		50	40 825	42 297	57 703	98 528	10
51	37 909	39 190	60 810	98 719	9		51	40 873	42 348	57 652	98 525	9
52	37 960	39 245	60 755	98 715	8		52	40 921	42 399	57 601	98 521	8
53	38 011	39 299	60 701	98 712	7		53	40 968	42 450	57 550	98 518	7
54	38 062	39 353	60 647	98 709	6		54	41 016	42 501	57 499	98 515	6
55	38 113	39 407	60 593	98 706	5		55	41 063	42 552	57 448	98 511	5
56	38 164	39 461	60 539	98 703	4		56	41 111	42 603	57 397	98 508	4
57	38 215	39 515	60 485	98 700	3		57	41 158	42 653	57 347	98 505	3
58	38 266	39 569	60 431	98 697	2		58	41 205	42 704	57 296	98 501	2
59	38 317	39 623	60 377	98 694	1		59	41 252	42 755	57 245	98 498	1
60	38 368	39 677	60 323	98 690	0		60	41 300	42 805	57 195	98 494	0
	9	9	10	9				9	9	10	9	
′	log cos	log cot	log tan	log sin	′		′	log cos	log cot	log tan	log sin	′

′	log sin 9	log tan 9	log cot 10	log cos 9	′
0	41 300	42 805	57 195	98 494	60
1	41 347	42 856	57 144	98 491	59
2	41 394	42 906	57 094	98 488	58
3	41 441	42 957	57 043	98 484	57
4	41 488	43 007	56 993	98 481	56
5	41 535	43 057	56 943	98 477	55
6	41 582	43 108	56 892	98 474	54
7	41 628	43 158	56 842	98 471	53
8	41 675	43 208	56 792	98 467	52
9	41 722	43 258	56 742	98 464	51
10	41 768	43 308	56 692	98 460	50
11	41 815	43 358	56 642	98 457	49
12	41 861	43 408	56 592	98 453	48
13	41 908	43 458	56 542	98 450	47
14	41 954	43 508	56 492	98 447	46
15	42 001	43 558	56 442	98 443	45
16	42 047	43 607	56 393	98 440	44
17	42 093	43 657	56 343	98 436	43
18	42 140	43 707	56 293	98 433	42
19	42 186	43 756	56 244	98 429	41
20	42 232	43 806	56 194	98 426	40
21	42 278	43 855	56 145	98 422	39
22	42 324	43 905	56 095	98 419	38
23	42 370	43 954	56 046	98 415	37
24	42 416	44 004	55 996	98 412	36
25	42 461	44 053	55 947	98 409	35
26	42 507	44 102	55 898	98 405	34
27	42 553	44 151	55 849	98 402	33
28	42 599	44 201	55 799	98 398	32
29	42 644	44 250	55 750	98 395	31
30	42 690	44 299	55 701	98 391	30
31	42 735	44 348	55 652	98 388	29
32	42 781	44 397	55 603	98 384	28
33	42 826	44 446	55 554	98 381	27
34	42 872	44 495	55 505	98 377	26
35	42 917	44 544	55 456	98 373	25
36	42 962	44 592	55 408	98 370	24
37	43 008	44 641	55 359	98 366	23
38	43 053	44 690	55 310	98 363	22
39	43 098	44 738	55 262	98 359	21
40	43 143	44 787	55 213	98 356	20
41	43 188	44 836	55 164	98 352	19
42	43 233	44 884	55 116	98 349	18
43	43 278	44 933	55 067	98 345	17
44	43 323	44 981	55 019	98 342	16
45	43 367	45 029	54 971	98 338	15
46	43 412	45 078	54 922	98 334	14
47	43 457	45 126	54 874	98 331	13
48	43 502	45 174	54 826	98 327	12
49	43 546	45 222	54 778	98 324	11
50	43 591	45 271	54 729	98 320	10
51	43 635	45 319	54 681	98 317	9
52	43 680	45 367	54 633	98 313	8
53	43 724	45 415	54 585	98 309	7
54	43 769	45 463	54 537	98 306	6
55	43 813	45 511	54 489	98 302	5
56	43 857	45 559	54 441	98 299	4
57	43 901	45 606	54 394	98 295	3
58	43 946	45 654	54 346	98 291	2
59	43 990	45 702	54 298	98 288	1
60	44 034	45 750	54 250	98 284	0
′	log cos 9	log cot 9	log tan 10	log sin 9	′

′	log sin 9	log tan 9	log cot 10	log cos 9	′
0	44 034	45 750	54 250	98 284	60
1	44 078	45 797	54 203	98 281	59
2	44 122	45 845	54 155	98 277	58
3	44 166	45 892	54 108	98 273	57
4	44 210	45 940	54 060	98 270	56
5	44 253	45 987	54 013	98 266	55
6	44 297	46 035	53 965	98 262	54
7	44 341	46 082	53 918	98 259	53
8	44 385	46 130	53 870	98 255	52
9	44 428	46 177	53 823	98 251	51
10	44 472	46 224	53 776	98 248	50
11	44 516	46 271	53 729	98 244	49
12	44 559	46 319	53 681	98 240	48
13	44 602	46 366	53 634	98 237	47
14	44 646	46 413	53 587	98 233	46
15	44 689	46 460	53 540	98 229	45
16	44 733	46 507	53 493	98 226	44
17	44 776	46 554	53 446	98 222	43
18	44 819	46 601	53 399	98 218	42
19	44 862	46 648	53 352	98 215	41
20	44 905	46 694	53 306	98 211	40
21	44 948	46 741	53 259	98 207	39
22	44 992	46 788	53 212	98 204	38
23	45 035	46 835	53 165	98 200	37
24	45 077	46 881	53 119	98 196	36
25	45 120	46 928	53 072	98 192	35
26	45 163	46 975	53 025	98 189	34
27	45 206	47 021	52 979	98 185	33
28	45 249	47 068	52 932	98 181	32
29	45 292	47 114	52 886	98 177	31
30	45 334	47 160	52 840	98 174	30
31	45 377	47 207	52 793	98 170	29
32	45 419	47 253	52 747	98 166	28
33	45 462	47 299	52 701	98 162	27
34	45 504	47 346	52 654	98 159	26
35	45 547	47 392	52 608	98 155	25
36	45 589	47 438	52 562	98 151	24
37	45 632	47 484	52 516	98 147	23
38	45 674	47 530	52 470	98 144	22
39	45 716	47 576	52 424	98 140	21
40	45 758	47 622	52 378	98 136	20
41	45 801	47 668	52 332	98 132	19
42	45 843	47 714	52 286	98 129	18
43	45 885	47 760	52 240	98 125	17
44	45 927	47 806	52 194	98 121	16
45	45 969	47 852	52 148	98 117	15
46	46 011	47 897	52 103	98 113	14
47	46 053	47 943	52 057	98 110	13
48	46 095	47 989	52 011	98 106	12
49	46 136	48 035	51 965	98 102	11
50	46 178	48 080	51 920	98 098	10
51	46 220	48 126	51 874	98 094	9
52	46 262	48 171	51 829	98 090	8
53	46 303	48 217	51 783	98 087	7
54	46 345	48 262	51 738	98 083	6
55	46 386	48 307	51 693	98 079	5
56	46 428	48 353	51 647	98 075	4
57	46 469	48 398	51 602	98 071	3
58	46 511	48 443	51 557	98 067	2
59	46 552	48 489	51 511	98 063	1
60	46 594	48 534	51 466	98 060	0
′	log cos 9	log cot 9	log tan 10	log sin 9	′

′	log sin 9	log tan 9	log cot 10	log cos 9	′
0	46 594	48 534	51 466	98 060	60
1	46 635	48 579	51 421	98 056	59
2	46 676	48 624	51 376	98 052	58
3	46 717	48 669	51 331	98 048	57
4	46 758	48 714	51 286	98 044	56
5	46 800	48 759	51 241	98 040	55
6	46 841	48 804	51 196	98 036	54
7	46 882	48 849	51 151	98 032	53
8	46 923	48 894	51 106	98 029	52
9	46 964	48 939	51 061	98 025	51
10	47 005	48 984	51 016	98 021	50
11	47 045	49 029	50 971	98 017	49
12	47 086	49 073	50 927	98 013	48
13	47 127	49 118	50 882	98 009	47
14	47 168	49 163	50 837	98 005	46
15	47 209	49 207	50 793	98 001	45
16	47 249	49 252	50 748	97 997	44
17	47 290	49 296	50 704	97 993	43
18	47 330	49 341	50 659	97 989	42
19	47 371	49 385	50 615	97 986	41
20	47 411	49 430	50 570	97 982	40
21	47 452	49 474	50 526	97 978	39
22	47 492	49 519	50 481	97 974	38
23	47 533	49 563	50 437	97 970	37
24	47 573	49 607	50 393	97 966	36
25	47 613	49 652	50 348	97 962	35
26	47 654	49 696	50 304	97 958	34
27	47 694	49 740	50 260	97 954	33
28	47 734	49 784	50 216	97 950	32
29	47 774	49 828	50 172	97 946	31
30	47 814	49 872	50 128	97 942	30
31	47 854	49 916	50 084	97 938	29
32	47 894	49 960	50 040	97 934	28
33	47 934	50 004	49 996	97 930	27
34	47 974	50 048	49 952	97 926	26
35	48 014	50 092	49 908	97 922	25
36	48 054	50 136	49 864	97 918	24
37	48 094	50 180	49 820	97 914	23
38	48 133	50 223	49 777	97 910	22
39	48 173	50 267	49 733	97 906	21
40	48 213	50 311	49 689	97 902	20
41	48 252	50 355	49 645	97 898	19
42	48 292	50 398	49 602	97 894	18
43	48 332	50 442	49 558	97 890	17
44	48 371	50 485	49 515	97 886	16
45	48 411	50 529	49 471	97 882	15
46	48 450	50 572	49 428	97 878	14
47	48 490	50 616	49 384	97 874	13
48	48 529	50 659	49 341	97 870	12
49	48 568	50 703	49 297	97 866	11
50	48 607	50 746	49 254	97 861	10
51	48 647	50 789	49 211	97 857	9
52	48 686	50 833	49 167	97 853	8
53	48 725	50 876	49 124	97 849	7
54	48 764	50 919	49 081	97 845	6
55	48 803	50 962	49 038	97 841	5
56	48 842	51 005	48 995	97 837	4
57	48 881	51 048	48 952	97 833	3
58	48 920	51 092	48 908	97 829	2
59	48 959	51 135	48 865	97 825	1
60	48 998	51 178	48 822	97 821	0
′	log cos 9	log cot 10	log tan 9	log sin	′

72°

′	log sin 9	log tan 9	log cot 10	log cos 9	′
0	48 998	51 178	48 822	97 821	60
1	49 037	51 221	48 779	97 817	59
2	49 076	51 264	48 736	97 812	58
3	49 115	51 306	48 694	97 808	57
4	49 153	51 349	48 651	97 804	56
5	49 192	51 392	48 608	97 800	55
6	49 231	51 435	48 565	97 796	54
7	49 269	51 478	48 522	97 792	53
8	49 308	51 520	48 480	97 788	52
9	49 347	51 563	48 437	97 784	51
10	49 385	51 606	48 394	97 779	50
11	49 424	51 648	48 352	97 775	49
12	49 462	51 691	48 309	97 771	48
13	49 500	51 734	48 266	97 767	47
14	49 539	51 776	48 224	97 763	46
15	49 577	51 819	48 181	97 759	45
16	49 615	51 861	48 139	97 754	44
17	49 654	51 903	48 097	97 750	43
18	49 692	51 946	48 054	97 746	42
19	49 730	51 988	48 012	97 742	41
20	49 768	52 031	47 969	97 738	40
21	49 806	52 073	47 927	97 734	39
22	49 844	52 115	47 885	97 729	38
23	49 882	52 157	47 843	97 725	37
24	49 920	52 200	47 800	97 721	36
25	49 958	52 242	47 758	97 717	35
26	49 996	52 284	47 716	97 713	34
27	50 034	52 326	47 674	97 708	33
28	50 072	52 368	47 632	97 704	32
29	50 110	52 410	47 590	97 700	31
30	50 148	52 452	47 548	97 696	30
31	50 185	52 494	47 506	97 691	29
32	50 223	52 536	47 464	97 687	28
33	50 261	52 578	47 422	97 683	27
34	50 298	52 620	47 380	97 679	26
35	50 336	52 661	47 339	97 674	25
36	50 374	52 703	47 297	97 670	24
37	50 411	52 745	47 255	97 666	23
38	50 449	52 787	47 213	97 662	22
39	50 486	52 829	47 171	97 657	21
40	50 523	52 870	47 130	97 653	20
41	50 561	52 912	47 088	97 649	19
42	50 598	52 953	47 047	97 645	18
43	50 635	52 995	47 005	97 640	17
44	50 673	53 037	46 963	97 636	16
45	50 710	53 078	46 922	97 632	15
46	50 747	53 120	46 880	97 628	14
47	50 784	53 161	46 839	97 623	13
48	50 821	53 202	46 798	97 619	12
49	50 858	53 244	46 756	97 615	11
50	50 896	53 285	46 715	97 610	10
51	50 933	53 327	46 673	97 606	9
52	50 970	53 368	46 632	97 602	8
53	51 007	53 409	46 591	97 597	7
54	51 043	53 450	46 550	97 593	6
55	51 080	53 492	46 508	97 589	5
56	51 117	53 533	46 467	97 584	4
57	51 154	53 574	46 426	97 580	3
58	51 191	53 615	46 385	97 576	2
59	51 227	53 656	46 344	97 571	1
60	51 264	53 697	46 303	97 567	0
′	log cos 9	log cot 10	log tan 9	log sin	′

71°

′	log sin	log tan	log cot	log cos	′
	9	9	10	9	
0	51 264	53 697	46 303	97 567	60
1	51 301	53 738	46 262	97 563	59
2	51 338	53 779	46 221	97 558	58
3	51 374	53 820	46 180	97 554	57
4	51 411	53 861	46 139	97 550	56
5	51 447	53 902	46 098	97 545	55
6	51 484	53 943	46 057	97 541	54
7	51 520	53 984	46 016	97 536	53
8	51 557	54 025	45 975	97 532	52
9	51 593	54 065	45 935	97 528	51
10	51 629	54 106	45 894	97 523	50
11	51 666	54 147	45 853	97 519	49
12	51 702	54 187	45 813	97 515	48
13	51 738	54 228	45 772	97 510	47
14	51 774	54 269	45 731	97 506	46
15	51 811	54 309	45 691	97 501	45
16	51 847	54 350	45 650	97 497	44
17	51 883	54 390	45 610	97 492	43
18	51 919	54 431	45 569	97 488	42
19	51 955	54 471	45 529	97 484	41
20	51 991	54 512	45 488	97 479	40
21	52 027	54 552	45 448	97 475	39
22	52 063	54 593	45 407	97 470	38
23	52 099	54 633	45 367	97 466	37
24	52 135	54 673	45 327	97 461	36
25	52 171	54 714	45 286	97 457	35
26	52 207	54 754	45 246	97 453	34
27	52 242	54 794	45 206	97 448	33
28	52 278	54 835	45 165	97 444	32
29	52 314	54 875	45 125	97 439	31
30	52 350	54 915	45 085	97 435	30
31	52 385	54 955	45 045	97 430	29
32	52 421	54 995	45 005	97 426	28
33	52 456	55 035	44 965	97 421	27
34	52 492	55 075	44 925	97 417	26
35	52 527	55 115	44 885	97 412	25
36	52 563	55 155	44 845	97 408	24
37	52 598	55 195	44 805	97 403	23
38	52 634	55 235	44 765	97 399	22
39	52 669	55 275	44 725	97 394	21
40	52 705	55 315	44 685	97 390	20
41	52 740	55 355	44 645	97 385	19
42	52 775	55 395	44 605	97 381	18
43	52 811	55 434	44 566	97 376	17
44	52 846	55 474	44 526	97 372	16
45	52 881	55 514	44 486	97 367	15
46	52 916	55 554	44 446	97 363	14
47	52 951	55 593	44 407	97 358	13
48	52 986	55 633	44 367	97 353	12
49	53 021	55 673	44 327	97 349	11
50	53 056	55 712	44 288	97 344	10
51	53 092	55 752	44 248	97 340	9
52	53 126	55 791	44 209	97 335	8
53	53 161	55 831	44 169	97 331	7
54	53 196	55 870	44 130	97 326	6
55	53 231	55 910	44 090	97 322	5
56	53 266	55 949	44 051	97 317	4
57	53 301	55 989	44 011	97 312	3
58	53 336	56 028	43 972	97 308	2
59	53 370	56 067	43 933	97 303	1
60	53 405	56 107	43 893	97 299	0
	9	9	10	9	
′	log cos	log cot	log tan	log sin	′

70°

′	log sin	log tan	log cot	log cos	′
	9	9	10	9	
0	53 405	56 107	43 893	97 299	60
1	53 440	56 146	43 854	97 294	59
2	53 475	56 185	43 815	97 289	58
3	53 509	56 224	43 776	97 285	57
4	53 544	56 264	43 736	97 280	56
5	53 578	56 303	43 697	97 276	55
6	53 613	56 342	43 658	97 271	54
7	53 647	56 381	43 619	97 266	53
8	53 682	56 420	43 580	97 262	52
9	53 716	56 459	43 541	97 257	51
10	53 751	56 498	43 502	97 252	50
11	53 785	56 537	43 463	97 248	49
12	53 819	56 576	43 424	97 243	48
13	53 854	56 615	43 385	97 238	47
14	53 888	56 654	43 346	97 234	46
15	53 922	56 693	43 307	97 229	45
16	53 957	56 732	43 268	97 224	44
17	53 991	56 771	43 229	97 220	43
18	54 025	56 810	43 190	97 215	42
19	54 059	56 849	43 151	97 210	41
20	54 093	56 887	43 113	97 206	40
21	54 127	56 926	43 074	97 201	39
22	54 161	56 965	43 035	97 196	38
23	54 195	57 004	42 996	97 192	37
24	54 229	57 042	42 958	97 187	36
25	54 263	57 081	42 919	97 182	35
26	54 297	57 120	42 880	97 178	34
27	54 331	57 158	42 842	97 173	33
28	54 365	57 197	42 803	97 168	32
29	54 399	57 235	42 765	97 163	31
30	54 433	57 274	42 726	97 159	30
31	54 466	57 312	42 688	97 154	29
32	54 500	57 351	42 649	97 149	28
33	54 534	57 389	42 611	97 145	27
34	54 567	57 428	42 572	97 140	26
35	54 601	57 466	42 534	97 135	25
36	54 635	57 504	42 496	97 130	24
37	54 668	57 543	42 457	97 126	23
38	54 702	57 581	42 419	97 121	22
39	54 735	57 619	42 381	97 116	21
40	54 769	57 658	42 342	97 111	20
41	54 802	57 696	42 304	97 107	19
42	54 836	57 734	42 266	97 102	18
43	54 869	57 772	42 228	97 097	17
44	54 903	57 810	42 190	97 092	16
45	54 936	57 849	42 151	97 087	15
46	54 969	57 887	42 113	97 083	14
47	55 003	57 925	42 075	97 078	13
48	55 036	57 963	42 037	97 073	12
49	55 069	58 001	41 999	97 068	11
50	55 102	58 039	41 961	97 063	10
51	55 136	58 077	41 923	97 059	9
52	55 169	58 115	41 885	97 054	8
53	55 202	58 153	41 847	97 049	7
54	55 235	58 191	41 809	97 044	6
55	55 268	58 229	41 771	97 039	5
56	55 301	58 267	41 733	97 035	4
57	55 334	58 304	41 696	97 030	3
58	55 367	58 342	41 658	97 025	2
59	55 400	58 380	41 620	97 020	1
60	55 433	58 418	41 582	97 015	0
	9	9	·10	9	
′	log cos	log cot	log tan	log sin	′

69°

′	log sin	log tan	log cot	log cos	′
	9	**9**	**10**	**9**	
0	55 433	58 418	41 582	97 015	**60**
1	55 466	58 455	41 545	97 010	59
2	55 499	58 493	41 507	97 005	58
3	55 532	58 531	41 469	97 001	57
4	55 564	58 569	41 431	96 996	56
5	55 597	58 606	41 394	96 991	**55**
6	55 630	58 644	41 356	96 986	54
7	55 663	58 681	41 319	96 981	53
8	55 695	58 719	41 281	96 976	52
9	55 728	58 757	41 243	96 971	51
10	55 761	58 794	41 206	96 966	**50**
11	55 793	58 832	41 168	96 962	49
12	55 826	58 869	41 131	96 957	48
13	55 858	58 907	41 093	96 952	47
14	55 891	58 944	41 056	96 947	46
15	55 923	58 981	41 019	96 942	**45**
16	55 956	59 019	40 981	96 937	44
17	55 988	59 056	40 944	96 932	43
18	56 021	59 094	40 906	96 927	42
19	56 053	59 131	40 869	96 922	41
20	56 085	59 168	40 832	96 917	**40**
21	56 118	59 205	40 795	96 912	39
22	56 150	59 243	40 757	96 907	38
23	56 182	59 280	40 720	96 903	37
24	56 215	59 317	40 683	96 898	36
25	56 247	59 354	40 646	96 893	**35**
26	56 279	59 391	40 609	96 888	34
27	56 311	59 429	40 571	96 883	33
28	56 343	59 466	40 534	96 878	32
29	56 375	59 503	40 497	96 873	31
30	56 408	59 540	40 460	96 868	**30**
31	56 440	59 577	40 423	96 863	29
32	56 472	59 614	40 386	96 858	28
33	56 504	59 651	40 349	96 853	27
34	56 536	59 688	40 312	96 848	26
35	56 568	59 725	40 275	96 843	**25**
36	56 599	59 762	40 238	96 838	24
37	56 631	59 799	40 201	96 833	23
38	56 663	59 835	40 165	96 828	22
39	56 695	59 872	40 128	96 823	21
40	56 727	59 909	40 091	96 818	**20**
41	56 759	59 946	40 054	96 813	19
42	56 790	59 983	40 017	96 808	18
43	56 822	60 019	39 981	96 803	17
44	56 854	60 056	39 944	96 798	16
45	56 886	60 093	39 907	96 793	**15**
46	56 917	60 130	39 870	96 788	14
47	56 949	60 166	39 834	96 783	13
48	56 980	60 203	39 797	96 778	12
49	57 012	60 240	39 760	96 772	11
50	57 044	60 276	39 724	96 767	**10**
51	57 075	60 313	39 687	96 762	9
52	57 107	60 349	39 651	96 757	8
53	57 138	60 386	39 614	96 752	7
54	57 169	60 422	39 578	96 747	6
55	57 201	60 459	39 541	96 742	**5**
56	57 232	60 495	39 505	96 737	4
57	57 264	60 532	39 468	96 732	3
58	57 295	60 568	39 432	96 727	2
59	57 326	60 605	39 395	96 722	1
60	57 358	60 641	39 359	96 717	**0**
	9	**9**	**10**	**9**	
′	log cos	log cot	log tan	log sin	′

68°

′	log sin	log tan	log cot	log cos	′
	9	**9**	**10**	**9**	
0	57 358	60 641	39 359	96 717	**60**
1	57 389	60 677	39 323	96 711	59
2	57 420	60 714	39 286	96 706	58
3	57 451	60 750	39 250	96 701	57
4	57 482	60 786	39 214	96 696	56
5	57 514	60 823	39 177	96 691	**55**
6	57 545	60 859	39 141	96 686	54
7	57 576	60 895	39 105	96 681	53
8	57 607	60 931	39 069	96 676	52
9	57 638	60 967	39 033	96 670	51
10	57 669	61 004	38 996	96 665	**50**
11	57 700	61 040	38 960	96 660	49
12	57 731	61 076	38 924	96 655	48
13	57 762	61 112	38 888	96 650	47
14	57 793	61 148	38 852	96 645	46
15	57 824	61 184	38 816	96 640	**45**
16	57 855	61 220	38 780	96 634	44
17	57 885	61 256	38 744	96 629	43
18	57 916	61 292	38 708	96 624	42
19	57 947	61 328	38 672	96 619	41
20	57 978	61 364	38 636	96 614	**40**
21	58 008	61 400	38 600	96 608	39
22	58 039	61 436	38 564	96 603	38
23	58 070	61 472	38 528	96 598	37
24	58 101	61 508	38 492	96 593	36
25	58 131	61 544	38 456	96 588	**35**
26	58 162	61 579	38 421	96 582	34
27	58 192	61 615	38 385	96 577	33
28	58 223	61 651	38 349	96 572	32
29	58 253	61 687	38 313	96 567	31
30	58 284	61 722	38 278	96 562	**30**
31	58 314	61 758	38 242	96 556	29
32	58 345	61 794	38 206	96 551	28
33	58 375	61 830	38 170	96 546	27
34	58 406	61 865	38 135	96 541	26
35	58 436	61 901	38 099	96 535	**25**
36	58 467	61 936	38 064	96 530	24
37	58 497	61 972	38 028	96 525	23
38	58 527	62 008	37 992	96 520	22
39	58 557	62 043	37 957	96 514	21
40	58 588	62 079	37 921	96 509	**20**
41	58 618	62 114	37 886	96 504	19
42	58 648	62 150	37 850	96 498	18
43	58 678	62 185	37 815	96 493	17
44	58 709	62 221	37 779	96 488	16
45	58 739	62 256	37 744	96 483	**15**
46	58 769	62 292	37 708	96 477	14
47	58 799	62 327	37 673	96 472	13
48	58 829	62 362	37 638	96 467	12
49	58 859	62 398	37 602	96 461	11
50	58 889	62 433	37 567	96 456	**10**
51	58 919	62 468	37 532	96 451	9
52	58 949	62 504	37 496	96 445	8
53	58 979	62 539	37 461	96 440	7
54	59 009	62 574	37 426	96 435	6
55	59 039	62 609	37 391	96 429	**5**
56	59 069	62 645	37 355	96 424	4
57	59 098	62 680	37 320	96 419	3
58	59 128	62 715	37 285	96 413	2
59	59 158	62 750	37 250	96 408	1
60	59 188	62 785	37 215	96 403	**0**
	9	**9**	**10**	**9**	
′	log cos	log cot	log tan	log sin	′

67°

′	log sin	log tan	log cot	log cos	′
	9	9	10	9	
0	59 188	62 785	37 215	96 403	60
1	59 218	62 820	37 180	96 397	59
2	59 247	62 855	37 145	96 392	58
3	59 277	62 890	37 110	96 387	57
4	59 307	62 926	37 074	96 381	56
5	59 336	62 961	37 039	96 376	55
6	59 366	62 996	37 004	96 370	54
7	59 396	63 031	36 969	96 365	53
8	59 425	63 066	36 934	96 360	52
9	59 455	63 101	36 899	96 354	51
10	59 484	63 135	36 865	96 349	50
11	59 514	63 170	36 830	96 343	49
12	59 543	63 205	36 795	96 338	48
13	59 573	63 240	36 760	96 333	47
14	59 602	63 275	36 725	96 327	46
15	59 632	63 310	36 690	96 322	45
16	59 661	63 345	36 655	96 316	44
17	59 690	63 379	36 621	96 311	43
18	59 720	63 414	36 586	96 305	42
19	59 749	63 449	36 551	96 300	41
20	59 778	63 484	36 516	96 294	40
21	59 808	63 519	36 481	96 289	39
22	59 837	63 553	36 447	96 284	38
23	59 866	63 588	36 412	96 278	37
24	59 895	63 623	36 377	96 273	36
25	59 924	63 657	36 343	96 267	35
26	59 954	63 692	36 308	96 262	34
27	59 983	63 726	36 274	96 256	33
28	60 012	63 761	36 239	96 251	32
29	60 041	63 796	36 204	96 245	31
30	60 070	63 830	36 170	96 240	30
31	60 099	63 865	36 135	96 234	29
32	60 128	63 899	36 101	96 229	28
33	60 157	63 934	36 066	96 223	27
34	60 186	63 968	36 032	96 218	26
35	60 215	64 003	35 997	96 212	25
36	60 244	64 037	35 963	96 207	24
37	60 273	64 072	35 928	96 201	23
38	60 302	64 106	35 894	96 196	22
39	60 331	64 140	35 860	96 190	21
40	60 359	64 175	35 825	96 185	20
41	60 388	64 209	35 791	96 179	19
42	60 417	64 243	35 757	96 174	18
43	60 446	64 278	35 722	96 168	17
44	60 474	64 312	35 688	96 162	16
45	60 503	64 346	35 654	96 157	15
46	60 532	64 381	35 619	96 151	14
47	60 561	64 415	35 585	96 146	13
48	60 589	64 449	35 551	96 140	12
49	60 618	64 483	35 517	96 135	11
50	60 646	64 517	35 483	96 129	10
51	60 675	64 552	35 448	96 123	9
52	60 704	64 586	35 414	96 118	8
53	60 732	64 620	35 380	96 112	7
54	60 761	64 654	35 346	96 107	6
55	60 789	64 688	35 312	96 101	5
56	60 818	64 722	35 278	96 095	4
57	60 846	64 756	35 244	96 090	3
58	60 875	64 790	35 210	96 084	2
59	60 903	64 824	35 176	96 079	1
60	60 931	64 858	35 142	96 073	0
	9	9	10	9	
′	log cos	log cot	log tan	log sin	′

66°

′	log sin	log tan	log cot	log cos	′
	9	9	10	9	
0	60 931	64 858	35 142	96 073	60
1	60 960	64 892	35 108	96 067	59
2	60 988	64 926	35 074	96 062	58
3	61 016	64 960	35 040	96 056	57
4	61 045	64 994	35 006	96 050	56
5	61 073	65 028	34 972	96 045	55
6	61 101	65 062	34 938	96 039	54
7	61 129	65 096	34 904	96 034	53
8	61 158	65 130	34 870	96 028	52
9	61 186	65 164	34 836	96 022	51
10	61 214	65 197	34 803	96 017	50
11	61 242	65 231	34 769	96 011	49
12	61 270	65 265	34 735	96 005	48
13	61 298	65 299	34 701	96 000	47
14	61 326	65 333	34 667	95 994	46
15	61 354	65 366	34 634	95 988	45
16	61 382	65 400	34 600	95 982	44
17	61 411	65 434	34 566	95 977	43
18	61 438	65 467	34 533	95 971	42
19	61 466	65 501	34 499	95 965	41
20	61 494	65 535	34 465	95 960	40
21	61 522	65 568	34 432	95 954	39
22	61 550	65 602	34 398	95 948	38
23	61 578	65 636	34 364	95 942	37
24	61 606	65 669	34 331	95 937	36
25	61 634	65 703	34 297	95 931	35
26	61 662	65 736	34 264	95 925	34
27	61 689	65 770	34 230	95 920	33
28	61 717	65 803	34 197	95 914	32
29	61 745	65 837	34 163	95 908	31
30	61 773	65 870	34 130	95 902	30
31	61 800	65 904	34 096	95 897	29
32	61 828	65 937	34 063	95 891	28
33	61 856	65 971	34 029	95 885	27
34	61 883	66 004	33 996	95 879	26
35	61 911	66 038	33 962	95 873	25
36	61 939	66 071	33 929	95 868	24
37	61 966	66 104	33 896	95 862	23
38	61 994	66 138	33 862	95 856	22
39	62 021	66 171	33 829	95 850	21
40	62 049	66 204	33 796	95 844	20
41	62 076	66 238	33 762	95 839	19
42	62 104	66 271	33 729	95 833	18
43	62 131	66 304	33 696	95 827	17
44	62 159	66 337	33 663	95 821	16
45	62 186	66 371	33 629	95 815	15
46	62 214	66 404	33 596	95 810	14
47	62 241	66 437	33 563	95 804	13
48	62 268	66 470	33 530	95 798	12
49	62 296	66 503	33 497	95 792	11
50	62 323	66 537	33 463	95 786	10
51	62 350	66 570	33 430	95 780	9
52	62 377	66 603	33 397	95 775	8
53	62 405	66 636	33 364	95 769	7
54	62 432	66 669	33 331	95 763	6
55	62 459	66 702	33 298	95 757	5
56	62 486	66 735	33 265	95 751	4
57	62 513	66 768	33 232	95 745	3
58	62 541	66 801	33 199	95 739	2
59	62 568	66 834	33 166	95 733	1
60	62 595	66 867	33 133	95 728	0
	9	9	10	9	
′	log cos	log cot	log tan	log sin	′

65°

'	log sin 9	log tan 9	log cot 10	log cos 9	'
0	62 595	66 867	33 133	95 728	60
1	62 622	66 900	33 100	95 722	59
2	62 649	66 933	33 067	95 716	58
3	62 676	66 966	33 034	95 710	57
4	62 703	66 999	33 001	95 704	56
5	62 730	67 032	32 968	95 698	55
6	62 757	67 065	32 935	95 692	54
7	62 784	67 098	32 902	95 686	53
8	62 811	67 131	32 869	95 680	52
9	62 838	67 163	32 837	95 674	51
10	62 865	67 196	32 804	95 668	50
11	62 892	67 229	32 771	95 663	49
12	62 918	67 262	32 738	95 657	48
13	62 945	67 295	32 705	95 651	47
14	62 972	67 327	32 673	95 645	46
15	62 999	67 360	32 640	95 639	45
16	63 026	67 393	32 607	95 633	44
17	63 052	67 426	32 574	95 627	43
18	63 079	67 458	32 542	95 621	42
19	63 106	67 491	32 509	95 615	41
20	63 133	67 524	32 476	95 609	40
21	63 159	67 556	32 444	95 603	39
22	63 186	67 589	32 411	95 597	38
23	63 213	67 622	32 378	95 591	37
24	63 239	67 654	32 346	95 585	36
25	63 266	67 687	32 313	95 579	35
26	63 292	67 719	32 281	95 573	34
27	63 319	67 752	32 248	95 567	33
28	63 345	67 785	32 215	95 561	32
29	63 372	67 817	32 183	95 555	31
30	63 398	67 850	32 150	95 549	30
31	63 425	67 882	32 118	95 543	29
32	63 451	67 915	32 085	95 537	28
33	63 478	67 947	32 053	95 531	27
34	63 504	67 980	32 020	95 525	26
35	63 531	68 012	31 988	95 519	25
36	63 557	68 044	31 956	95 513	24
37	63 583	68 077	31 923	95 507	23
38	63 610	68 109	31 891	95 500	22
39	63 636	68 142	31 858	95 494	21
40	63 662	68 174	31 826	95 488	20
41	63 689	68 206	31 794	95 482	19
42	63 715	68 239	31 761	95 476	18
43	63 741	68 271	31 729	95 470	17
44	63 767	68 303	31 697	95 464	16
45	63 794	68 336	31 664	95 458	15
46	63 820	68 368	31 632	95 452	14
47	63 846	68 400	31 600	95 446	13
48	63 872	68 432	31 568	95 440	12
49	63 898	68 465	31 535	95 434	11
50	63 924	68 497	31 503	95 427	10
51	63 950	68 529	31 471	95 421	9
52	63 976	68 561	31 439	95 415	8
53	64 002	68 593	31 407	95 409	7
54	64 028	68 626	31 374	95 403	6
55	64 054	68 658	31 342	95 397	5
56	64 080	68 690	31 310	95 391	4
57	64 106	68 722	31 278	95 384	3
58	64 132	68 754	31 246	95 378	2
59	64 158	68 786	31 214	95 372	1
60	64 184	68 818	31 182	95 366	0
'	log cos 9	log cot 9	log tan 10	log sin 9	'

'	log sin 9	log tan 9	log cot 10	log cos 9	'
0	64 184	68 818	31 182	95 366	60
1	64 210	68 850	31 150	95 360	59
2	64 236	68 882	31 118	95 354	58
3	64 262	68 914	31 086	95 348	57
4	64 288	68 946	31 054	95 341	56
5	64 313	68 978	31 022	95 335	55
6	64 339	69 010	30 990	95 329	54
7	64 365	69 042	30 958	95 323	53
8	64 391	69 074	30 926	95 317	52
9	64 417	69 106	30 894	95 310	51
10	64 442	69 138	30 862	95 304	50
11	64 468	69 170	30 830	95 298	49
12	64 494	69 202	30 798	95 292	48
13	64 519	69 234	30 766	95 286	47
14	64 545	69 266	30 734	95 279	46
15	64 571	69 298	30 702	95 273	45
16	64 596	69 329	30 671	95 267	44
17	64 622	69 361	30 639	95 261	43
18	64 647	69 393	30 607	95 254	42
19	64 673	69 425	30 575	95 248	41
20	64 698	69 457	30 543	95 242	40
21	64 724	69 488	30 512	95 236	39
22	64 749	69 520	30 480	95 229	38
23	64 775	69 552	30 448	95 223	37
24	64 800	69 584	30 416	95 217	36
25	64 826	69 615	30 385	95 211	35
26	64 851	69 647	30 353	95 204	34
27	64 877	69 679	30 321	95 198	33
28	64 902	69 710	30 290	95 192	32
29	64 927	69 742	30 258	95 185	31
30	64 953	69 774	30 226	95 179	30
31	64 978	69 805	30 195	95 173	29
32	65 003	69 837	30 163	95 167	28
33	65 029	69 868	30 132	95 160	27
34	65 054	69 900	30 100	95 154	26
35	65 079	69 932	30 068	95 148	25
36	65 104	69 963	30 037	95 141	24
37	65 130	69 995	30 005	95 135	23
38	65 155	70 026	29 974	95 129	22
39	65 180	70 058	29 942	95 122	21
40	65 205	70 089	29 911	95 116	20
41	65 230	70 121	29 879	95 110	19
42	65 255	70 152	29 848	95 103	18
43	65 281	70 184	29 816	95 097	17
44	65 306	70 215	29 785	95 090	16
45	65 331	70 247	29 753	95 084	15
46	65 356	70 278	29 722	95 078	14
47	65 381	70 309	29 691	95 071	13
48	65 406	70 341	29 659	95 065	12
49	65 431	70 372	29 628	95 059	11
50	65 456	70 404	29 596	95 052	10
51	65 481	70 435	29 565	95 046	9
52	65 506	70 466	29 534	95 039	8
53	65 531	70 498	29 502	95 033	7
54	65 556	70 529	29 471	95 027	6
55	65 580	70 560	29 440	95 020	5
56	65 605	70 592	29 408	95 014	4
57	65 630	70 623	29 377	95 007	3
58	65 655	70 654	29 346	95 001	2
59	65 680	70 685	29 315	94 995	1
60	65 705	70 717	29 283	94 988	0
'	log cos 9	log cot 9	log tan 10	log sin 9	'

64°   **63°**

′	log sin 9	log tan 9	log cot 10	log cos 9	′
0	65 705	70 717	29 283	94 988	60
1	65 729	70 748	29 252	94 982	59
2	65 754	70 779	29 221	94 975	58
3	65 779	70 810	29 190	94 969	57
4	65 804	70 841	29 159	94 962	56
5	65 828	70 873	29 127	94 956	55
6	65 853	70 904	29 096	94 949	54
7	65 878	70 935	29 065	94 943	53
8	65 902	70 966	29 034	94 936	52
9	65 927	70 997	29 003	94 930	51
10	65 952	71 028	28 972	94 923	50
11	65 976	71 059	28 941	94 917	49
12	66 001	71 090	28 910	94 911	48
13	66 025	71 121	28 879	94 904	47
14	66 050	71 153	28 847	94 898	46
15	66 075	71 184	28 816	94 891	45
16	66 099	71 215	28 785	94 885	44
17	66 124	71 246	28 754	94 878	43
18	66 148	71 277	28 723	94 871	42
19	66 173	71 308	28 692	94 865	41
20	66 197	71 339	28 661	94 858	40
21	66 221	71 370	28 630	94 852	39
22	66 246	71 401	28 599	94 845	38
23	66 270	71 431	28 569	94 839	37
24	66 295	71 462	28 538	94 832	36
25	66 319	71 493	28 507	94 826	35
26	66 343	71 524	28 476	94 819	34
27	66 368	71 555	28 445	94 813	33
28	66 392	71 586	28 414	94 806	32
29	66 416	71 617	28 383	94 799	31
30	66 441	71 648	28 352	94 793	30
31	66 465	71 679	28 321	94 786	29
32	66 489	71 709	28 291	94 780	28
33	66 513	71 740	28 260	94 773	27
34	66 537	71 771	28 229	94 767	26
35	66 562	71 802	28 198	94 760	25
36	66 586	71 833	28 167	94 753	24
37	66 610	71 863	28 137	94 747	23
38	66 634	71 894	28 106	94 740	22
39	66 658	71 925	28 075	94 734	21
40	66 682	71 955	28 045	94 727	20
41	66 706	71 986	28 014	94 720	19
42	66 731	72 017	27 983	94 714	18
43	66 755	72 048	27 952	94 707	17
44	66 779	72 078	27 922	94 700	16
45	66 803	72 109	27 891	94 694	15
46	66 827	72 140	27 860	94 687	14
47	66 851	72 170	27 830	94 680	13
48	66 875	72 201	27 799	94 674	12
49	66 899	72 231	27 769	94 667	11
50	66 922	72 262	27 738	94 660	10
51	66 946	72 293	27 707	94 654	9
52	66 970	72 323	27 677	94 647	8
53	66 994	72 354	27 646	94 640	7
54	67 018	72 384	27 616	94 634	6
55	67 042	72 415	27 585	94 627	5
56	67 066	72 445	27 555	94 620	4
57	67 090	72 476	27 524	94 614	3
58	67 113	72 506	27 494	94 607	2
59	67 137	72 537	27 463	94 600	1
60	67 161	72 567	27 433	94 593	0
′	log cos 9	log cot 9	log tan 10	log sin 9	′

62°

′	log sin 9	log tan 9	log cot 10	log cos 9	′
0	67 161	72 567	27 433	94 593	60
1	67 185	72 598	27 402	94 587	59
2	67 208	72 628	27 372	94 580	58
3	67 232	72 659	27 341	94 573	57
4	67 256	72 689	27 311	94 567	56
5	67 280	72 720	27 280	94 560	55
6	67 303	72 750	27 250	94 553	54
7	67 327	72 780	27 220	94 546	53
8	67 350	72 811	27 189	94 540	52
9	67 374	72 841	27 159	94 533	51
10	67 398	72 872	27 128	94 526	50
11	67 421	72 902	27 098	94 519	49
12	67 445	72 932	27 068	94 513	48
13	67 468	72 963	27 037	94 506	47
14	67 492	72 993	27 007	94 499	46
15	67 515	73 023	26 977	94 492	45
16	67 539	73 054	26 946	94 485	44
17	67 562	73 084	26 916	94 479	43
18	67 586	73 114	26 886	94 472	42
19	67 609	73 144	26 856	94 465	41
20	67 633	73 175	26 825	94 458	40
21	67 656	73 205	26 795	94 451	39
22	67 680	73 235	26 765	94 445	38
23	67 703	73 265	26 735	94 438	37
24	67 726	73 295	26 705	94 431	36
25	67 750	73 326	26 674	94 424	35
26	67 773	73 356	26 644	94 417	34
27	67 796	73 386	26 614	94 410	33
28	67 820	73 416	26 584	94 404	32
29	67 843	73 446	26 554	94 397	31
30	67 866	73 476	26 524	94 390	30
31	67 890	73 507	26 493	94 383	29
32	67 913	73 537	26 463	94 376	28
33	67 936	73 567	26 433	94 369	27
34	67 959	73 597	26 403	94 362	26
35	67 982	73 627	26 373	94 355	25
36	68 006	73 657	26 343	94 349	24
37	68 029	73 687	26 313	94 342	23
38	68 052	73 717	26 283	94 335	22
39	68 075	73 747	26 253	94 328	21
40	68 098	73 777	26 223	94 321	20
41	68 121	73 807	26 193	94 314	19
42	68 144	73 837	26 163	94 307	18
43	68 167	73 867	26 133	94 300	17
44	68 190	73 897	26 103	94 293	16
45	68 213	73 927	26 073	94 286	15
46	68 237	73 957	26 043	94 279	14
47	68 260	73 987	26 013	94 273	13
48	68 283	74 017	25 983	94 266	12
49	68 305	74 047	25 953	94 259	11
50	68 328	74 077	25 923	94 252	10
51	68 351	74 107	25 893	94 245	9
52	68 374	74 137	25 863	94 238	8
53	68 397	74 166	25 834	94 231	7
54	68 420	74 196	25 804	94 224	6
55	68 443	74 226	25 774	94 217	5
56	68 466	74 256	25 744	94 210	4
57	68 489	74 286	25 714	94 203	3
58	68 512	74 316	25 684	94 196	2
59	68 534	74 345	25 655	94 189	1
60	68 557	74 375	25 625	94 182	0
′	log cos 9	log cot 9	log tan 10	log sin 9	′

61°

′	log sin 9	log tan 9	log cot 10	log cos 9	′
0	68 557	74 375	25 625	94 182	60
1	68 580	74 405	25 595	94 175	59
2	68 603	74 435	25 565	94 168	58
3	68 625	74 465	25 535	94 161	57
4	68 648	74 494	25 506	94 154	56
5	68 671	74 524	25 476	94 147	55
6	68 694	74 554	25 446	94 140	54
7	68 716	74 583	25 417	94 133	53
8	68 739	74 613	25 387	94 126	52
9	68 762	74 643	25 357	94 119	51
10	68 784	74 673	25 327	94 112	50
11	68 807	74 702	25 298	94 105	49
12	68 829	74 732	25 268	94 098	48
13	68 852	74 762	25 238	94 090	47
14	68 875	74 791	25 209	94 083	46
15	68 897	74 821	25 179	94 076	45
16	68 920	74 851	25 149	94 069	44
17	68 942	74 880	25 120	94 062	43
18	68 965	74 910	25 090	94 055	42
19	68 987	74 939	25 061	94 048	41
20	69 010	74 969	25 031	94 041	40
21	69 032	74 998	25 002	94 034	39
22	69 055	75 028	24 972	94 027	38
23	69 077	75 058	24 942	94 020	37
24	69 100	75 087	24 913	94 012	36
25	69 122	75 117	24 883	94 005	35
26	69 144	75 146	24 854	93 998	34
27	69 167	75 176	24 824	93 991	33
28	69 189	75 205	24 795	93 984	32
29	69 212	75 235	24 765	93 977	31
30	69 234	75 264	24 736	93 970	30
31	69 256	75 294	24 706	93 963	29
32	69 279	75 323	24 677	93 955	28
33	69 301	75 353	24 647	93 948	27
34	69 323	75 382	24 618	93 941	26
35	69 345	75 411	24 589	93 934	25
36	69 368	75 441	24 559	93 927	24
37	69 390	75 470	24 530	93 920	23
38	69 412	75 500	24 500	93 912	22
39	69 434	75 529	24 471	93 905	21
40	69 456	75 558	24 442	93 898	20
41	69 479	75 588	24 412	93 891	19
42	69 501	75 617	24 383	93 884	18
43	69 523	75 647	24 353	93 876	17
44	69 545	75 676	24 324	93 869	16
45	69 567	75 705	24 295	93 862	15
46	69 589	75 735	24 265	93 855	14
47	69 611	75 764	24 236	93 847	13
48	69 633	75 793	24 207	93 840	12
49	69 655	75 822	24 178	93 833	11
50	69 677	75 852	24 148	93 826	10
51	69 699	75 881	24 119	93 819	9
52	69 721	75 910	24 090	93 811	8
53	69 743	75 939	24 061	93 804	7
54	69 765	75 969	24 031	93 797	6
55	69 787	75 998	24 002	93 789	5
56	69 809	76 027	23 973	93 782	4
57	69 831	76 056	23 944	93 775	3
58	69 853	76 085	23 914	93 768	2
59	69 875	76 115	23 885	93 760	1
60	69 897	76 144	23 856	93 753	0
′	log cos 9	log cot 9	log tan 10	log sin 9	′

60°

′	log sin 9	log tan 9	log cot 10	log cos 9	′
0	69 897	76 144	23 856	93 753	60
1	69 919	76 173	23 827	93 746	59
2	69 941	76 202	23 798	93 738	58
3	69 963	76 231	23 769	93 731	57
4	69 984	76 261	23 739	93 724	56
5	70 006	76 290	23 710	93 717	55
6	70 028	76 319	23 681	93 709	54
7	70 050	76 348	23 652	93 702	53
8	70 072	76 377	23 623	93 695	52
9	70 093	76 406	23 594	93 687	51
10	70 115	76 435	23 565	93 680	50
11	70 137	76 464	23 536	93 673	49
12	70 159	76 493	23 507	93 665	48
13	70 180	76 522	23 478	93 658	47
14	70 202	76 551	23 449	93 650	46
15	70 224	76 580	23 420	93 643	45
16	70 245	76 609	23 391	93 636	44
17	70 267	76 639	23 361	93 628	43
18	70 288	76 668	23 332	93 621	42
19	70 310	76 697	23 303	93 614	41
20	70 332	76 725	23 275	93 606	40
21	70 353	76 754	23 246	93 599	39
22	70 375	76 783	23 217	93 591	38
23	70 396	76 812	23 188	93 584	37
24	70 418	76 841	23 159	93 577	36
25	70 439	76 870	23 130	93 569	35
26	70 461	76 899	23 101	93 562	34
27	70 482	76 928	23 072	93 554	33
28	70 504	76 957	23 043	93 547	32
29	70 525	76 986	23 014	93 539	31
30	70 547	77 015	22 985	93 532	30
31	70 568	77 044	22 956	93 525	29
32	70 590	77 073	22 927	93 517	28
33	70 611	77 101	22 899	93 510	27
34	70 633	77 130	22 870	93 502	26
35	70 654	77 159	22 841	93 495	25
36	70 675	77 188	22 812	93 487	24
37	70 697	77 217	22 783	93 480	23
38	70 718	77 246	22 754	93 472	22
39	70 739	77 274	22 726	93 465	21
40	70 761	77 303	22 697	93 457	20
41	70 782	77 332	22 668	93 450	19
42	70 803	77 361	22 639	93 442	18
43	70 824	77 390	22 610	93 435	17
44	70 846	77 418	22 582	93 427	16
45	70 867	77 447	22 553	93 420	15
46	70 888	77 476	22 524	93 412	14
47	70 909	77 505	22 495	93 405	13
48	70 931	77 533	22 467	93 397	12
49	70 952	77 562	22 438	93 390	11
50	70 973	77 591	22 409	93 382	10
51	70 994	77 619	22 381	93 375	9
52	71 015	77 648	22 352	93 367	8
53	71 036	77 677	22 323	93 360	7
54	71 058	77 706	22 294	93 352	6
55	71 079	77 734	22 266	93 344	5
56	71 100	77 763	22 237	93 337	4
57	71 121	77 791	22 209	93 329	3
58	71 142	77 820	22 180	93 322	2
59	71 163	77 849	22 151	93 314	1
60	71 184	77 877	22 123	93 307	0
′	log cos 9	log cot 9	log tan 10	log sin 9	′

59°

′	log sin	log tan	log cot	log cos	′
	9	9	10	9	
0	71 184	77 877	22 123	93 307	60
1	71 205	77 906	22 094	93 299	59
2	71 226	77 935	22 065	93 291	58
3	71 247	77 963	22 037	93 284	57
4	71 268	77 992	22 008	93 276	56
5	71 289	78 020	21 980	93 269	55
6	71 310	78 049	21 951	93 261	54
7	71 331	78 077	21 923	93 253	53
8	71 352	78 106	21 894	93 246	52
9	71 373	78 135	21 865	93 238	51
10	71 393	78 163	21 837	93 230	50
11	71 414	78 192	21 808	93 223	49
12	71 435	78 220	21 780	93 215	48
13	71 456	78 249	21 751	93 207	47
14	71 477	78 277	21 723	93 200	46
15	71 498	78 306	21 694	93 192	45
16	71 519	78 334	21 666	93 184	44
17	71 539	78 363	21 637	93 177	43
18	71 560	78 391	21 609	93 169	42
19	71 581	78 419	21 581	93 161	41
20	71 602	78 448	21 552	93 154	40
21	71 622	78 476	21 524	93 146	39
22	71 643	78 505	21 495	93 138	38
23	71 664	78 533	21 467	93 131	37
24	71 685	78 562	21 438	93 123	36
25	71 705	78 590	21 410	93 115	35
26	71 726	78 618	21 382	93 108	34
27	71 747	78 647	21 353	93 100	33
28	71 767	78 675	21 325	93 092	32
29	71 788	78 704	21 296	93 084	31
30	71 809	78 732	21 268	93 077	30
31	71 829	78 760	21 240	93 069	29
32	71 850	78 789	21 211	93 061	28
33	71 870	78 817	21 183	93 053	27
34	71 891	78 845	21 155	93 046	26
35	71 911	78 874	21 126	93 038	25
36	71 932	78 902	21 098	93 030	24
37	71 952	78 930	21 070	93 022	23
38	71 973	78 959	21 041	93 014	22
39	71 994	78 987	21 013	93 007	21
40	72 014	79 015	20 985	92 999	20
41	72 034	79 043	20 957	92 991	19
42	72 055	79 072	20 928	92 983	18
43	72 075	79 100	20 900	92 976	17
44	72 096	79 128	20 872	92 968	16
45	72 116	79 156	20 844	92 960	15
46	72 137	79 185	20 815	92 952	14
47	72 157	79 213	20 787	92 944	13
48	72 177	79 241	20 759	92 936	12
49	72 198	79 269	20 731	92 929	11
50	72 218	79 297	20 703	92 921	10
51	72 238	79 326	20 674	92 913	9
52	72 259	79 354	20 646	92 905	8
53	72 279	79 382	20 618	92 897	7
54	72 299	79 410	20 590	92 889	6
55	72 320	79 438	20 562	92 881	5
56	72 340	79 466	20 534	92 874	4
57	72 360	79 495	20 505	92 866	3
58	72 381	79 523	20 477	92 858	2
59	72 401	79 551	20 449	92 850	1
60	72 421	79 579	20 421	92 842	0
	9	9	10	9	
′	log cos	log cot	log tan	log sin	′

′	log sin	log tan	log cot	log cos	′
	9	9	10	9	
0	72 421	79 579	20 421	92 842	60
1	72 441	79 607	20 393	92 834	59
2	72 461	79 635	20 365	92 826	58
3	72 482	79 663	20 337	92 818	57
4	72 502	79 691	20 309	92 810	56
5	72 522	79 719	20 281	92 803	55
6	72 542	79 747	20 253	92 795	54
7	72 562	79 776	20 224	92 787	53
8	72 582	79 804	20 196	92 779	52
9	72 602	79 832	20 168	92 771	51
10	72 622	79 860	20 140	92 763	50
11	72 643	79 888	20 112	92 755	49
12	72 663	79 916	20 084	92 747	48
13	72 683	79 944	20 056	92 739	47
14	72 703	79 972	20 028	92 731	46
15	72 723	80 000	20 000	92 723	45
16	72 743	80 028	19 972	92 715	44
17	72 763	80 056	19 944	92 707	43
18	72 783	80 084	19 916	92 699	42
19	72 803	80 112	19 888	92 691	41
20	72 823	80 140	19 860	92 683	40
21	72 843	80 168	19 832	92 675	39
22	72 863	80 195	19 805	92 667	38
23	72 883	80 223	19 777	92 659	37
24	72 902	80 251	19 749	92 651	36
25	72 922	80 279	19 721	92 643	35
26	72 942	80 307	19 693	92 635	34
27	72 962	80 335	19 665	92 627	33
28	72 982	80 363	19 637	92 619	32
29	73 002	80 391	19 609	92 611	31
30	73 022	80 419	19 581	92 603	30
31	73 041	80 447	19 553	92 595	29
32	73 061	80 474	19 526	92 587	28
33	73 081	80 502	19 498	92 579	27
34	73 101	80 530	19 470	92 571	26
35	73 121	80 558	19 442	92 563	25
36	73 140	80 586	19 414	92 555	24
37	73 160	80 614	19 386	92 546	23
38	73 180	80 642	19 358	92 538	22
39	73 200	80 669	19 331	92 530	21
40	73 219	80 697	19 303	92 522	20
41	73 239	80 725	19 275	92 514	19
42	73 259	80 753	19 247	92 506	18
43	73 278	80 781	19 219	92 498	17
44	73 298	80 808	19 192	92 490	16
45	73 318	80 836	19 164	92 482	15
46	73 337	80 864	19 136	92 473	14
47	73 357	80 892	19 108	92 465	13
48	73 377	80 919	19 081	92 457	12
49	73 396	80 947	19 053	92 449	11
50	73 416	80 975	19 025	92 441	10
51	73 435	81 003	18 997	92 433	9
52	73 455	81 030	18 970	92 425	8
53	73 474	81 058	18 942	92 416	7
54	73 494	81 086	18 914	92 408	6
55	73 513	81 113	18 887	92 400	5
56	73 533	81 141	18 859	92 392	4
57	73 552	81 169	18 831	92 384	3
58	73 572	81 196	18 804	92 376	2
59	73 591	81 224	18 776	92 367	1
60	73 611	81 252	18 748	92 359	0
	9	9	10	9	
′	log cos	log cot	log tan	log sin	′

58° **57°**

′	log sin	log tan	log cot	log cos	′
	9	9	10	9	
0	73 611	81 252	18 748	92 359	60
1	73 630	81 279	18 721	92 351	59
2	73 650	81 307	18 693	92 343	58
3	73 669	81 335	18 665	92 335	57
4	73 689	81 362	18 638	92 326	56
5	73 708	81 390	18 610	92 318	55
6	73 727	81 418	18 582	92 310	54
7	73 747	81 445	18 555	92 302	53
8	73 766	81 473	18 527	92 293	52
9	73 785	81 500	18 500	92 285	51
10	73 805	81 528	18 472	92 277	50
11	73 824	81 556	18 444	92 269	49
12	73 843	81 583	18 417	92 260	48
13	73 863	81 611	18 389	92 252	47
14	73 882	81 638	18 362	92 244	46
15	73 901	81 666	18 334	92 235	45
16	73 921	81 693	18 307	92 227	44
17	73 940	81 721	18 279	92 219	43
18	73 959	81 748	18 252	92 211	42
19	73 978	81 776	18 224	92 202	41
20	73 997	81 803	18 197	92 194	40
21	74 017	81 831	18 169	92 186	39
22	74 036	81 858	18 142	92 177	38
23	74 055	81 886	18 114	92 169	37
24	74 074	81 913	18 087	92 161	36
25	74 093	81 941	18 059	92 152	35
26	74 113	81 968	18 032	92 144	34
27	74 132	81 996	18 004	92 136	33
28	74 151	82 023	17 977	92 127	32
29	74 170	82 051	17 949	92 119	31
30	74 189	82 078	17 922	92 111	30
31	74 208	82 106	17 894	92 102	29
32	74 227	82 133	17 867	92 094	28
33	74 246	82 161	17 839	92 086	27
34	74 265	82 188	17 812	92 077	26
35	74 284	82 215	17 785	92 069	25
36	74 303	82 243	17 757	92 060	24
37	74 322	82 270	17 730	92 052	23
38	74 341	82 298	17 702	92 044	22
39	74 360	82 325	17 675	92 035	21
40	74 379	82 352	17 648	92 027	20
41	74 398	82 380	17 620	92 018	19
42	74 417	82 407	17 593	92 010	18
43	74 436	82 435	17 565	92 002	17
44	74 455	82 462	17 538	91 993	16
45	74 474	82 489	17 511	91 985	15
46	74 493	82 517	17 483	91 976	14
47	74 512	82 544	17 456	91 968	13
48	74 531	82 571	17 429	91 959	12
49	74 549	82 599	17 401	91 951	11
50	74 568	82 626	17 374	91 942	10
51	74 587	82 653	17 347	91 934	9
52	74 606	82 681	17 319	91 925	8
53	74 625	82 708	17 292	91 917	7
54	74 644	82 735	17 265	91 908	6
55	74 662	82 762	17 238	91 900	5
56	74 681	82 790	17 210	91 891	4
57	74 700	82 817	17 183	91 883	3
58	74 719	82 844	17 156	91 874	2
59	74 737	82 871	17 129	91 866	1
60	74 756	82 899	17 101	91 857	0
	9	9	10	9	
′	log cos	log cot	log tan	log sin	′

56°

′	log sin	log tan	log cot	log cos	′
	9	9	10	9	
0	74 756	82 899	17 101	91 857	60
1	74 775	82 926	17 074	91 849	59
2	74 794	82 953	17 047	91 840	58
3	74 812	82 980	17 020	91 832	57
4	74 831	83 008	16 992	91 823	56
5	74 850	83 035	16 965	91 815	55
6	74 868	83 062	16 938	91 806	54
7	74 887	83 089	16 911	91 798	53
8	74 906	83 117	16 883	91 789	52
9	74 924	83 144	16 856	91 781	51
10	74 943	83 171	16 829	91 772	50
11	74 961	83 198	16 802	91 763	49
12	74 980	83 225	16 775	91 755	48
13	74 999	83 252	16 748	91 746	47
14	75 017	83 280	16 720	91 738	46
15	75 036	83 307	16 693	91 729	45
16	75 054	83 334	16 666	91 720	44
17	75 073	83 361	16 639	91 712	43
18	75 091	83 388	16 612	91 703	42
19	75 110	83 415	16 585	91 695	41
20	75 128	83 442	16 558	91 686	40
21	75 147	83 470	16 530	91 677	39
22	75 165	83 497	16 503	91 669	38
23	75 184	83 524	16 476	91 660	37
24	75 202	83 551	16 449	91 651	36
25	75 221	83 578	16 422	91 643	35
26	75 239	83 605	16 395	91 634	34
27	75 258	83 632	16 368	91 625	33
28	75 276	83 659	16 341	91 617	32
29	75 294	83 686	16 314	91 608	31
30	75 313	83 713	16 287	91 599	30
31	75 331	83 740	16 260	91 591	29
32	75 350	83 768	16 232	91 582	28
33	75 368	83 795	16 205	91 573	27
34	75 386	83 822	16 178	91 565	26
35	75 405	83 849	16 151	91 556	25
36	75 423	83 876	16 124	91 547	24
37	75 441	83 903	16 097	91 538	23
38	75 459	83 930	16 070	91 530	22
39	75 478	83 957	16 043	91 521	21
40	75 496	83 984	16 016	91 512	20
41	75 514	84 011	15 989	91 504	19
42	75 533	84 038	15 962	91 495	18
43	75 551	84 065	15 935	91 486	17
44	75 569	84 092	15 908	91 477	16
45	75 587	84 119	15 881	91 469	15
46	75 605	84 146	15 854	91 460	14
47	75 624	84 173	15 827	91 451	13
48	75 642	84 200	15 800	91 442	12
49	75 660	84 227	15 773	91 433	11
50	75 678	84 254	15 746	91 425	10
51	75 696	84 280	15 720	91 416	9
52	75 714	84 307	15 693	91 407	8
53	75 733	84 334	15 666	91 398	7
54	75 751	84 361	15 639	91 389	6
55	75 769	84 388	15 612	91 381	5
56	75 787	84 415	15 585	91 372	4
57	75 805	84 442	15 558	91 363	3
58	75 823	84 469	15 531	91 354	2
59	75 841	84 496	15 504	91 345	1
60	75 859	84 523	15 477	91 336	0
	9	9	10	9	
′	log cos	log cot	log tan	log sin	′

55°

′	log sin	log tan	log cot	log cos	′
	9	9	10	9	
0	75 859	84 523	15 477	91 336	60
1	75 877	84 550	15 450	91 328	59
2	75 895	84 576	15 424	91 319	58
3	75 913	84 603	15 397	91 310	57
4	75 931	84 630	15 370	91 301	56
5	75 949	84 657	15 343	91 292	55
6	75 967	84 684	15 316	91 283	54
7	75 985	84 711	15 289	91 274	53
8	76 003	84 738	15 262	91 266	52
9	76 021	84 764	15 236	91 257	51
10	76 039	84 791	15 209	91 248	50
11	76 057	84 818	15 182	91 239	49
12	76 075	84 845	15 155	91 230	48
13	76 093	84 872	15 128	91 221	47
14	76 111	84 899	15 101	91 212	46
15	76 129	84 925	15 075	91 203	45
16	76 146	84 952	15 048	91 194	44
17	76 164	84 979	15 021	91 185	43
18	76 182	85 006	14 994	91 176	42
19	76 200	85 033	14 967	91 167	41
20	76 218	85 059	14 941	91 158	40
21	76 236	85 086	14 914	91 149	39
22	76 253	85 113	14 887	91 141	38
23	76 271	85 140	14 860	91 132	37
24	76 289	85 166	14 834	91 123	36
25	76 307	85 193	14 807	91 114	35
26	76 324	85 220	14 780	91 105	34
27	76 342	85 247	14 753	91 096	33
28	76 360	85 273	14 727	91 087	32
29	76 378	85 300	14 700	91 078	31
30	76 395	85 327	14 673	91 069	30
31	76 413	85 354	14 646	91 060	29
32	76 431	85 380	14 620	91 051	28
33	76 448	85 407	14 593	91 042	27
34	76 466	85 434	14 566	91 033	26
35	76 484	85 460	14 540	91 023	25
36	76 501	85 487	14 513	91 014	24
37	76 519	85 514	14 486	91 005	23
38	76 537	85 540	14 460	90 996	22
39	76 554	85 567	14 433	90 987	21
40	76 572	85 594	14 406	90 978	20
41	76 590	85 620	14 380	90 969	19
42	76 607	85 647	14 353	90 960	18
43	76 625	85 674	14 326	90 951	17
44	76 642	85 700	14 300	90 942	16
45	76 660	85 727	14 273	90 933	15
46	76 677	85 754	14 246	90 924	14
47	76 695	85 780	14 220	90 915	13
48	76 712	85 807	14 193	90 906	12
49	76 730	85 834	14 166	90 896	11
50	76 747	85 860	14 140	90 887	10
51	76 765	85 887	14 113	90 878	9
52	76 782	85 913	14 087	90 869	8
53	76 800	85 940	14 060	90 860	7
54	76 817	85 967	14 033	90 851	6
55	76 835	85 993	14 007	90 842	5
56	76 852	86 020	13 980	90 832	4
57	76 870	86 046	13 954	90 823	3
58	76 887	86 073	13 927	90 814	2
59	76 904	86 100	13 900	90 805	1
60	76 922	86 126	13 874	90 796	0
	9	9	10	9	
′	log cos	log cot	log tan	log sin	′

54°

′	log sin	log tan	log cot	log cos	′
	9	9	10	9	
0	76 922	86 126	13 874	90 796	60
1	76 939	86 153	13 847	90 787	59
2	76 957	86 179	13 821	90 777	58
3	76 974	86 206	13 794	90 768	57
4	76 991	86 232	13 768	90 759	56
5	77 009	86 259	13 741	90 750	55
6	77 026	86 285	13 715	90 741	54
7	77 043	86 312	13 688	90 731	53
8	77 061	86 338	13 662	90 722	52
9	77 078	86 365	13 635	90 713	51
10	77 095	86 392	13 608	90 704	50
11	77 112	86 418	13 582	90 694	49
12	77 130	86 445	13 555	90 685	48
13	77 147	86 471	13 529	90 676	47
14	77 164	86 498	13 502	90 667	46
15	77 181	86 524	13 476	90 657	45
16	77 199	86 551	13 449	90 648	44
17	77 216	86 577	13 423	90 639	43
18	77 233	86 603	13 397	90 630	42
19	77 250	86 630	13 370	90 620	41
20	77 268	86 656	13 344	90 611	40
21	77 285	86 683	13 317	90 602	39
22	77 302	86 709	13 291	90 592	38
23	77 319	86 736	13 264	90 583	37
24	77 336	86 762	13 238	90 574	36
25	77 353	86 789	13 211	90 565	35
26	77 370	86 815	13 185	90 555	34
27	77 387	86 842	13 158	90 546	33
28	77 405	86 868	13 132	90 537	32
29	77 422	86 894	13 106	90 527	31
30	77 439	86 921	13 079	90 518	30
31	77 456	86 947	13 053	90 509	29
32	77 473	86 974	13 026	90 499	28
33	77 490	87 000	13 000	90 490	27
34	77 507	87 027	12 973	90 480	26
35	77 524	87 053	12 947	90 471	25
36	77 541	87 079	12 921	90 462	24
37	77 558	87 106	12 894	90 452	23
38	77 575	87 132	12 868	90 443	22
39	77 592	87 158	12 842	90 434	21
40	77 609	87 185	12 815	90 424	20
41	77 626	87 211	12 789	90 415	19
42	77 643	87 238	12 762	90 405	18
43	77 660	87 264	12 736	90 396	17
44	77 677	87 290	12 710	90 386	16
45	77 694	87 317	12 683	90 377	15
46	77 711	87 343	12 657	90 368	14
47	77 728	87 369	12 631	90 358	13
48	77 744	87 396	12 604	90 349	12
49	77 761	87 422	12 578	90 339	11
50	77 778	87 448	12 552	90 330	10
51	77 795	87 475	12 525	90 320	9
52	77 812	87 501	12 499	90 311	8
53	77 829	87 527	12 473	90 301	7
54	77 846	87 554	12 446	90 292	6
55	77 862	87 580	12 420	90 282	5
56	77 879	87 606	12 394	90 273	4
57	77 896	87 633	12 367	90 263	3
58	77 913	87 659	12 341	90 254	2
59	77 930	87 685	12 315	90 244	1
60	77 946	87 711	12 289	90 235	0
	9	9	10	9	
′	log cos	log cot	log tan	log sin	′

53°

′	log sin 9	log tan 9	log cot 10	log cos 9	′
0	77 946	87 711	12 289	90 235	60
1	77 963	87 738	12 262	90 225	59
2	77 980	87 764	12 236	90 216	58
3	77 997	87 790	12 210	90 206	57
4	78 013	87 817	12 183	90 197	56
5	78 030	87 843	12 157	90 187	55
6	78 047	87 869	12 131	90 178	54
7	78 063	87 895	12 105	90 168	53
8	78 080	87 922	12 078	90 159	52
9	78 097	87 948	12 052	90 149	51
10	78 113	87 974	12 026	90 139	50
11	78 130	88 000	12 000	90 130	49
12	78 147	88 027	11 973	90 120	48
13	78 163	88 053	11 947	90 111	47
14	78 180	88 079	11 921	90 101	46
15	78 197	88 105	11 895	90 091	45
16	78 213	88 131	11 869	90 082	44
17	78 230	88 158	11 842	90 072	43
18	78 246	88 184	11 816	90 063	42
19	78 263	88 210	11 790	90 053	41
20	78 280	88 236	11 764	90 043	40
21	78 296	88 262	11 738	90 034	39
22	78 313	88 289	11 711	90 024	38
23	78 329	88 315	11 685	90 014	37
24	78 346	88 341	11 659	90 005	36
25	78 362	88 367	11 633	89 995	35
26	78 379	88 393	11 607	89 985	34
27	78 395	88 420	11 580	89 976	33
28	78 412	88 446	11 554	89 966	32
29	78 428	88 472	11 528	89 956	31
30	78 445	88 498	11 502	89 947	30
31	78 461	88 524	11 476	89 937	29
32	78 478	88 550	11 450	89 927	28
33	78 494	88 577	11 423	89 918	27
34	78 510	88 603	11 397	89 908	26
35	78 527	88 629	11 371	89 898	25
36	78 543	88 655	11 345	89 888	24
37	78 560	88 681	11 319	89 879	23
38	78 576	88 707	11 293	89 869	22
39	78 592	88 733	11 267	89 859	21
40	78 609	88 759	11 241	89 849	20
41	78 625	88 786	11 214	89 840	19
42	78 642	88 812	11 188	89 830	18
43	78 658	88 838	11 162	89 820	17
44	78 674	88 864	11 136	89 810	16
45	78 691	88 890	11 110	89 801	15
46	78 707	88 916	11 084	89 791	14
47	78 723	88 942	11 058	89 781	13
48	78 739	88 968	11 032	89 771	12
49	78 756	88 994	11 006	89 761	11
50	78 772	89 020	10 980	89 752	10
51	78 788	89 046	10 954	89 742	9
52	78 805	89 073	10 927	89 732	8
53	78 821	89 099	10 901	89 722	7
54	78 837	89 125	10 875	89 712	6
55	78 853	89 151	10 849	89 702	5
56	78 869	89 177	10 823	89 693	4
57	78 886	89 203	10 797	89 683	3
58	78 902	89 229	10 771	89 673	2
59	78 918	89 255	10 745	89 663	1
60	78 934	89 281	10 719	89 653	0
′	log cos 9	log cot 9	log tan 10	log sin 9	′

′	log sin 9	log tan 9	log cot 10	log cos 9	′
0	78 934	89 281	10 719	89 653	60
1	78 950	89 307	10 693	89 643	59
2	78 967	89 333	10 667	89 633	58
3	78 983	89 359	10 641	89 624	57
4	78 999	89 385	10 615	89 614	56
5	79 015	89 411	10 589	89 604	55
6	79 031	89 437	10 563	89 594	54
7	79 047	89 463	10 537	89 584	53
8	79 063	89 489	10 511	89 574	52
9	79 079	89 515	10 485	89 564	51
10	79 095	89 541	10 459	89 554	50
11	79 111	89 567	10 433	89 544	49
12	79 128	89 593	10 407	89 534	48
13	79 144	89 619	10 381	89 524	47
14	79 160	89 645	10 355	89 514	46
15	79 176	89 671	10 329	89 504	45
16	79 192	89 697	10 303	89 495	44
17	79 208	89 723	10 277	89 485	43
18	79 224	89 749	10 251	89 475	42
19	79 240	89 775	10 225	89 465	41
20	79 256	89 801	10 199	89 455	40
21	79 272	89 827	10 173	89 445	39
22	79 288	89 853	10 147	89 435	38
23	79 304	89 879	10 121	89 425	37
24	79 319	89 905	10 095	89 415	36
25	79 335	89 931	10 069	89 405	35
26	79 351	89 957	10 043	89 395	34
27	79 367	89 983	10 017	89 385	33
28	79 383	90 009	09 991	89 375	32
29	79 399	90 035	09 965	89 364	31
30	79 415	90 061	09 939	89 354	30
31	79 431	90 086	09 914	89 344	29
32	79 447	90 112	09 888	89 334	28
33	79 463	90 138	09 862	89 324	27
34	79 478	90 164	09 836	89 314	26
35	79 494	90 190	09 810	89 304	25
36	79 510	90 216	09 784	89 294	24
37	79 526	90 242	09 758	89 284	23
38	79 542	90 268	09 732	89 274	22
39	79 558	90 294	09 706	89 264	21
40	79 573	90 320	09 680	89 254	20
41	79 589	90 346	09 654	89 244	19
42	79 605	90 371	09 629	89 233	18
43	79 621	90 397	09 603	89 223	17
44	79 636	90 423	09 577	89 213	16
45	79 652	90 449	09 551	89 203	15
46	79 668	90 475	09 525	89 193	14
47	79 684	90 501	09 499	89 183	13
48	79 699	90 527	09 473	89 173	12
49	79 715	90 553	09 447	89 162	11
50	79 731	90 578	09 422	89 152	10
51	79 746	90 604	09 396	89 142	9
52	79 762	90 630	09 370	89 132	8
53	79 778	90 656	09 344	89 122	7
54	79 793	90 682	09 318	89 112	6
55	79 809	90 708	09 292	89 101	5
56	79 825	90 734	09 266	89 091	4
57	79 840	90 759	09 241	89 081	3
58	79 856	90 785	09 215	89 071	2
59	79 872	90 811	09 189	89 060	1
60	79 887	90 837	09 163	89 050	0
′	log cos 9	log cot 9	log tan 10	log sin 9	′

′	log sin	log tan	log cot	log cos	′
	9	9	10	9	
0	79 887	90 837	09 163	89 050	60
1	79 903	90 863	09 137	89 040	59
2	79 918	90 889	09 111	89 030	58
3	79 934	90 914	09 086	89 020	57
4	79 950	90 940	09 060	89 009	56
5	79 965	90 966	09 034	88 999	55
6	79 981	90 992	09 008	88 989	54
7	79 996	91 018	08 982	88 978	53
8	80 012	91 043	08 957	88 968	52
9	80 027	91 069	08 931	88 958	51
10	80 043	91 095	08 905	88 948	50
11	80 058	91 121	08 879	88 937	49
12	80 074	91 147	08 853	88 927	48
13	80 089	91 172	08 828	88 917	47
14	80 105	91 198	08 802	88 906	46
15	80 120	91 224	08 776	88 896	45
16	80 136	91 250	08 750	88 886	44
17	80 151	91 276	08 724	88 875	43
18	80 166	91 301	08 699	88 865	42
19	80 182	91 327	08 673	88 855	41
20	80 197	91 353	08 647	88 844	40
21	80 213	91 379	08 621	88 834	39
22	80 228	91 404	08 596	88 824	38
23	80 244	91 430	08 570	88 813	37
24	80 259	91 456	08 544	88 803	36
25	80 274	91 482	08 518	88 793	35
26	80 290	91 507	08 493	88 782	34
27	80 305	91 533	08 467	88 772	33
28	80 320	91 559	08 441	88 761	32
29	80 336	91 585	08 415	88 751	31
30	80 351	91 610	08 390	88 741	30
31	80 366	91 636	08 364	88 730	29
32	80 382	91 662	08 338	88 720	28
33	80 397	91 688	08 312	88 709	27
34	80 412	91 713	08 287	88 699	26
35	80 428	91 739	08 261	88 688	25
36	80 443	91 765	08 235	88 678	24
37	80 458	91 791	08 209	88 668	23
38	80 473	91 816	08 184	88 657	22
39	80 489	91 842	08 158	88 647	21
40	80 504	91 868	08 132	88 636	20
41	80 519	91 893	08 107	88 626	19
42	80 534	91 919	08 081	88 615	18
43	80 550	91 945	08 055	88 605	17
44	80 565	91 971	08 029	88 594	16
45	80 580	91 996	08 004	88 584	15
46	80 595	92 022	07 978	88 573	14
47	80 610	92 048	07 952	88 563	13
48	80 625	92 073	07 927	88 552	12
49	80 641	92 099	07 901	88 542	11
50	80 656	92 125	07 875	88 531	10
51	80 671	92 150	07 850	88 521	9
52	80 686	92 176	07 824	88 510	8
53	80 701	92 202	07 798	88 499	7
54	80 716	92 227	07 773	88 489	6
55	80 731	92 253	07 747	88 478	5
56	80 746	92 279	07 721	88 468	4
57	80 762	92 304	07 696	88 457	3
58	80 777	92 330	07 670	88 447	2
59	80 792	92 356	07 644	88 436	1
60	80 807	92 381	07 619	88 425	0
	9	9	10	9	
′	log cos	log cot	log tan	log sin	′

50°

′	log sin	log tan	log cot	log cos	′
	9	9	10	9	
0	80 807	92 381	07 619	88 425	60
1	80 822	92 407	07 593	88 415	59
2	80 837	92 433	07 567	88 404	58
3	80 852	92 458	07 542	88 394	57
4	80 867	92 484	07 516	88 383	56
5	80 882	92 510	07 490	88 372	55
6	80 897	92 535	07 465	88 362	54
7	80 912	92 561	07 439	88 351	53
8	80 927	92 587	07 413	88 340	52
9	80 942	92 612	07 388	88 330	51
10	80 957	92 638	07 362	88 319	50
11	80 972	92 663	07 337	88 308	49
12	80 987	92 689	07 311	88 298	48
13	81 002	92 715	07 285	88 287	47
14	81 017	92 740	07 260	88 276	46
15	81 032	92 766	07 234	88 266	45
16	81 047	92 792	07 208	88 255	44
17	81 061	92 817	07 183	88 244	43
18	81 076	92 843	07 157	88 234	42
19	81 091	92 868	07 132	88 223	41
20	81 106	92 894	07 106	88 212	40
21	81 121	92 920	07 080	88 201	39
22	81 136	92 945	07 055	88 191	38
23	81 151	92 971	07 029	88 180	37
24	81 166	92 996	07 004	88 169	36
25	81 180	93 022	06 978	88 158	35
26	81 195	93 048	06 952	88 148	34
27	81 210	93 073	06 927	88 137	33
28	81 225	93 099	06 901	88 126	32
29	81 240	93 124	06 876	88 115	31
30	81 254	93 150	06 850	88 105	30
31	81 269	93 175	06 825	88 094	29
32	81 284	93 201	06 799	88 083	28
33	81 299	93 227	06 773	88 072	27
34	81 314	93 252	06 748	88 061	26
35	81 328	93 278	06 722	88 051	25
36	81 343	93 303	06 697	88 040	24
37	81 358	93 329	06 671	88 029	23
38	81 372	93 354	06 646	88 018	22
39	81 387	93 380	06 620	88 007	21
40	81 402	93 406	06 594	87 996	20
41	81 417	93 431	06 569	87 985	19
42	81 431	93 457	06 543	87 975	18
43	81 446	93 482	06 518	87 964	17
44	81 461	93 508	06 492	87 953	16
45	81 475	93 533	06 467	87 942	15
46	81 490	93 559	06 441	87 931	14
47	81 505	93 584	06 416	87 920	13
48	81 519	93 610	06 390	87 909	12
49	81 534	93 636	06 364	87 898	11
50	81 549	93 661	06 339	87 887	10
51	81 563	93 687	06 313	87 877	9
52	81 578	93 712	06 288	87 866	8
53	81 592	93 738	06 262	87 855	7
54	81 607	93 763	06 237	87 844	6
55	81 622	93 789	06 211	87 833	5
56	81 636	93 814	06 186	87 822	4
57	81 651	93 840	06 160	87 811	3
58	81 665	93 865	06 135	87 800	2
59	81 680	93 891	06 109	87 789	1
60	81 694	93 916	06 084	87 778	0
	9	9	10	9	
′	log cos	log cot	log tan	log sin	′

49°

′	log sin 9	log tan 9	log cot 10	log cos 9	′
0	81 694	93 916	06 084	87 778	**60**
1	81 709	93 942	06 058	87 767	59
2	81 723	93 967	06 033	87 756	58
3	81 738	93 993	06 007	87 745	57
4	81 752	94 018	05 982	87 734	56
5	81 767	94 044	05 956	87 723	**55**
6	81 781	94 069	05 931	87 712	54
7	81 796	94 095	05 905	87 701	53
8	81 810	94 120	05 880	87 690	52
9	81 825	94 146	05 854	87 679	51
10	81 839	94 171	05 829	87 668	**50**
11	81 854	94 197	05 803	87 657	49
12	81 868	94 222	05 778	87 646	48
13	81 882	94 248	05 752	87 635	47
14	81 897	94 273	05 727	87 624	46
15	81 911	94 299	05 701	87 613	**45**
16	81 926	94 324	05 676	87 601	44
17	81 940	94 350	05 650	87 590	43
18	81 955	94 375	05 625	87 579	42
19	81 969	94 401	05 599	87 568	41
20	81 983	94 426	05 574	87 557	**40**
21	81 998	94 452	05 548	87 546	39
22	82 012	94 477	05 523	87 535	38
23	82 026	94 503	05 497	87 524	37
24	82 041	94 528	05 472	87 513	36
25	82 055	94 554	05 446	87 501	**35**
26	82 069	94 579	05 421	87 490	34
27	82 084	94 604	05 396	87 479	33
28	82 098	94 630	05 370	87 468	32
29	82 112	94 655	05 345	87 457	31
30	82 126	94 681	05 319	87 446	**30**
31	82 141	94 706	05 294	87 434	29
32	82 155	94 732	05 268	87 423	28
33	82 169	94 757	05 243	87 412	27
34	82 184	94 783	05 217	87 401	26
35	82 198	94 808	05 192	87 390	**25**
36	82 212	94 834	05 166	87 378	24
37	82 226	94 859	05 141	87 367	23
38	82 240	94 884	05 116	87 356	22
39	82 255	94 910	05 090	87 345	21
40	82 269	94 935	05 065	87 334	**20**
41	82 283	94 961	05 039	87 322	19
42	82 297	94 986	05 014	87 311	18
43	82 311	95 012	04 988	87 300	17
44	82 326	95 037	04 963	87 288	16
45	82 340	95 062	04 938	87 277	**15**
46	82 354	95 088	04 912	87 266	14
47	82 368	95 113	04 887	87 255	13
48	82 382	95 139	04 861	87 243	12
49	82 396	95 164	04 836	87 232	11
50	82 410	95 190	04 810	87 221	**10**
51	82 424	95 215	04 785	87 209	9
52	82 439	95 240	04 760	87 198	8
53	82 453	95 266	04 734	87 187	7
54	82 467	95 291	04 709	87 175	6
55	82 481	95 317	04 683	87 164	**5**
56	82 495	95 342	04 658	87 153	4
57	82 509	95 368	04 632	87 141	3
58	82 523	95 393	04 607	87 130	2
59	82 537	95 418	04 582	87 119	1
60	82 551	95 444	04 556	87 107	**0**
′	log cos 9	log cot 9	log tan 10	log sin 9	′

48°

′	log sin 9	log tan 9	log cot 10	log cos 9	′
0	82 551	95 444	04 556	87 107	**60**
1	82 565	95 469	04 531	87 096	59
2	82 579	95 495	04 505	87 085	58
3	82 593	95 520	04 480	87 073	57
4	82 607	95 545	04 455	87 062	56
5	82 621	95 571	04 429	87 050	**55**
6	82 635	95 596	04 404	87 039	54
7	82 649	95 622	04 378	87 028	53
8	82 663	95 647	04 353	87 016	52
9	82 677	95 672	04 328	87 005	51
10	82 691	95 698	04 302	86 993	**50**
11	82 705	95 723	04 277	86 982	49
12	82 719	95 748	04 252	86 970	48
13	82 733	95 774	04 226	86 959	47
14	82 747	95 799	04 201	86 947	46
15	82 761	95 825	04 175	86 936	**45**
16	82 775	95 850	04 150	86 924	44
17	82 788	95 875	04 125	86 913	43
18	82 802	95 901	04 099	86 902	42
19	82 816	95 926	04 074	86 890	41
20	82 830	95 952	04 048	86 879	**40**
21	82 844	95 977	04 023	86 867	39
22	82 858	96 002	03 998	86 855	38
23	82 872	96 028	03 972	86 844	37
24	82 885	96 053	03 947	86 832	36
25	82 899	96 078	03 922	86 821	**35**
26	82 913	96 104	03 896	86 809	34
27	82 927	96 129	03 871	86 798	33
28	82 941	96 155	03 845	86 786	32
29	82 955	96 180	03 820	86 775	31
30	82 968	96 205	03 795	86 763	**30**
31	82 982	96 231	03 769	86 752	29
32	82 996	96 256	03 744	86 740	28
33	83 010	96 281	03 719	86 728	27
34	83 023	96 307	03 693	86 717	26
35	83 037	96 332	03 668	86 705	**25**
36	83 051	96 357	03 643	86 694	24
37	83 065	96 383	03 617	86 682	23
38	83 078	96 408	03 592	86 670	22
39	83 092	96 433	03 567	86 659	21
40	83 106	96 459	03 541	86 647	**20**
41	83 120	96 484	03 516	86 635	19
42	83 133	96 510	03 490	86 624	18
43	83 147	96 535	03 465	86 612	17
44	83 161	96 560	03 440	86 600	16
45	83 174	96 586	03 414	86 589	**15**
46	83 188	96 611	03 389	86 577	14
47	83 202	96 636	03 364	86 565	13
48	83 215	96 662	03 338	86 554	12
49	83 229	96 687	03 313	86 542	11
50	83 242	96 712	03 288	86 530	**10**
51	83 256	96 738	03 262	86 518	9
52	83 270	96 763	03 237	86 507	8
53	83 283	96 788	03 212	86 495	7
54	83 297	96 814	03 186	86 483	6
55	83 310	96 839	03 161	86 472	**5**
56	83 324	96 864	03 136	86 460	4
57	83 338	96 890	03 110	86 448	3
58	83 351	96 915	03 085	86 436	2
59	83 365	96 940	03 060	86 425	1
60	83 378	96 966	03 034	86 413	**0**
′	log cos 9	log cot 9	log tan 10	log sin 9	′

47°

'	log sin	log tan	log cot	log cos	'
	9	9	10	9	
0	83 378	96 966	03 034	86 413	60
1	83 392	96 991	03 009	86 401	59
2	83 405	97 016	02 984	86 389	58
3	83 419	97 042	02 958	86 377	57
4	83 432	97 067	02 933	86 366	56
5	83 446	97 092	02 908	86 354	55
6	83 459	97 118	02 882	86 342	54
7	83 473	97 143	02 857	86 330	53
8	83 486	97 168	02 832	86 318	52
9	83 500	97 193	02 807	86 306	51
10	83 513	97 219	02 781	86 295	50
11	83 527	97 244	02 756	86 283	49
12	83 540	97 269	02 731	86 271	48
13	83 554	97 295	02 705	86 259	47
14	83 567	97 320	02 680	86 247	46
15	83 581	97 345	02 655	86 235	45
16	83 594	97 371	02 629	86 223	44
17	83 608	97 396	02 604	86 211	43
18	83 621	97 421	02 579	86 200	42
19	83 634	97 447	02 553	86 188	41
20	83 648	97 472	02 528	86 176	40
21	83 661	97 497	02 503	86 164	39
22	83 674	97 523	02 477	86 152	38
23	83 688	97 548	02 452	86 140	37
24	83 701	97 573	02 427	86 128	36
25	83 715	97 598	02 402	86 116	35
26	83 728	97 624	02 376	86 104	34
27	83 741	97 649	02 351	86 092	33
28	83 755	97 674	02 326	86 080	32
29	83 768	97 700	02 300	86 068	31
30	83 781	97 725	02 275	86 056	30
31	83 795	97 750	02 250	86 044	29
32	83 808	97 776	02 224	86 032	28
33	83 821	97 801	02 199	86 020	27
34	83 834	97 826	02 174	86 008	26
35	83 848	97 851	02 149	85 996	25
36	83 861	97 877	02 123	85 984	24
37	83 874	97 902	02 098	85 972	23
38	83 887	97 927	02 073	85 960	22
39	83 901	97 953	02 047	85 948	21
40	83 914	97 978	02 022	85 936	20
41	83 927	98 003	01 997	85 924	19
42	83 940	98 029	01 971	85 912	18
43	83 954	98 054	01 946	85 900	17
44	83 967	98 079	01 921	85 888	16
45	83 980	98 104	01 896	85 876	15
46	83 993	98 130	01 870	85 864	14
47	84 006	98 155	01 845	85 851	13
48	84 020	98 180	01 820	85 839	12
49	84 033	98 206	01 794	85 827	11
50	84 046	98 231	01 769	85 815	10
51	84 059	98 256	01 744	85 803	9
52	84 072	98 281	01 719	85 791	8
53	84 085	98 307	01 693	85 779	7
54	84 098	98 332	01 668	85 766	6
55	84 112	98 357	01 643	85 754	5
56	84 125	98 383	01 617	85 742	4
57	84 138	98 408	01 592	85 730	3
58	84 151	98 433	01 567	85 718	2
59	84 164	98 458	01 542	85 706	1
60	84 177	98 484	01 516	85 693	0
'	log cos	log cot	log tan	log sin	'
	9	10	10	9	

'	log sin	log tan	log cot	log cos	'
	9	9	10	9	
0	84 177	98 484	01 516	85 693	60
1	84 190	98 509	01 491	85 681	59
2	84 203	98 534	01 466	85 669	58
3	84 216	98 560	01 440	85 657	57
4	84 229	98 585	01 415	85 645	56
5	84 242	98 610	01 390	85 632	55
6	84 255	98 635	01 365	85 620	54
7	84 269	98 661	01 339	85 608	53
8	84 282	98 686	01 314	85 596	52
9	84 295	98 711	01 289	85 583	51
10	84 308	98 737	01 263	85 571	50
11	84 321	98 762	01 238	85 559	49
12	84 334	98 787	01 213	85 547	48
13	84 347	98 812	01 188	85 534	47
14	84 360	98 838	01 162	85 522	46
15	84 373	98 863	01 137	85 510	45
16	84 385	98 888	01 112	85 497	44
17	84 398	98 913	01 087	85 485	43
18	84 411	98 939	01 061	85 473	42
19	84 424	98 964	01 036	85 460	41
20	84 437	98 989	01 011	85 448	40
21	84 450	99 015	00 985	85 436	39
22	84 463	99 040	00 960	85 423	38
23	84 476	99 065	00 935	85 411	37
24	84 489	99 090	00 910	85 399	36
25	84 502	99 116	00 884	85 386	35
26	84 515	99 141	00 859	85 374	34
27	84 528	99 166	00 834	85 361	33
28	84 540	99 191	00 809	85 349	32
29	84 553	99 217	00 783	85 337	31
30	84 566	99 242	00 758	85 324	30
31	84 579	99 267	00 733	85 312	29
32	84 592	99 293	00 707	85 299	28
33	84 605	99 318	00 682	85 287	27
34	84 618	99 343	00 657	85 274	26
35	84 630	99 368	00 632	85 262	25
36	84 643	99 394	00 606	85 250	24
37	84 656	99 419	00 581	85 237	23
38	84 669	99 444	00 556	85 225	22
39	84 682	99 469	00 531	85 212	21
40	84 694	99 495	00 505	85 200	20
41	84 707	99 520	00 480	85 187	19
42	84 720	99 545	00 455	85 175	18
43	84 733	99 570	00 430	85 162	17
44	84 745	99 596	00 404	85 150	16
45	84 758	99 621	00 379	85 137	15
46	84 771	99 646	00 354	85 125	14
47	84 784	99 672	00 328	85 112	13
48	84 796	99 697	00 303	85 100	12
49	84 809	99 722	00 278	85 087	11
50	84 822	99 747	00 253	85 074	10
51	84 835	99 773	00 227	85 062	9
52	84 847	99 798	00 202	85 049	8
53	84 860	99 823	00 177	85 037	7
54	84 873	99 848	00 152	85 024	6
55	84 885	99 874	00 126	85 012	5
56	84 898	99 899	00 101	84 999	4
57	84 911	99 924	00 076	84 986	3
58	84 923	99 949	00 051	84 974	2
59	84 936	99 975	00 025	84 961	1
60	84 949	00 000	00 000	84 949	0
'	log cos	log cot	log tan	log sin	'
	9	10	10	9	

TABLE IV.

—◆—

For Determining with Greater Accuracy than can be done by means of Table III.:

1. *log sin*, *log tan*, and *log cot*, when the angle is between 0° and 2°;
2. *log cos*, *log tan*, and *log cot*, when the angle is between 88° and 90°;
3. The value of the angle when the logarithm of the function does *not* lie between the limits **8.54 684** and **11.45 316**.

—◆—

FORMULAS FOR THE USE OF THE NUMBERS S AND T.

I. When the angle α is between 0° and 2°:

$\log \sin \alpha = \log \alpha'' + S.$	$\log \alpha'' = \log \sin \alpha - S,$
$\log \tan \alpha = \log \alpha'' + T.$	$= \log \tan \alpha - T,$
$\log \cot \alpha = \operatorname{colog} \tan \alpha.$	$= \operatorname{colog} \cot \alpha - T.$

II. When the angle α is between 88° and 90°:

$\log \cos \alpha = \log (90° - \alpha)'' + S.$	$\log (90° - \alpha)'' = \log \cos \alpha - S,$
$\log \cot \alpha = \log (90° - \alpha)'' + T.$	$= \log \cot \alpha - T,$
$\log \tan \alpha = \operatorname{colog} \cot \alpha.$	$= \operatorname{colog} \tan \alpha - T,$
	$\text{and } \alpha = 90° - (90° - \alpha).$

—∘◦⊗◦∘—

VALUES OF S AND T.

α''	S	log sin α	α''	T	log tan α	α	T	log tan α
0		—	0		—	5 146		8.39 713
	4.68 557			4.68 557			4.68 567	
2 409		8.06 740	200		6.98 660	5 424		8.41 999
	4.68 556			4.68 558			4.68 568	
3 417		8.21 920	1 726		7.92 263	5 689		8.44 072
	4.68 555			4.68 559			4.68 569	
3 823		8.26 795	2 432		8.07 156	5 941		8.45 955
	4.68 555			4.68 560			4.68 570	
4 190		8.30 776	2 976		8.15 924	6 184		8.47 697
	4.68 554			4.68 561			4.68 571	
4 840		8.37 038	3 434		8.22 142	6 417		8.49 305
	4.68 553			4.68 562			4.68 572	
5 414		8.41 904	3 838		8.26 973	6 642		8.50 802
	4.68 552			4.68 563			4.68 573	
5 932		8.45 872	4 204		8.30 930	6 859		8.52 200
	4.68 551			4.68 564			4.68 574	
6 408		8.49 223	4 540		8.34 270	7 070		8.53 516
	4.68 550			4.68 565			4.68 575	
6 633		8.50 721	4 699		8.35 766	7 173		8.54 145
	4.68 550			4.68 565			4.68 575	
6 851		8.52 125	4 853		8.37 167	7 274		8.54 753
	4.68 549			4.68 566				
7 267		8.54 684	5 146		8.39 713			
α''	S	log sin α	α''	T	log tan α	α	T	log tan α

TABLE V.—CIRCUMFERENCES AND AREAS OF CIRCLES. 51

If N = the radius of the circle, the circumference = $2\pi N$.									
If N = the radius of the circle, the area $\quad\quad = \pi N^2$.									
If N = the circumference of the circle, the radius $= \dfrac{1}{2\pi} N$.									
If N = the circumference of the circle, the area $\quad = \dfrac{1}{4\pi} N^2$.									

N	$2\pi N$	πN^2	$\frac{1}{2\pi}N$	$\frac{1}{4\pi}N^2$	N	$2\pi N$	πN^2	$\frac{1}{2\pi}N$	$\frac{1}{4\pi}N^2$
0	0.00	0.0	0.000	0.00	50	314.16	7 854	7.96	198.94
1	6.28	3.1	0.159	0.08	51	320.44	8 171	8.12	206.98
2	12.57	12.6	0.318	0.32	52	326.73	8 495	8.28	215.18
3	18.85	28.3	0.477	0.72	53	333.01	8 825	8.44	223.53
4	25.13	50.3	0.637	1.27	54	339.29	9 161	8.59	232.05
5	31.42	78.5	0.796	1.99	55	345.58	9 503	8.75	240.72
6	37.70	113.1	0.955	2.86	56	351.86	9 852	8.91	249.55
7	43.98	153.9	1.114	3.90	57	358.14	10 207	9.07	258.55
8	50.27	201.1	1.273	5.09	58	364.42	10 568	9.23	267.70
9	56.55	254.5	1.432	6.45	59	370.71	10 936	9.39	277.01
10	62.83	314.2	1.592	7.96	60	376.99	11 310	9.55	286.48
11	69.12	380.1	1.751	9.63	61	383.27	11 690	9.71	296.11
12	75.40	452.4	1.910	11.46	62	389.56	12 076	9.87	305.90
13	81.68	530.9	2.069	13.45	63	395.84	12 469	10.03	315.84
14	87.96	615.8	2.228	15.60	64	402.12	12 868	10.19	325.95
15	94.25	706.9	2.387	17.90	65	408.41	13 273	10.35	336.21
16	100.53	804.2	2.546	20.37	66	414.69	13 685	10.50	346.64
17	106.81	907.9	2.706	23.00	67	420.97	14 103	10.66	357.22
18	113.10	1 017.9	2.865	25.78	68	427.26	14 527	10.82	367.97
19	119.38	1 134.1	3.024	28.73	69	433.54	14 957	10.98	378.87
20	125.66	1 256.6	3.183	31.83	70	439.82	15 394	11.14	389.93
21	131.95	1 385.4	3.342	35.09	71	446.11	15 837	11.30	401.15
22	138.23	1 520.5	3.501	38.52	72	452.39	16 286	11.46	412.53
23	144.51	1 661.9	3.661	42.10	73	458.67	16 742	11.62	424.07
24	150.80	1 809.6	3.820	45.84	74	464.96	17 203	11.78	435.77
25	157.08	1 963.5	3.979	49.74	75	471.24	17 671	11.94	447.62
26	163.36	2 123.7	4.138	53.79	76	477.52	18 146	12.10	459.64
27	169.65	2 290.2	4.297	58.01	77	483.81	18 627	12.25	471.81
28	175.93	2 463.0	4.456	62.39	78	490.09	19 113	12.41	484.15
29	182.21	2 642.1	4.615	66.92	79	496.37	19 607	12.57	496.64
30	188.50	2 827.4	4.775	71.62	80	502.65	20 106	12.73	509.30
31	194.78	3 019.1	4.934	76.47	81	508.94	20 612	12.89	522.11
32	201.06	3 217.0	5.093	81.49	82	515.22	21 124	13.05	535.08
33	207.35	3 421.2	5.252	86.66	83	521.50	21 642	13.21	548.21
34	213.63	3 631.7	5.411	91.99	84	527.79	22 167	13.37	561.50
35	219.91	3 848.5	5.570	97.48	85	534.07	22 698	13.53	574.95
36	226.19	4 071.5	5.730	103.13	86	540.35	23 235	13.69	588.55
37	232.48	4 300.8	5.889	108.94	87	546.64	23 779	13.85	602.32
38	238.76	4 536.5	6.048	114.91	88	552.92	24 328	14.01	616.25
39	245.04	4 778.4	6.207	121.04	89	559.20	24 885	14.16	630.33
40	251.33	5 026.5	6.366	127.32	90	565.49	25 447	14.32	644.58
41	257.61	5 281.0	6.525	133.77	91	571.77	26 016	14.48	658.98
42	263.89	5 541.8	6.685	140.37	92	578.05	26 590	14.64	673.54
43	270.18	5 808.8	6.844	147.14	93	584.34	27 172	14.80	688.27
44	276.46	6 082.1	7.003	154.06	94	590.62	27 759	14.96	703.15
45	282.74	6 361.7	7.162	161.14	95	596.90	28 353	15.12	718.19
46	289.03	6 647.6	7.321	168.39	96	603.19	28 953	15.28	733.39
47	295.31	6 939.8	7.480	175.79	97	609.47	29 559	15.44	748.74
48	301.59	7 238.2	7.639	183.35	98	615.75	30 172	15.60	764.26
49	307.88	7 543.0	7.799	191.07	99	622.04	30 791	15.76	779.94
50	314.16	7 854.0	7.958	198.94	100	628.32	31 416	15.92	795.77
N	$2\pi N$	πN^2	$\frac{1}{2\pi}N$	$\frac{1}{4\pi}N^2$	N	$2\pi N$	πN^2	$\frac{1}{2\pi}N$	$\frac{1}{4\pi}N^2$

′	0°		1°		2°		3°		4°		′
	sin	cos	sin	cos	sin	cos	sin	cos	sin	cos	
0	0000	1.000	0175	9998	0349	9994	0523	9986	0698	9976	60
1	0003	1.000	0177	9998	0352	9994	0526	9986	0700	9975	59
2	0006	1.000	0180	9998	0355	9994	0529	9986	0703	9975	58
3	0009	1.000	0183	9998	0358	9994	0532	9986	0706	9975	57
4	0012	1.000	0186	9998	0361	9993	0535	9986	0709	9975	56
5	0015	1.000	0189	9998	0364	9993	0538	9986	0712	9975	55
6	0017	1.000	0192	9998	0366	9993	0541	9985	0715	9974	54
7	0020	1.000	0195	9998	0369	9993	0544	9985	0718	9974	53
8	0023	1.000	0198	9998	0372	9993	0547	9985	0721	9974	52
9	0026	1.000	0201	9998	0375	9993	0550	9985	0724	9974	51
10	0029	1.000	0204	9998	0378	9993	0552	9985	0727	9974	50
11	0032	1.000	0207	9998	0381	9993	0555	9985	0729	9973	49
12	0035	1.000	0209	9998	0384	9993	0558	9984	0732	9973	48
13	0038	1.000	0212	9998	0387	9993	0561	9984	0735	9973	47
14	0041	1.000	0215	9998	0390	9992	0564	9984	0738	9973	46
15	0044	1.000	0218	9998	0393	9992	0567	9984	0741	9973	45
16	0047	1.000	0221	9998	0396	9992	0570	9984	0744	9972	44
17	0049	1.000	0224	9997	0398	9992	0573	9984	0747	9972	43
18	0052	1.000	0227	9997	0401	9992	0576	9983	0750	9972	42
19	0055	1.000	0230	9997	0404	9992	0579	9983	0753	9972	41
20	0058	1.000	0233	9997	0407	9992	0581	9983	0756	9971	40
21	0061	1.000	0236	9997	0410	9992	0584	9983	0758	9971	39
22	0064	1.000	0239	9997	0413	9991	0587	9983	0761	9971	38
23	0067	1.000	0241	9997	0416	9991	0590	9983	0764	9971	37
24	0070	1.000	0244	9997	0419	9991	0593	9982	0767	9971	36
25	0073	1.000	0247	9997	0422	9991	0596	9982	0770	9970	35
26	0076	1.000	0250	9997	0425	9991	0599	9982	0773	9970	34
27	0079	1.000	0253	9997	0427	9991	0602	9982	0776	9970	33
28	0081	1.000	0256	9997	0430	9991	0605	9982	0779	9970	32
29	0084	1.000	0259	9997	0433	9991	0608	9982	0782	9969	31
30	0087	1.000	0262	9997	0436	9990	0610	9981	0785	9969	30
31	0090	1.000	0265	9996	0439	9990	0613	9981	0787	9969	29
32	0093	1.000	0268	9996	0442	9990	0616	9981	0790	9969	28
33	0096	1.000	0270	9996	0445	9990	0619	9981	0793	9968	27
34	0099	1.000	0273	9996	0448	9990	0622	9981	0796	9968	26
35	0102	9999	0276	9996	0451	9990	0625	9980	0799	9968	25
36	0105	9999	0279	9996	0454	9990	0628	9980	0802	9968	24
37	0108	9999	0282	9996	0457	9990	0631	9980	0805	9968	23
38	0111	9999	0285	9996	0459	9989	0634	9980	0808	9967	22
39	0113	9999	0288	9996	0462	9989	0637	9980	0811	9967	21
40	0116	9999	0291	9996	0465	9989	0640	9980	0814	9967	20
41	0119	9999	0294	9996	0468	9989	0642	9979	0816	9967	19
42	0122	9999	0297	9996	0471	9989	0645	9979	0819	9966	18
43	0125	9999	0300	9996	0474	9989	0648	9979	0822	9966	17
44	0128	9999	0302	9995	0477	9989	0651	9979	0825	9966	16
45	0131	9999	0305	9995	0480	9988	0654	9979	0828	9966	15
46	0134	9999	0308	9995	0483	9988	0657	9978	0831	9965	14
47	0137	9999	0311	9995	0486	9988	0660	9978	0834	9965	13
48	0140	9999	0314	9995	0488	9988	0663	9978	0837	9965	12
49	0143	9999	0317	9995	0491	9988	0666	9978	0840	9965	11
50	0145	9999	0320	9995	0494	9988	0669	9978	0843	9964	10
51	0148	9999	0323	9995	0497	9988	0671	9977	0845	9964	9
52	0151	9999	0326	9995	0500	9987	0674	9977	0848	9964	8
53	0154	9999	0329	9995	0503	9987	0677	9977	0851	9964	7
54	0157	9999	0332	9995	0506	9987	0680	9977	0854	9963	6
55	0160	9999	0334	9994	0509	9987	0683	9977	0857	9963	5
56	0163	9999	0337	9994	0512	9987	0686	9976	0860	9963	4
57	0166	9999	0340	9994	0515	9987	0689	9976	0863	9963	3
58	0169	9999	0343	9994	0518	9987	0692	9976	0866	9962	2
59	0172	9999	0346	9994	0520	9986	0695	9976	0869	9962	1
60	0175	9999	0349	9994	0523	9986	0698	9976	0872	9962	0
	cos	sin	cos	sin	cos	sin	cos	sin	cos	sin	
′	89°		88°		87°		86°		85°		′

′	5° sin	cos	6° sin	cos	7° sin	cos	8° sin	cos	9° sin	cos	′
0	0872	9962	1045	9945	1219	9925	1392	9903	1564	9877	60
1	0874	9962	1048	9945	1222	9925	1395	9902	1567	9876	59
2	0877	9961	1051	9945	1224	9925	1397	9902	1570	9876	58
3	0880	9961	1054	9944	1227	9924	1400	9901	1573	9876	57
4	0883	9961	1057	9944	1230	9924	1403	9901	1576	9875	56
5	0886	9961	1060	9944	1233	9924	1406	9901	1579	9875	55
6	0889	9960	1063	9943	1236	9923	1409	9900	1582	9874	54
7	0892	9960	1066	9943	1239	9923	1412	9900	1584	9874	53
8	0895	9960	1068	9943	1241	9923	1415	9899	1587	9873	52
9	0898	9960	1071	9942	1245	9922	1418	9899	1590	9873	51
10	0901	9959	1074	9942	1248	9922	1421	9899	1593	9872	50
11	0903	9959	1077	9942	1250	9922	1423	9898	1596	9872	49
12	0906	9959	1080	9942	1253	9921	1426	9898	1599	9871	48
13	0909	9959	1083	9941	1256	9921	1429	9897	1602	9871	47
14	0912	9958	1086	9941	1259	9920	1432	9897	1605	9870	46
15	0915	9958	1089	9941	1262	9920	1435	9897	1607	9870	45
16	0918	9958	1092	9940	1265	9920	1438	9896	1610	9869	44
17	0921	9958	1094	9940	1268	9919	1441	9896	1613	9869	43
18	0924	9957	1097	9940	1271	9919	1444	9895	1616	9869	42
19	0927	9957	1100	9939	1274	9919	1446	9895	1619	9868	41
20	0929	9957	1103	9939	1276	9918	1449	9894	1622	9868	40
21	0932	9956	1106	9939	1279	9918	1452	9894	1625	9867	39
22	0935	9956	1109	9938	1282	9917	1455	9894	1628	9867	38
23	0938	9956	1112	9938	1285	9917	1458	9893	1630	9866	37
24	0941	9956	1115	9938	1288	9917	1461	9893	1633	9866	36
25	0944	9955	1118	9937	1291	9916	1464	9892	1636	9865	35
26	0947	9955	1120	9937	1294	9916	1467	9892	1639	9865	34
27	0950	9955	1123	9937	1297	9916	1469	9891	1642	9864	33
28	0953	9955	1126	9936	1299	9915	1472	9891	1645	9864	32
29	0956	9954	1129	9936	1302	9915	1475	9891	1648	9863	31
30	0958	9954	1132	9936	1305	9914	1478	9890	1650	9863	30
31	0961	9954	1135	9935	1308	9914	1481	9890	1653	9862	29
32	0964	9953	1138	9935	1311	9914	1484	9889	1656	9862	28
33	0967	9953	1141	9935	1314	9913	1487	9889	1659	9861	27
34	0970	9953	1144	9934	1317	9913	1490	9888	1662	9861	26
35	0973	9953	1146	9934	1320	9913	1492	9888	1665	9860	25
36	0976	9952	1149	9934	1323	9912	1495	9888	1668	9860	24
37	0979	9952	1152	9933	1325	9912	1498	9887	1671	9859	23
38	0982	9952	1155	9933	1328	9911	1501	9887	1673	9859	22
39	0985	9951	1158	9933	1331	9911	1504	9886	1676	9859	21
40	0987	9951	1161	9932	1334	9911	1507	9886	1679	9858	20
41	0990	9951	1164	9932	1337	9910	1510	9885	1682	9858	19
42	0993	9951	1167	9932	1340	9910	1513	9885	1685	9857	18
43	0996	9950	1170	9931	1343	9909	1515	9884	1688	9857	17
44	0999	9950	1172	9931	1346	9909	1518	9884	1691	9856	16
45	1002	9950	1175	9931	1349	9909	1521	9884	1693	9856	15
46	1005	9949	1178	9930	1351	9908	1524	9883	1696	9855	14
47	1008	9949	1181	9930	1354	9908	1527	9883	1699	9855	13
48	1011	9949	1184	9930	1357	9907	1530	9882	1702	9854	12
49	1013	9949	1187	9929	1360	9907	1533	9882	1705	9854	11
50	1016	9948	1190	9929	1363	9907	1536	9881	1708	9853	10
51	1019	9948	1193	9929	1366	9906	1538	9881	1711	9853	9
52	1022	9948	1196	9928	1369	9906	1541	9880	1714	9852	8
53	1025	9947	1198	9928	1372	9905	1544	9880	1716	9852	7
54	1028	9947	1201	9928	1374	9905	1547	9880	1719	9851	6
55	1031	9947	1204	9927	1377	9905	1550	9879	1722	9851	5
56	1034	9946	1207	9927	1380	9904	1553	9879	1725	9850	4
57	1037	9946	1210	9927	1383	9904	1556	9878	1728	9850	3
58	1039	9946	1213	9926	1386	9903	1559	9878	1731	9849	2
59	1042	9946	1216	9926	1389	9903	1561	9877	1734	9849	1
60	1045	9945	1219	9925	1392	9903	1564	9877	1736	9848	0
	cos	sin	cos	sin	cos	sin	cos	sin	cos	sin	
′	84°		83°		82°		81°		80°		′

′	10° sin	cos	11° sin	cos	12° sin	cos	13° sin	cos	14° sin	cos	′
0	1736	9848	1908	9816	2079	9781	2250	9744	2419	9703	60
1	1739	9848	1911	9816	2082	9781	2252	9743	2422	9702	59
2	1742	9847	1914	9815	2085	9780	2255	9742	2425	9702	58
3	1745	9847	1917	9815	2088	9780	2258	9742	2428	9701	57
4	1748	9846	1920	9814	2090	9779	2261	9741	2431	9700	56
5	1751	9846	1922	9813	2093	9778	2264	9740	2433	9699	55
6	1754	9845	1925	9813	2096	9778	2267	9740	2436	9699	54
7	1757	9845	1928	9812	2099	9777	2269	9739	2439	9698	53
8	1759	9844	1931	9812	2102	9777	2272	9738	2442	9697	52
9	1762	9843	1934	9811	2105	9776	2275	9738	2445	9697	51
10	1765	9843	1937	9811	2108	9775	2278	9737	2447	9696	50
11	1768	9842	1939	9810	2110	9775	2281	9736	2450	9695	49
12	1771	9842	1942	9810	2113	9774	2284	9736	2453	9694	48
13	1774	9841	1945	9809	2116	9774	2286	9735	2456	9694	47
14	1777	9841	1948	9808	2119	9773	2289	9734	2459	9693	46
15	1779	9840	1951	9808	2122	9772	2292	9734	2462	9692	45
16	1782	9840	1954	9807	2125	9772	2295	9733	2464	9692	44
17	1785	9839	1957	9807	2127	9771	2298	9732	2467	9691	43
18	1788	9839	1959	9806	2130	9770	2300	9732	2470	9690	42
19	1791	9838	1962	9806	2133	9770	2303	9731	2473	9689	41
20	1794	9838	1965	9805	2136	9769	2306	9730	2476	9689	40
21	1797	9837	1968	9804	2139	9769	2309	9730	2478	9688	39
22	1799	9837	1971	9804	2142	9768	2312	9729	2481	9687	38
23	1802	9836	1974	9803	2145	9767	2315	9728	2484	9687	37
24	1805	9836	1977	9803	2147	9767	2317	9728	2487	9686	36
25	1808	9835	1979	9802	2150	9766	2320	9727	2490	9685	35
26	1811	9835	1982	9802	2153	9765	2323	9726	2493	9684	34
27	1814	9834	1985	9801	2156	9765	2326	9726	2495	9684	33
28	1817	9834	1988	9800	2159	9764	2329	9725	2498	9683	32
29	1819	9833	1991	9800	2162	9764	2332	9724	2501	9682	31
30	1822	9833	1994	9799	2164	9763	2334	9724	2504	9681	30
31	1825	9832	1997	9799	2167	9762	2337	9723	2507	9681	29
32	1828	9831	1999	9798	2170	9762	2340	9722	2509	9680	28
33	1831	9831	2002	9798	2173	9761	2343	9722	2512	9679	27
34	1834	9830	2005	9797	2176	9760	2346	9721	2515	9679	26
35	1837	9830	2008	9796	2179	9760	2349	9720	2518	9678	25
36	1840	9829	2011	9796	2181	9759	2351	9720	2521	9677	24
37	1842	9829	2014	9795	2184	9759	2354	9719	2524	9676	23
38	1845	9828	2016	9795	2187	9758	2357	9718	2526	9676	22
39	1848	9828	2019	9794	2190	9757	2360	9718	2529	9675	21
40	1851	9827	2022	9793	2193	9757	2363	9717	2532	9674	20
41	1854	9827	2025	9793	2196	9756	2366	9716	2535	9673	19
42	1857	9826	2028	9792	2198	9755	2368	9715	2538	9673	18
43	1860	9826	2031	9792	2201	9755	2371	9715	2540	9672	17
44	1862	9825	2034	9791	2204	9754	2374	9714	2543	9671	16
45	1865	9825	2036	9790	2207	9753	2377	9713	2546	9670	15
46	1868	9824	2039	9790	2210	9753	2380	9713	2549	9670	14
47	1871	9823	2042	9789	2213	9752	2383	9712	2552	9669	13
48	1874	9823	2045	9789	2215	9751	2385	9711	2554	9668	12
49	1877	9822	2048	9788	2218	9751	2388	9711	2557	9667	11
50	1880	9822	2051	9787	2221	9750	2391	9710	2560	9667	10
51	1882	9821	2054	9787	2224	9750	2394	9709	2563	9666	9
52	1885	9821	2056	9786	2227	9749	2397	9709	2566	9665	8
53	1888	9820	2059	9786	2230	9748	2399	9708	2569	9665	7
54	1891	9820	2062	9785	2233	9748	2402	9707	2571	9664	6
55	1894	9819	2065	9784	2235	9747	2405	9706	2574	9663	5
56	1897	9818	2068	9784	2238	9746	2408	9706	2577	9662	4
57	1900	9818	2071	9783	2241	9746	2411	9705	2580	9662	3
58	1902	9817	2073	9783	2244	9745	2414	9704	2583	9661	2
59	1905	9817	2076	9782	2247	9744	2416	9704	2585	9660	1
60	1908	9816	2079	9781	2250	9744	2419	9703	2588	9659	0
	cos	sin	cos	sin	cos	sin	cos	sin	cos	sin	
′	79°		78°		77°		76°		75°		′

′	15° sin	cos	16° sin	cos	17° sin	cos	18° sin	cos	19° sin	cos	′
0	2588	9659	2756	9613	2924	9563	3090	9511	3256	9455	60
1	2591	9659	2759	9612	2926	9562	3093	9510	3258	9454	59
2	2594	9658	2762	9611	2929	9561	3096	9509	3261	9453	58
3	2597	9657	2165	9610	2932	9560	3098	9508	3264	9452	57
4	2599	9656	2768	9609	2935	9560	3101	9507	3267	9451	56
5	2602	9655	2770	9609	2938	9559	3104	9506	3269	9450	55
6	2605	9655	2773	9608	2940	9558	.3107	9505	3272	9449	54
7	2608	9654	2776	9607	2943	9557	3110	9504	3275	9449	53
8	2611	9653	2779	9606	2946	9556	3112	9503	3278	9448	52
9	2613	9652	2782	9605	2949	9555	3115	9502	3280	9447	51
10	2616	9652	2784	9605	2952	9555	3118	9502	3283	9446	50
11	2619	9651	2787	9604	2954	9554	3121	9501	3286	9445	49
12	2622	9650	2790	9603	2957	9553	3123	9500	3289	9444	48
13	2625	9649	2793	9602	2960	9552	3126	9499	3291	9443	47
14	2628	9649	2795	9601	2963	9551	3129	9498	3294	9442	46
15	2630	9648	2798	9600	2965	9550	3132	9497	3297	9441	45
16	2633	9647	2801	9600	2968	9549	3134	9496	3300	9440	44
17	2636	9646	2804	9599	2971	9548	3137	9495	3302	9439	43
18	2639	9646	2807	9598	2974	9548	3140	9494	3305	9438	42
19	2642	9645	2809	9597	2977	9547	3143	9493	3308	9437	41
20	2644	9644	2812	9596	2979	9546	3145	9492	3311	9436	40
21	2647	9643	2815	9596	2982	9545	3148	9492	3313	9435	39
22	2650	9642	2818	9595	2985	9544	3151	9491	3316	9434	38
23	2653	9642	2821	9594	2988	9543	3154	9490	3319	9433	37
24	2656	9641	2823	9593	2990	9542	3156	9489	3322	9432	36
25	2658	9640	2826	9592	2993	9542	3159	9488	3324	9431	35
26	2661	9639	2829	9591	2996	9541	3162	9487	3327	9430	34
27	2664	9639	2832	9591	2999	9540	3165	9486	3330	9429	33
28	2667	9638	2835	9590	3002	9539	3168	9485	3333	9428	32
29	2670	9637	2837	9589	3004	9538	3170	9484	3335	9427	31
30	2672	9636	2840	9588	3007	9537	3173	9483	3338	9426	30
31	2675	9636	2843	9587	3010	9536	3176	9482	3341	9425	29
32	2678	9635	2846	9587	3013	9535	3179	9481	3344	9424	28
33	2681	9634	2849	9586	3015	9535	3181	9480	3346	9423	27
34	2684	9633	2851	9585	3018	9534	3184	9480	3349	9423	26
35	2686	9632	2854	9584	3021	9533	3187	9479	3352	9422	25
36	2689	9632	2857	9583	3024	9532	3190	9478	3355	9421	24
37	2692	9631	2860	9582	3026	9531	3192	9477	3357	9420	23
38	2695	9630	2862	9582	3029	9530	3195	9476	3360	9419	22
39	2698	9629	2865	9581	3032	9529	3198	9475	3363	9418	21
40	2700	9628	2868	9580	3035	9528	3201	9474	3365	9417	20
41	2703	9628	2871	9579	3038	9527	3203	9473	3368	9416	19
42	2706	9627	2874	9578	3040	9527	3206	9472	3371	9415	18
43	2709	9626	2876	9577	3043	9526	3209	9471	3374	9414	17
44	2712	9625	2879	9577	3046	9525	3212	9470	3376	9413	16
45	2714	9625	2882	9576	3049	9524	3214	9469	3379	9412	15
46	2717	9624	2885	9575	3051	9523	3217	9468	3382	9411	14
47	2720	9623	2888	9574	3054	9522	3220	9467	3385	9410	13
48	2723	9622	2890	9573	3057	9521	3223	9466	3387	9409	12
49	2726	9621	2893	9572	3060	9520	3225	9466	3390	9408	ᵢ 11
50	2728	9621	2896	9572	3062	9520	3228	9465	3393	9407	10
51	2731	9620	2899	9571	3065	9519	3231	9464	3396	9406	9
52	2734	9619	2901	9570	3068	9518	3234	9463	3398	9405	8
53	2737	9618	2904	9569	3071	9517	3236	9462	3401	9404	7
54	2740	9617	2907	9568	3074	9516	3239	9461	3404	9403	6
55	2742	9617	2910	9567	3076	9515	3242	9460	3407	9402	5
56	2745	9616	2913	9566	3079	9514	3245	9459	3409	9401	4
57	2748	9615	2915	9566	3082	9513	3247	9458	3412	9400	3
58	2751	9614	2918	9565	3085	9512	3250	9457	3415	9399	2
59	2754	9613	2921	9564	3087	9511	3253	9456	3417	9398	1
60	2756	9613	2924	9563	3090	9511	3256	9455	3420	9397	0
	cos	sin	cos	sin	cos	sin	cos	sin	cos	sin	
′	74°		73°		72°		71°		70°		′

′	20° sin	cos	21° sin	cos	22° sin	cos	23° sin	cos	24° sin	cos	′
0	3420	9397	3584	9336	3746	9272	3907	9205	4067	9135	60
1	3423	9396	3586	9335	3749	9271	3910	9204	4070	9134	59
2	3426	9395	3589	9334	3751	9270	3913	9203	4073	9133	58
3	3428	9394	3592	9333	3754	9269	3915	9202	4075	9132	57
4	3431	9393	3595	9332	3757	9267	3918	9200	4078	9131	56
5	3434	9392	3597	9331	3760	9266	3921	9199	4081	9130	55
6	3437	9391	3600	9330	3762	9265	3923	9198	4083	9128	54
7	3439	9390	3603	9328	3765	9264	3926	9197	4086	9127	53
8	3442	9389	3605	9327	3768	9263	3929	9196	4089	9126	52
9	3445	9388	3608	9326	3770	9262	3931	9195	4091	9125	51
10	3448	9387	3611	9325	3773	9261	3934	9194	4094	9124	50
11	3450	9386	3614	9324	3776	9260	3937	9192	4097	9122	49
12	3453	9385	3616	9323	3778	9259	3939	9191	4099	9121	48
13	3456	9384	3619	9322	3781	9258	3942	9190	4102	9120	47
14	3458	9383	3622	9321	3784	9257	3945	9189	4105	9119	46
15	3461	9382	3624	9320	3786	9255	3947	9188	4107	9118	45
16	3464	9381	3627	9319	3789	9254	3950	9187	4110	9116	44
17	3467	9380	3630	9318	3792	9253	3953	9186	4112	9115	43
18	3469	9379	3633	9317	3795	9252	3955	9184	4115	9114	42
19	3472	9378	3635	9316	3797	9251	3958	9183	4118	9113	41
20	3475	9377	3638	9315	3800	9250	3961	9182	4120	9112	40
21	3478	9376	3641	9314	3803	9249	3963	9181	4123	9110	39
22	3480	9375	3643	9313	3805	9248	3966	9180	4126	9109	38
23	3483	9374	3646	9312	3808	9247	3969	9179	4128	9108	37
24	3486	9373	3649	9311	3811	9245	3971	9178	4131	9107	36
25	3488	9372	3651	9309	3813	9244	3974	9176	4134	9106	35
26	3491	9371	3654	9308	3816	9243	3977	9175	4136	9104	34
27	3494	9370	3657	9307	3819	9242	3979	9174	4139	9103	33
28	3497	9369	3660	9306	3821	9241	3982	9173	4142	9102	32
29	3499	9368	3662	9305	3824	9240	3985	9172	4144	9101	31
30	3502	9367	3665	9304	3827	9239	3987	9171	4147	9100	30
31	3505	9366	3668	9303	3830	9238	3990	9169	4150	9098	29
32	3508	9365	3670	9302	3832	9237	3993	9168	4152	9097	28
33	3510	9364	3673	9301	3835	9235	3995	9167	4155	9096	27
34	3513	9363	3676	9300	3838	9234	3998	9166	4158	9095	26
35	3516	9362	3679	9299	3840	9233	4001	9165	4160	9094	25
36	3518	9361	3681	9298	3843	9232	4003	9164	4163	9092	24
37	3521	9360	3684	9297	3846	9231	4006	9162	4165	9091	23
38	3524	9359	3687	9296	3848	9230	4009	9161	4168	9090	22
39	3527	9358	3689	9295	3851	9229	4011	9160	4171	9088	21
40	3529	9356	3692	9293	3854	9228	4014	9159	4173	9088	20
41	3532	9355	3695	9292	3856	9227	4017	9158	4176	9086	19
42	3535	9354	3697	9291	3859	9225	4019	9157	4179	9085	18
43	3537	9353	3700	9290	3862	9224	4022	9155	4181	9084	17
44	3540	9352	3703	9289	3864	9223	4025	9154	4184	9083	16
45	3543	9351	3706	9288	3867	9222	4027	9153	4187	9081	15
46	3546	9350	3708	9287	3870	9221	4030	9152	4189	9080	14
47	3548	9349	3711	9286	3872	9220	4033	9151	4192	9079	13
48	3551	9348	3714	9285	3875	9219	4035	9150	4195	9078	12
49	3554	9347	3716	9284	3878	9218	4038	9148	4197	9077	11
50	3557	9346	3719	9283	3881	9216	4041	9147	4200	9075	10
51	3559	9345	3722	9282	3883	9215	4043	9146	4202	9074	9
52	3562	9344	3724	9281	3886	9214	4046	9145	4205	9073	8
53	3565	9343	3727	9279	3889	9213	4049	9143	4208	9072	7
54	3567	9342	3730	9278	3891	9212	4051	9143	4210	9070	6
55	3570	9341	3733	9277	3894	9211	4054	9141	4213	9069	5
56	3573	9340	3735	9276	3897	9210	4057	9140	4216	9068	4
57	3576	9339	3738	9275	3899	9208	4059	9139	4218	9067	3
58	3578	9338	3741	9274	3902	9207	4062	9138	4221	9066	2
59	3581	9337	3743	9273	3905	9206	4065	9137	4224	9064	1
60	3584	9336	3746	9272	3907	9205	4067	9135	4226	9063	0
	cos	sin	cos	sin	cos	sin	cos	sin	cos	sin	′
	69°		**68°**		**67°**		**66°**		**65°**		

′	25°		26°		27°		28°		29°		′
	sin	cos	sin	cos	sin	cos	sin	cos	sin	cos	
0	4226	9063	4384	8988	4540	8910	4695	8829	4848	8746	60
1	4229	9062	4386	8987	4542	8909	4697	8828	4851	8745	59
2	4231	9061	4389	8985	4545	8907	4700	8827	4853	8743	58
3	4234	9059	4392	8984	4548	8906	4702	8825	4856	8742	57
4	4237	9058	4394	8983	4550	8905	4705	8824	4858	8741	56
5	4239	9057	4397	8982	4553	8903	4708	8823	4861	8739	55
6	4242	9056	4399	8980	4555	8902	4710	8821	4863	8738	54
7	4245	9054	4402	8979	4558	8901	4713	8820	4866	8736	53
8	4247	9053	4405	8978	4561	8899	4715	8819	4868	8735	52
9	4250	9052	4407	8976	4563	8898	4718	8817	4871	8733	51
10	4253	9051	4410	8975	4566	8897	4720	8816	4874	8732	50
11	4255	9050	4412	8974	4568	8895	4723	8814	4876	8731	49
12	4258	9048	4415	8973	4571	8894	4726	8813	4879	8729	48
13	4260	9047	4418	8971	4574	8893	4728	8812	4881	8728	47
14	4263	9046	4420	8970	4576	8892	4731	8810	4884	8726	46
15	4266	9045	4423	8969	4579	8890	4733	8809	4886	8725	45
16	4268	9043	4425	8967	4581	8889	4736	8808	4889	8724	44
17	4271	9042	4428	8966	4584	8888	4738	8806	4891	8722	43
18	4274	9041	4431	8965	4586	8886	4741	8805	4894	8721	42
19	4276	9040	4433	8964	4589	8885	4743	8803	4896	8719	41
20	4279	9038	4436	8962	4592	8884	4746	8802	4899	8718	40
21	4281	9037	4439	8961	4594	8882	4749	8801	4901	8716	39
22	4284	9036	4441	8960	4597	8881	4751	8799	4904	8715	38
23	4287	9035	4444	8958	4599	8879	4754	8798	4907	8714	37
24	4289	9033	4446	8957	4602	8878	4756	8796	4909	8712	36
25	4292	9032	4449	8956	4605	8877	4759	8795	4912	8711	35
26	4295	9031	4452	8955	4607	8875	4761	8794	4914	8709	34
27	4297	9030	4454	8953	4610	8874	4764	8792	4917	8708	33
28	4300	9028	4457	8952	4612	8873	4766	8791	4919	8706	32
29	4302	9027	4459	8951	4615	8871	4769	8790	4922	8705	31
30	4305	9026	4462	8949	4617	8870	4772	8788	4924	8704	30
31	4308	9025	4465	8948	4620	8869	4774	8787	4927	8702	29
32	4310	9023	4467	8947	4623	8867	4777	8785	4929	8701	28
33	4313	9022	4470	8945	4625	8866	4779	8784	4932	8699	27
34	4316	9021	4472	8944	4628	8865	4782	8783	4934	8698	26
35	4318	9020	4475	8943	4630	8863	4784	8781	4937	8696	25
36	4321	9018	4478	8942	4633	8862	4787	8780	4939	8695	24
37	4323	9017	4480	8940	4636	8861	4789	8778	4942	8694	23
38	4326	9016	4483	8939	4638	8859	4792	8777	4944	8692	22
39	4329	9015	4485	8938	4641	8858	4795	8776	4947	8691	21
40	4331	9013	4488	8936	4643	8857	4797	8774	4950	8689	20
41	4334	9012	4491	8935	4646	8855	4800	8773	4952	8688	19
42	4337	9011	4493	8934	4648	8854	4802	8771	4955	8686	18
43	4339	9010	4496	8932	4651	8853	4805	8770	4957	8685	17
44	4342	9008	4498	8931	4654	8851	4807	8769	4960	8683	16
45	4344	9007	4501	8930	4656	8850	4810	8767	4962	8682	15
46	4347	9006	4504	8928	4659	8849	4812	8766	4965	8681	14
47	4350	9004	4506	8927	4661	8847	4815	8764	4967	8679	13
48	4352	9003	4509	8926	4664	8846	4818	8763	4970	8678	12
49	4355	9002	4511	8925	4666	8844	4820	8762	4972	8676	11
50	4358	9001	4514	8923	4669	8843	4823	8760	4975	8675	10
51	4360	8999	4517	8922	4672	8842	4825	8759	4977	8673	9
52	4363	8998	4519	8921	4674	8840	4828	8757	4980	8672	8
53	4365	8997	4522	8919	4677	8839	4830	8756	4982	8670	7
54	4368	8996	4524	8918	4679	8838	4833	8755	4985	8669	6
55	4371	8994	4527	8917	4682	8836	4835	8753	4987	8668	5
56	4373	8993	4530	8915	4684	8835	4838	8752	4990	8666	4
57	4376	8992	4532	8914	4687	8834	4840	8750	4992	8665	3
58	4378	8990	4535	8913	4690	8832	4843	8749	4995	8663	2
59	4381	8989	4537	8911	4692	8831	4846	8748	4997	8662	1
60	4384	8988	4540	8910	4695	8829	4848	8746	5000	8660	0
	cos	sin	cos	sin	cos	sin	cos	sin	cos	sin	
′	64°		63°		62°		61°		60°		′

′	30° sin	cos	31° sin	cos	32° sin	cos	33° sin	cos	34° sin	cos	′
0	5000	8660	5150	8572	5299	8480	5446	8387	5592	8290	60
1	5003	8659	5153	8570	5302	8479	5449	8385	5594	8289	59
2	5005	8657	5155	8569	5304	8477	5451	8384	5597	8287	58
3	5008	8656	5158	8567	5307	8476	5454	8382	5599	8285	57
4	5010	8654	5160	8566	5309	8474	5456	8380	5602	8284	56
5	5013	8653	5163	8564	5312	8473	5459	8379	5604	8282	55
6	5015	8652	5165	8563	5314	8471	5461	8377	5606	8281	54
7	5018	8650	5168	8561	5316	8470	5463	8376	5609	8279	53
8	5020	8649	5170	8560	5319	8468	5466	8374	5611	8277	52
9	5023	8647	5173	8558	5321	8467	5468	8372	5614	8276	51
10	5025	8646	5175	8557	5324	8465	5471	8371	5616	8274	50
11	5028	8644	5178	8555	5326	8463	5473	8369	5618	8272	49
12	5030	8643	5180	8554	5329	8462	5476	8368	5621	8271	48
13	5033	8641	5183	8552	5331	8460	5478	8366	5623	8269	47
14	5035	8640	5185	8551	5334	8459	5480	8364	5626	8268	46
15	5038	8638	5188	8549	5336	8457	5483	8363	5628	8266	45
16	5040	8637	5190	8548	5339	8456	5485	8361	5630	8264	44
17	5043	8635	5193	8546	5341	8454	5488	8360	5633	8263	43
18	5045	8634	5195	8545	5344	8453	5490	8358	5635	8261	42
19	5048	8632	5198	8543	5346	8451	5493	8356	5638	8259	41
20	5050	8631	5200	8542	5348	8450	5495	8355	5640	8258	40
21	5053	8630	5203	8540	5351	8448	5498	8353	5642	8256	39
22	5055	8628	5205	8539	5353	8446	5500	8352	5645	8254	38
23	5058	8627	5208	8537	5356	8445	5502	8350	5647	8253	37
24	5060	8625	5210	8536	5358	8443	5505	8348	5650	8251	36
25	5063	8624	5213	8534	5361	8442	5507	8347	5652	8249	35
26	5065	8622	5215	8532	5363	8440	5510	8345	5654	8248	34
27	5068	8621	5218	8531	5366	8439	5512	8344	5657	8246	33
28	5070	8619	5220	8529	5368	8437	5515	8342	5659	8245	32
29	5073	8618	5223	8528	5371	8435	5517	8340	5662	8243	31
30	5075	8616	5225	8526	5373	8434	5519	8339	5664	8241	30
31	5078	8615	5227	8525	5375	8432	5522	8337	5666	8240	29
32	5080	8613	5230	8523	5378	8431	5524	8336	5669	8238	28
33	5083	8612	5232	8522	5380	8429	5527	8334	5671	8236	27
34	5085	8610	5235	8520	5383	8428	5529	8332	5674	8235	26
35	5088	8609	5237	8519	5385	8426	5531	8331	5676	8233	25
36	5090	8607	5240	8517	5388	8425	5534	8329	5678	8231	24
37	5093	8606	5242	8516	5390	8423	5536	8328	5681	8230	23
38	5095	8604	5245	8514	5393	8421	5539	8326	5683	8228	22
39	5098	8603	5247	8513	5395	8420	5541	8324	5686	8226	21
40	5100	8601	5250	8511	5398	8418	5544	8323	5688	8225	20
41	5103	8600	5252	8510	5400	8417	5546	8321	5690	8223	19
42	5105	8599	5255	8508	5402	8415	5548	8320	5693	8221	18
43	5108	8597	5257	8507	5405	8414	5551	8318	5695	8220	17
44	5110	8596	5260	8505	5407	8412	5553	8316	5698	8218	16
45	5113	8594	5262	8504	5410	8410	5556	8315	5700	8216	15
46	5115	8593	5265	8502	5412	8409	5558	8313	5702	8215	14
47	5118	8591	5267	8500	5415	8407	5561	8311	5705	8213	13
48	5120	8590	5270	8499	5417	8406	5563	8310	5707	8211	12
49	5123	8588	5272	8497	5420	8404	5565	8308	5710	8210	11
50	5125	8587	5275	8496	5422	8403	5568	8307	5712	8208	10
51	5128	8585	5277	8494	5424	8401	5570	8305	5714	8207	9
52	5130	8584	5279	8493	5427	8399	5573	8303	5717	8205	8
53	5133	8582	5282	8491	5429	8398	5575	8302	5719	8203	7
54	5135	8581	5284	8490	5432	8396	5577	8300	5721	8202	6
55	5138	8579	5287	8488	5434	8395	5580	8299	5724	8200	5
56	5140	8578	5289	8487	5437	8393	5582	8297	5726	8198	4
57	5143	8576	5292	8485	5439	8391	5585	8295	5729	8197	3
58	5145	8575	5294	8484	5442	8390	5587	8294	5731	8195	2
59	5148	8573	5297	8482	5444	8388	5590	8292	5733	8193	1
60	5150	8572	5299	8480	5446	8387	5592	8290	5736	8192	0
′	cos	sin	cos	sin	cos	sin	cos	sin	cos	sin	′
	59°		58°		57°		56°		55°		

′	35° sin	cos	36° sin	cos	37° sin	cos	38° sin	cos	39° sin	cos	′
0	5736	8192	5878	8090	6018	7986	6157	7880	6293	7771	60
1	5738	8190	5880	8088	6020	7985	6159	7878	6295	7770	59
2	5741	8188	5883	8087	6023	7983	6161	7877	6298	7768	58
3	5743	8187	5885	8085	6025	7981	6163	7875	6300	7766	57
4	5745	8185	5887	8083	6027	7979	6166	7873	6302	7764	56
5	5748	8183	5890	8082	6030	7978	6168	7871	6305	7762	55
6	5750	8181	5892	8080	6032	7976	6170	7869	6307	7760	54
7	5752	8180	5894	8078	6034	7974	6173	7868	6309	7759	53
8	5755	8178	5897	8076	6037	7972	6175	7866	6311	7757	52
9	5757	8176	5899	8075	6039	7971	6177	7864	6314	7755	51
10	5760	8175	5901	8073	6041	7969	6180	7862	6316	7753	50
11	5762	8173	5904	8071	6044	7967	6182	7860	6318	7751	49
12	5764	8171	5906	8070	6046	7965	6184	7859	6320	7749	48
13	5767	8170	5908	8068	6048	7964	6186	7857	6323	7748	47
14	5769	8168	5911	8066	6051	7962	6189	7855	6325	7746	46
15	5771	8166	5913	8064	6053	7960	6191	7853	6327	7744	45
16	5774	8165	5915	8063	6055	7958	6193	7851	6329	7742	44
17	5776	8163	5918	8061	6058	7956	6196	7850	6332	7740	43
18	5779	8161	5920	8059	6060	7955	6198	7848	6334	7738	42
19	5781	8160	5922	8058	6062	7953	6200	7346	6336	7737	41
20	5783	8158	5925	8056	6065	7951	6202	7844	6338	7735	40
21	5786	8156	5927	8054	6067	7950	6205	7842	6341	7733	39
22	5788	8155	5930	8052	6069	7948	6207	7841	6343	7731	38
23	5790	8153	5932	8051	6071	7946	6209	7839	6345	7729	37
24	5793	8151	5934	8049	6074	7944	6211	7837	6347	7727	36
25	5795	8150	5937	8047	6076	7942	6214	7835	6350	7725	35
26	5798	8148	5939	8045	6078	7941	6216	7833	6352	7724	34
27	5800	8146	5941	8044	6081	7939	6218	7832	6354	7722	33
28	5802	8145	5944	8042	6083	7937	6221	7830	6356	7720	32
29	5805	8143	5946	8040	6085	7935	6223	7828	6359	7718	31
30	5807	8141	5948	8039	6088	7934	6225	7826	6361	7716	30
31	5809	8139	5951	8037	6090	7932	6227	7824	6363	7714	29
32	5812	8138	5953	8035	6092	7930	6230	7822	6365	7713	28
33	5814	8136	5955	8033	6095	7928	6232	7821	6368	7711	27
34	5816	8134	5958	8032	6097	7926	6234	7819	6370	7709	26
35	5819	8133	5960	8030	6099	7925	6237	7817	6372	7707	25
36	5821	8131	5962	8028	6101	7923	6239	7815	6374	7705	24
37	5824	8129	5965	8026	6104	7921	6241	7813	9376	7703	23
38	5826	8128	5967	8025	6106	7919	6243	7812	6379	7701	22
39	5828	8126	5969	8023	6108	7918	6246	7810	6381	7700	21
40	5831	8124	5972	8021	6111	7916	6248	7808	6383	7698	20
41	5833	8123	5974	8020	6113	7914	6250	7806	6385	7696	19
42	5835	8121	5976	8018	6115	7912	6252	7804	6388	7694	18
43	5838	8119	5979	8016	6118	7910	6255	7802	6390	7692	17
44	5840	8117	5981	8014	6120	7909	6257	7801	6392	7690	16
45	5842	8116	5983	8013	6122	7907	6259	7799	6394	7688	15
46	5845	8114	5986	8011	6124	7905	6262	7797	6397	7687	14
47	5847	8112	5988	8009	6127	7903	6264	7795	6399	7685	13
48	5850	8111	5990	8007	6129	7902	6266	7793	6401	7683	12
49	5852	8109	5993	8006	6131	7900	6268	7792	6403	7681	11
50	5854	8107	5995	8004	6134	7898	6271	7790	6406	7679	10
51	5857	8106	5997	8002	6136	7896	6273	7788	6408	7677	9
52	5859	8104	6000	8000	6138	7894	6275	7786	6410	7675	8
53	5861	8102	6002	7999	6141	7893	6277	7784	6412	7674	7
54	5864	8100	6004	7997	6143	7891	6280	7782	6414	7672	6
55	5866	8099	6007	7995	6145	7889	6282	7781	6417	7670	5
56	5868	8097	6009	7993	6147	7887	6284	7779	6419	7668	4
57	5871	8095	6011	7992	6150	7885	6286	7777	6421	7666	3
58	5873	8094	6014	7990	6152	7884	6289	7775	6423	7664	2
59	5875	8092	6016	7988	6154	7882	6291	7773	6426	7662	1
60	5878	8090	6018	7986	6157	7880	6293	7771	6428	7660	0
	cos	sin	cos	sin	cos	sin	cos	sin	cos	sin	
′	54°		53°		52°		51°		50°		′

′	40° sin cos	41° sin cos	42° sin cos	43° sin cos	44° sin cos	′
0	6428 7660	6561 7547	6691 7431	6820 7314	6947 7193	60
1	6430 7659	6563 7545	6693 7430	6822 7312	6949 7191	59
2	6432 7657	6565 7543	6696 7428	6824 7310	6951 7189	58
3	6435 7655	6567 7541	6698 7426	6826 7308	6953 7187	57
4	6437 7653	6569 7539	6700 7424	6828 7306	6955 7185	56
5	6439 7651	6572 7538	6702 7422	6831 7304	6957 7183	55
6	6441 7649	6574 7536	6704 7420	6833 7302	6959 7181	54
7	6443 7647	6576 7534	6706 7418	6835 7300	6961 7179	53
8	6446 7645	6578 7532	6709 7416	6837 7298	6963 7177	52
9	6448 7644	6580 7530	6711 7414	6839 7296	6965 7175	51
10	6450 7642	6583 7528	6713 7412	6841 7294	6967 7173	50
11	6452 7640	6585 7526	6715 7410	6843 7292	6970 7171	49
12	6455 7638	6587 7524	6717 7408	6845 7290	6972 7169	48
13	6457 7636	6589 7522	6719 7406	6848 7288	6974 7167	47
14	6459 7634	6591 7520	6722 7404	6850 7286	6976 7165	46
15	6461 7632	6593 7518	6724 7402	6852 7284	6978 7163	45
16	6463 7630	6596 7516	6726 7400	6854 7282	6980 7161	44
17	6466 7629	6598 7515	6728 7398	6856 7280	6982 7159	43
18	6468 7627	6600 7513	6730 7396	6858 7278	6984 7157	42
19	6470 7625	6602 7511	6732 7394	6860 7276	6986 7155	41
20	6472 7623	6604 7509	6734 7392	6862 7274	6988 7153	40
21	6475 7621	6607 7507	6737 7390	6865 7272	6990 7151	39
22	6477 7619	6609 7505	6739 7388	6867 7270	6992 7149	38
23	6479 7617	6611 7503	6741 7387	6869 7268	6995 7147	37
24	6481 7615	6613 7501	6743 7385	6871 7266	6997 7145	36
25	6483 7613	6615 7499	6745 7383	6873 7264	6999 7143	35
26	6486 7612	6617 7497	6747 7381	6875 7262	7001 7141	34
27	6488 7610	6620 7495	6749 7379	6877 7260	7003 7139	33
28	6490 7608	6622 7493	6752 7377	6879 7258	7005 7137	32
29	6492 7606	6624 7491	6754 7375	6881 7256	7007 7135	31
30	6494 7604	6626 7490	6756 7373	6884 7254	7009 7133	30
31	6497 7602	6628 7488	6758 7371	6886 7252	7011 7130	29
32	6499 7600	6631 7486	6760 7369	6888 7250	7013 7128	28
33	6501 7598	6633 7484	6762 7367	6890 7248	7015 7126	27
34	6503 7596	6635 7482	6764 7365	6892 7246	7017 7124	26
35	6506 7595	6637 7480	6767 7363	6894 7244	7019 7122	25
36	6508 7593	6639 7478	6769 7361	6896 7242	7022 7120	24
37	6510 7591	6641 7476	6771 7359	6898 7240	7024 7118	23
38	6512 7589	6644 7474	6773 7357	6900 7238	7026 7116	22
39	6514 7587	6646 7472	6775 7355	6903 7236	7028 7114	21
40	6517 7585	6648 7470	6777 7253	6905 7234	7030 7112	20
41	6519 7583	6650 7468	6779 7351	6907 7232	7032 7110	19
42	6521 7581	6652 7466	6782 7349	6909 7230	7034 7108	18
43	6523 7579	6654 7464	6784 7347	6911 7228	7036 7106	17
44	6525 7578	6657 7463	6786 7345	6913 7226	7038 7104	16
45	6528 7576	6659 7461	6788 7343	6915 7224	7040 7102	15
46	6530 7574	6661 7459	6790 7341	6917 7222	7042 7100	14
47	6532 7572	6663 7457	6792 7339	6919 7220	7044 7098	13
48	6534 7570	6665 7455	6794 7337	6921 7218	7046 7096	12
49	6536 7568	6667 7453	6797 7335	6924 7216	7048 7094	11
50	6539 7566	6670 7451	6799 7333	6926 7214	7050 7092	10
51	6541 7564	6672 7449	6801 7331	6928 7212	7053 7090	9
52	6543 7562	6674 7447	6803 7329	6930 7210	7055 7088	8
53	6545 7560	6676 7445	6805 7327	6932 7208	7057 7085	7
54	6547 7559	6678 7443	6807 7325	6934 7206	7059 7083	6
55	6550 7557	6680 7441	6809 7323	6936 7203	7061 7081	5
56	6552 7555	6683 7439	6811 7321	6938 7201	7063 7079	4
57	6554 7553	6685 7437	6814 7319	6940 7199	7065 7077	3
58	6556 7551	6687 7435	6816 7318	6942 7197	7067 7075	2
59	6558 7549	6689 7433	6818 7316	6944 7195	7069 7073	1
60	6561 7547	6691 7431	6820 7314	6947 7193	7071 7071	0
	cos sin	cos sin	cos sin	cos sin	cos sin	
′	49°	48°	47°	46°	45°	′

′	0°		1°		2°		3°		4°		′
	tan	cot	tan	cot	tan	cot	tan	cot	tan	cot	
0	0000	Infinite	0175	57.2900	0349	28.6363	0524	19.0811	0699	14.3007	60
1	0003	3437.75	0177	56.3506	0352	28.3994	0527	18.9755	0702	14.2411	59
2	0006	1718.87	0180	55.4415	0355	28.1664	0530	18 8711	0705	14.1821	58
3	0009	1145.92	0183	54.5613	0358	27.9372	0533	18.7678	0708	14.1235	57
4	0012	859.436	0186	53.7086	0361	27.7117	0536	18.6656	0711	14.0655	56
5	0015	687.549	0189	52.8821	0364	27.4899	0539	18.5645	0714	14 0079	55
6	0017	572.957	0192	52.0807	0367	27.2715	0542	18.4645	0717	13.9507	54
7	0020	491.106	0195	51.3032	0370	27.0566	0544	18.3655	0720	13.8940	53
8	0023	429.718	0198	50.5485	0373	26.8450	0547	18.2677	0723	13 8378	52
9	0026	381.971	0201	49.8157	0375	26.6367	0550	18.1708	0726	13.7821	51
10	0029	343.774	0204	49.1039	0378	26.4316	0553	18.0750	0729	13.7267	50
11	0032	312.521	0207	48.4121	0381	26.2296	0556	17.9802	0731	13.6719	49
12	0035	286.478	0209	47.7395	0384	26.0307	0559	17.8863	0734	13 6174	48
13	0038	264.441	0212	47.0853	0387	25.8348	0562	17.7934	0737	13.5634	47
14	0041	245.552	0215	46.4489	0390	25.6418	0565	17.7015	0740	13.5098	46
15	0044	229.182	0218	45.8294	0393	25.4517	0568	17.6106	0743	13.4566	45
16	0047	214.858	0221	45.2261	0396	25.2644	0571	17.5205	0746	13.4039	44
17	0049	202.219	0224	44.6386	0399	25.0798	0574	17.4314	0749.	13.3515	43
18	0052	190.984	0227	44.0661	0402	24.8978	0577	17.3432	0752	13.2996	42
19	0055	180.932	0230	43.5081	0405	24.7185	0580	17.2558	0755	13.2480	41
20	0058	171.885	0233	42.9641	0407	24.5418	0582	17.1693	0758	13.1969	40
21	0061	163.700	0236	42.4335	0410	24.3675	0585	17.0837	0761	13.1461	39
22	0064	156.259	0239	41.9158	0413	24.1957	0588	16.9990	0764	13 0958	38
23	0067	149.465	0241	41.4106	0416	24.0263	0591	16.9150	0767	13.0458	37
24	0070	143.237	0244	40.9174	0419	23.8593	0594	16.8319	0769	12.9962	36
25	0073	137.507	0247	40.4358	0422	23.6945	0597	16.7496	0772	12.9469	35
26	0076	132.219	0250	39.9655	0425	23.5321	0600	16.6681	0775	12.8981	34
27	0079	127.321	0253	39 5059	0428	23.3718	0603	16.5874	0778	12.8496	33
28	0081	122.774	0256	39.0568	0431	23.2137	0606	16.5075	0781	12.8014	32
29	0084	118.540	0259	38.6177	0434	23.0577	0609	16.4283	0784	12.7536	31
30	0087	114.589	0262	38.1885	0437	22.9038	0612	16 3499	0787	12.7062	30
31	0090	110.892	0265	37.7686	0440	22.7519	0615	16.2722	0790	12.6591	29
32	0093	107.426	0268	37.3579•	0442	22.6020	0617	16.1952	0793	12.6124	28
33	0096	104.171	0271	36.9560	0445	22.4541	0620	16.1190	0796	12.5660	27
34	0099	101.107	0274	36.5627	0448	22.3081	0623	16.0435	0799	12.5199	26
35	0102	98.2179	0276	36 1776	0451	22.1640	0626	15.9687	0802	12.4742	25
36	0105	95.4895	0279	35 8006	0454	22.0217	0629	15.8945	0805	12.4288	24
37	0108	92.9085	0282	35.4313	0457	21.8813	0632	15.8211	0808	12.3838	23
38	0111	90.4633	0285	35.0695	0460	21.7426	0635	15.7483	0810	12.3390	22
39	0113	88.1436	0288	34.7151	0463	21.6056	0638	15.6762	0813	12.2946	21
40	0116	85.9398	0291	34.3678	0466	21.4704	0641	15.6048	0816	12.2505	20
41	0119	83.8435	0294	34.0273	0469	21.3369	0644	15.5340	0819	12.2067	19
42	0122	81.8470	0297	33.6935	0472	21.2049	0647	15.4638	0822	12.1632	18
43	0125	79.9434	0300	33.3662	0475	21.0747	0650	15.3943	0825	12.1201	17
44	0128	78.1263	0303	33.0452	0477	20.9460	0653	15.3254	0828	12.0772	16
45	0131	76.3900	0306	32.7303	0480	20.8188	0656	15.2571	0831	12.0346	15
46	0134	74.7292	0308	32.4213	0483	20.6932	0658	15.1893	0834	11.9923	14
47	0137	73.1390	0311	32.1181	0486	20.5691	0661	15.1222	0837	11.9504	13
48	0140	71.6151	0314	31.8205	0489	20 4465	0664	15.0557	0840	11.9087	12
49	0143	70.1533	0317	31.5284	0492	20.3253	0667	14.9898	0843	11.8673	11
50	0146	68.7501	0320	31.2416	0495	20.2056	0670	14.9244	0846	11.8262	10
51	0148	67.4019	0323	30.9599	0498	20.0872	0673	14.8596	0849	11.7853	9
52	0151	66.1055	0326	30.6833	0501	19.9702	0676	14.7954	0851	11.7448	8
53	0154	64.8580	0329	30 4116	0504	19.8546	0679	14.7317	0854	11.7045	7
54	0157	63.6567	0332	30.1446	0507	19.7403	0682	14.6685	0857	11.6645	6
55	0160	62.4992	0335	29.8823	0509	19.6273	0685	14.6059	0860	11.6248	5
56	0163	61.3829	0338	29.6245	0512	19.5156	0688	14.5438	0863	11.5853	4
57	0166	60.3058	0340	29.3711	0515	19.4051	0690	14.4823	0866	11.5461	3
58	0169	59.2659	0343	29.1220	0518	19.2959	0693	14.4212	0869	11.5072	2
59	0172	58.2612	0346	28 8771	0521	19.1879	0696	14.3607	0872	11.4685	1
60	0175	57.2900	0349	28.6363	0524	19.0811	0699	14.3007	0875	11.4301	0
	cot	tan	cot	tan	cot	tan	cot	tan	cot	tan	′
	89°		88°		87°		86°		85°		

′	5°		6°		7°		8°		9°		′
	tan	cot	tan	cot	tan	cot	tan	cot	tan	cot	
0	0875	11.4301	1051	9.5144	1228	8.1443	1405	7.1154	1584	6.3138	**60**
1	0878	11.3919	1054	9.4878	1231	8.1248	1408	7.1004	1587	6.3019	59
2	0881	11.3540	1057	9.4614	1234	8.1054	1411	7.0855	1590	6.2901	58
3	0884	11.3163	1060	9.4352	1237	8.0860	1414	7.0706	1593	6.2783	57
4	0887	11.2789	1063	9.4090	1240	8.0667	1417	7.0558	1596	6.2666	56
5	0890	11.2417	1066	9.3831	1243	8.0476	1420	7.0410	1599	6.2549	**55**
6	0892	11.2048	1069	9.3572	1246	8.0285	1423	7.0264	1602	6.2432	54
7	0895	11.1681	1072	9.3315	1249	8.0095	1426	7.0117	1605	6.2316	53
8	0898	11.1316	1075	9.3060	1251	7.9906	1429	6.9972	1608	6.2200	52
9	0901	11.0954	1078	9.2806	1254	7.9718	1432	6.9827	1611	6.2085	51
10	0904	11.0594	1080	9.2553	1257	7.9530	1435	6.9682	1614	6.1970	**50**
11	0907	11.0237	1083	9.2302	1260	7.9344	1438	6.9538	1617	6.1856	49
12	0910	10.9882	1086	9.2052	1263	7.9158	1441	6.9395	1620	6.1742	48
13	0913	10.9529	1089	9.1803	1266	7.8973	1444	6.9252	1623	6.1628	47
14	0916	10.9178	1092	9.1555	1269	7.8789	1447	6.9110	1626	6.1515	46
15	0919	10.8829	1095	9.1309	1272	7.8606	1450	6.8969	1629	6.1402	**45**
16	0922	10.8483	1098	9.1065	1275	7.8424	1453	6.8828	1632	6.1290	44
17	0925	10.8139	1101	9.0821	1278	7.8243	1456	6.8687	1635	6.1178	43
18	0928	10.7797	1104	9.0579	1281	7.8062	1459	6.8548	1638	6.1066	42
19	0931	10.7457	1107	9.0338	1284	7.7883	1462	6.8408	1641	6.0955	41
20	0934	10.7119	1110	9.0098	1287	7.7704	1465	6.8269	1644	6.0844	**40**
21	0936	10.6783	1113	8.9860	1290	7.7525	1468	6.8131	1647	6.0734	39
22	0939	10.6450	1116	8.9623	1293	7.7348	1471	6.7994	1650	6.0624	38
23	0942	10.6118	1119	8.9387	1296	7.7171	1474	6.7856	1653	6.0514	37
24	0945	10.5789	1122	8.9152	1299	7.6996	1477	6.7720	1655	6.0405	36
25	0948	10.5462	1125	8.8919	1302	7.6821	1480	6.7584	1658	6.0296	**35**
26	0951	10.5136	1128	8.8686	1305	7.6647	1483	6.7448	1661	6.0188	34
27	0954	10.4813	1131	8.8455	1308	7.6473	1486	6.7313	1664	6.0080	33
28	0957	10.4491	1134	8.8225	1311	7.6301	1489	6.7179	1667	5.9972	32
29	0960	10.4172	1136	8.7996	1314	7.6129	1492	6.7045	1670	5.9865	31
30	0963	10.3854	1139	8.7769	1317	7.5958	1495	6.6912	1673	5.9758	**30**
31	0966	10.3538	1142	8.7542	1319	7.5787	1497	6.6779	1676	5.9651	29
32	0969	10.3224	1145	8.7317	1322	7.5618	1500	6.6646	1679	5.9545	28
33	0972	10.2913	1148	8.7093	1325	7.5449	1503	6.6514	1682	5.9439	27
34	0975	10.2602	1151	8.6870	1328	7.5281	1506	6.6383	1685	5.9333	26
35	0978	10.2294	1154	8.6648	1331	7.5113	1509	6.6252	1688	5.9228	**25**
36	0981	10.1988	1157	8.6427	1334	7.4947	1512	6.6122	1691	5.9124	24
37	0983	10.1683	1160	8.6208	1337	7.4781	1515	6.5992	1694	5.9019	23
38	0986	10.1381	1163	8.5989	1340	7.4615	1518	6.5863	1697	5.8915	22
39	0989	10.1080	1166	8.5772	1343	7.4451	1521	6.5734	1700	5.8811	21
40	0992	10.0780	1169	8.5555	1346	7.4287	1524	6.5606	1703	5.8708	**20**
41	0995	10.0483	1172	8.5340	1349	7.4124	1527	6.5478	1706	5.8605	19
42	0998	10.0187	1175	8.5126	1352	7.3962	1530	6.5350	1709	5.8502	18
43	1001	9.9893	1178	8.4913	1355	7.3800	1533	6.5223	1712	5.8400	17
44	1004	9.9601	1181	8.4701	1358	7.3639	1536	6.5097	1715	5.8298	16
45	1007	9.9310	1184	8.4490	1361	7.3479	1539	6.4971	1718	5.8197	**15**
46	1010	9.9021	1187	8.4280	1364	7.3319	1542	6.4846	1721	5.8095	14
47	1013	9.8734	1189	8.4071	1367	7.3160	1545	6.4721	1724	5.7994	13
48	1016	9.8448	1192	8.3863	1370	7.3002	1548	6.4596	1727	5.7894	12
49	1019	9.8164	1195	8.3656	1373	7.2844	1551	6.4472	1730	5.7794	11
50	1022	9.7882	1198	8.3450	1376	7.2687	1554	6.4348	1733	5.7694	**10**
51	1025	9.7601	1201	8.3245	1379	7.2531	1557	6.4225	1736	5.7594	9
52	1028	9.7322	1204	8.3041	1382	7.2375	1560	6.4103	1739	5.7495	8
53	1030	9.7044	1207	8.2838	1385	7.2220	1563	6.3980	1742	5.7396	7
54	1033	9.6768	1210	8.2636	1388	7.2066	1566	6.3859	1745	5.7297	6
55	1036	9.6499	1213	8.2434	1391	7.1912	1569	6.3737	1748	5.7199	**5**
56	1039	9.6220	1216	8.2234	1394	7.1759	1572	6.3617	1751	5.7101	4
57	1042	9.5949	1219	8.2035	1397	7.1607	1575	6.3496	1754	5.7004	3
58	1045	9.5679	1222	8.1837	1399	7.1455	1578	6.3376	1757	5.6906	2
59	1048	9.5411	1225	8.1640	1402	7.1304	1581	6.3257	1760	5.6809	1
60	1051	9.5144	1228	8.1443	1405	7.1154	1584	6.3138	1763	5.6713	**0**
	cot	tan	cot	tan	cot	tan	cot	tan	cot	tan	
′	84°		83°		82°		81°		80°		′

′	10°		11°		12°		13°		14°		′
	tan	cot	tan	cot	tan	cot	tan	cot	tan	cot	
0	1763	5.6713	1944	5.1446	2126	4.7046	2309	4.3315	2493	4.0108	**60**
1	1766	5.6617	1947	5.1366	2129	4.6979	2312	4.3257	2496	4.0058	59
2	1769	5.6521	1950	5.1286	2132	4.6912	2315	4.3200	2499	4.0009	58
3	1772	5.6425	1953	5.1207	2135	4.6845	2318	4.3143	2503	3.9959	57
4	1775	5.6330	1956	5.1128	2138	4.6779	2321	4.3086	2506	3.9910	56
5	1778	5.6234	1959	5.1049	2141	4.6712	2324	4.3029	2509	3.9861	**55**
6	1781	5.6140	1962	5.0970	2144	4.6646	2327	4.2972	2512	3.9812	54
7	1784	5.6045	1965	5.0892	2147	4.6580	2330	4.2916	2515	3.9763	53
8	1787	5.5951	1968	5.0814	2150	4.6514	2333	4.2859	2518	3.9714	52
9	1790	5.5857	1971	5.0736	2153	4.6448	2336	4.2803	2521	3.9665	51
10	1793	5.5764	1974	5.0658	2156	4.6382	2339	4.2747	2524	3.9617	**50**
11	1796	5.5671	1977	5.0581	2159	4.6317	2342	4.2691	2527	3.9568	49
12	1799	5.5578	1980	5.0504	2162	4.6252	2345	4.2635	2530	3.9520	48
13	1802	5.5485	1983	5.0427	2165	4.6187	2349	4.2580	2533	3.9471	47
14	1805	5.5393	1986	5.0350	2168	4.6122	2352	4.2524	2537	3.9423	46
15	1808	5.5301	1989	5.0273	2171	4.6057	2355	4.2468	2540	3.9375	**45**
16	1811	5.5209	1992	5.0197	2174	4.5993	2358	4.2413	2543	3.9327	44
17	1814	5.5118	1995	5.0121	2177	4.5928	2361	4.2358	2546	3.9279	43
18	1817	5.5026	1998	5.0045	2180	4.5864	2364	4.2303	2549	3.9232	42
19	1820	5.4936	2001	4.9969	2183	4.5800	2367	4.2248	2552	3.9184	41
20	1823	5.4845	2004	4.9894	2186	4.5736	2370	4.2193	2555	3.9136	**40**
21	1826	5.4755	2007	4.9819	2189	4.5673	2373	4.2139	2558	3.9089	39
22	1829	5.4665	2010	4.9744	2193	4.5609	2376	4.2084	2561	3.9042	38
23	1832	5.4575	2013	4.9669	2196	4.5546	2379	4.2030	2564	3.8995	37
24	1835	5.4486	2016	4.9594	2199	4.5483	2382	4.1976	2568	3.8947	36
25	1838	5.4397	2019	4.9520	2202	4.5420	2385	4.1922	2571	3.8900	**35**
26	1841	5.4308	2022	4.9446	2205	4.5357	2388	4.1868	2574	3.8854	34
27	1844	5.4219	2025	4.9372	2208	4.5294	2392	4.1814	2577	3.8807	33
28	1847	5.4131	2028	4.9298	2211	4.5232	2395	4.1760	2580	3.8760	32
29	1850	5.4043	2031	4.9225	2214	4.5169	2398	4.1706	2583	3.8714	31
30	1853	5.3955	2035	4.9152	2217	4.5107	2401	4.1653	2586	3.8667	**30**
31	1856	5.3868	2038	4.9078	2220	4.5045	2404	4.1600	2589	3.8621	29
32	1859	5.3781	2941	4.9006	2223	4.4983	2407	4.1547	2592	3.8575	28
33	1862	5.3694	2044	4.8933	2226	4.4922	2410	4.1493	2595	3.8528	27
34	1865	5.3607	2047	4.8860	2229	4.4860	2413	4.1441	2599	3.8482	26
35	1868	5.3521	2050	4.8788	2232	4.4799	2416	4.1388	2602	3.8436	**25**
36	1871	5.3435	2053	4.8716	2235	4.4737	2419	4.1335	2605	3.8391	24
37	1874	5.3349	2056	4.8644	2238	4.4676	2422	4.1282	2608	3.8345	23
38	1877	5.3263	2059	4.8573	2241	4.4615	2425	4.1230	2611	3.8299	22
39	1880	5.3178	2062	4.8501	2244	4.4555	2428	4.1178	2614	3.8254	21
40	1883	5.3093	2065	4.8430	2247	4.4494	2432	4.1126	2617	3.8208	**20**
41	1887	5.3008	2068	4.8359	2251	4.4434	2435	4.1074	2620	3.8163	19
42	1890	5.2924	2071	4.8288	2254	4.4374	2438	4.1022	2623	3.8118	18
43	1893	5.2839	2074	4.8218	2257	4.4313	2441	4.0970	2627	3.8073	17
44	1896	5.2755	2077	4.8147	2260	4.4253	2444	4.0918	2630	3.8028	16
45	1899	5.2672	2080	4.8077	2263	4.4194	2447	4.0867	2633	3.7983	**15**
46	1902	5.2588	2083	4.8007	2266	4.4134	2450	4.0815	2636	3.7938	14
47	1905	5.2505	2086	4.7937	2269	4.4075	2453	4.0764	2639	3.7893	13
48	1908	5.2422	2089	4.7867	2272	4.4015	2456	4.0713	2642	3.7848	12
49	1911	5.2339	2092	4.7798	2275	4.3956	2459	4.0662	2645	3.7804	11
50	1914	5.2257	2095	4.7729	2278	4.3897	2462	4.0611	2648	3.7760	**10**
51	1917	5.2174	2098	4.7659	2281	4.3838	2465	4.0560	2651	3.7715	9
52	1920	5.2092	2101	4.7591	2284	4.3779	2469	4.0509	2655	3.7671	8
53	1923	5.2011	2104	4.7522	2287	4.3721	2472	4.0459	2658	3.7627	7
54	1926	5.1929	2107	4.7453	2290	4.3662	2475	4.0408	2661	3.7583	6
55	1929	5.1848	2110	4.7385	2293	4.3604	2478	4.0358	2664	3.7539	**5**
56	1932	5.1767	2113	4.7317	2296	4.3546	2481	4.0308	2667	3.7495	4
57	1935	5.1686	2116	4.7249	2299	4.3488	2484	4.0257	2670	3.7451	3
58	1938	5.1606	2119	4.7181	2303	4.3430	2487	4.0207	2673	3.7408	2
59	1941	5.1526	2123	4.7114	2306	4.3372	2490	4.0158	2676	3.7364	1
60	1944	5.1446	2126	4.7046	2309	4.3315	2493	4.0108	2679	3.7321	0
	cot	tan	cot	tan	cot	tan	cot	tan	cot	tan	
′	**79°**		**78°**		**77°**		**76°**		**75°**		′

′	15°		16°		17°		18°		19°		′
	tan	cot	tan	cot	tan	cot	tan	cot	tan	cot	
0	2679	3.7321	2867	3.4874	3057	3.2709	3249	3.0777	3443	2.9042	**60**
1	2683	3.7277	2871	3.4836	3060	3.2675	3252	3.0746	3447	2.9015	59
2	2686	3.7234	2874	3.4798	3064	3.2641	3256	3.0716	3450	2.8987	58
3	2689	3.7191	2877	3.4760	3067	3.2607	3259	3.0686	3453	2.8960	57
4	2692	3.7148	2880	3.4722	3070	3.2573	3262	3.0655	3456	2.8933	56
5	2695	3.7105	2883	3.4684	3073	3.2539	3265	3.0625	3460	2.8905	**55**
6	2698	3.7062	2886	3.4646	3076	3.2506	3269	3.0595	3463	2.8878	54
7	2701	3.7019	2890	3.4608	3080	3.2472	3272	3.0565	3466	2.8851	53
8	2704	3.6976	2893	3.4570	3083	3.2438	3275	3.0535	3469	2.8824	52
9	2708	3.6933	2896	3.4533	3086	3.2405	3278	3.0505	3473	2.8797	51
10	2711	3.6891	2899	3.4495	3089	3.2371	3281	3.0475	3476	2.8770	**50**
11	2714	3.6848	2902	3.4458	3092	3.2338	3285	3.0445	3479	2.8743	49
12	2717	3.6806	2905	3.4420	3096	3.2305	3288	3.0415	3482	2.8716	48
13	2720	3.6764	2908	3.4383	3099	3.2272	3291	3.0385	3486	2.8689	47
14	2723	3.6722	2912	3.4346	3102	3.2238	3294	3.0356	3489	2.8662	46
15	2726	3.6680	2915	3.4308	3105	3.2205	3298	3.0326	3492	2.8636	**45**
16	2729	3.6638	2918	3.4271	3108	3.2172	3301	3.0296	3495	2.8609	44
17	2733	3.6596	2921	3.4234	3111	3.2139	3304	3.0267	3499	2.8582	43
18	2736	3.6554	2924	3.4197	3115	3.2106	3307	3.0237	3502	2.8556	42
19	2739	3.6512	2927	3.4160	3118	3.2073	3310	3.0208	3505	2.8529	41
20	2742	3.6470	2931	3.4124	3121	3.2041	3314	3.0178	3508	2.8502	**40**
21	2745	3.6429	2934	3.4087	3124	3.2008	3317	3.0149	3512	2.8476	39
22	2748	3.6387	2937	3.4050	3127	3.1975	3320	3.0120	3515	2.8449	38
23	2751	3.6346	2940	3.4014	3131	3.1943	3323	3.0090	3518	2.8423	37
24	2754	3.6305	2943	3.3977	3134	3.1910	3327	3.0061	3522	2.8397	36
25	2758	3.6264 ·	2946	3.3941	3137	3.1878	3330	3.0032	3525	2.8370	**35**
26	2761	3.6222	2949	3.3904	3140	3.1845	3333	3.0003	3528	2.8344	34
27	2764	3.6181	2953	3.3868	3143	3.1813	3336	2.9974	3531	2.8318	33
28	2767	3.6140	2956	3.3832	3147	3.1780	3339	2.9945	3535	2.8291	32
29	2770	3.6100	2959	3.3796	3150	3.1748	3343	2.9916	3538	2.8265	31
30	2773	3.6059	2962	3.3759	3153	3.1716	3346	2.9887	3541	2.8239	**30**
31	2776	3.6018	2965	3.3723	3156	3.1684	3349	2.9858	3544	2.8213	29
32	2780	3.5978	2968	3.3687	3159	3.1652	3352	2.9829	3548	2.8187	28
33	2783	3.5937	2972	3.3652	3163	3.1620	3356	2.9800	3551	2.8161	27
34	2786	3.5897	2975	3.3616	3166	3.1588	3359	2.9772	3554	2.8135	26
35	2789	3.5856	2978	3.3580	3169	3.1556	3362	2.9743	3558	2.8109	**25**
36	2792	3.5816	2981	3.3544	3172	3.1524	3365	2.9714	3561	2.8083	24
37	2795	3.5776	2984	3.3509	3175	3.1492	3369	2.9686	3564	2.8057	23
38	2798	3.5736	2987	3.3473	3179	3.1460	3372	2.9657	3567	2.8032	22
39	2801	3.5696	2991	3.3438	3182	3.1429	3375	2.9629	3571	2.8006	21
40	2805	3.5656	2994	3.3402	3185	3.1397	3378	2.9600	3574	2.7980	**20**
41	2808	3.5616	2997	3.3367	3188	3.1366	3382	2.9572	3577	2.7955	19
42	2811	3.5576	3000	3.3332	3191	3.1334	3385	2.9544	3581	2.7929	18
43	2814	3.5536	3003	3.3297	3195	3.1303	3388	2.9515	3584	2.7903	17
44	2817	3.5497	3006	3.3261	3198	3.1271	3391	2.9487	3587	2.7878	16
45	2820	3.5457	3010	3.3226	3201	3.1240	3395	2.9459	3590	2.7852	**15**
46	2823	3.5418	3013	3.3191	3204	3.1209	3398	2.9431	3594	2.7827	14
47	2827	3.5379	3016	3.3156	3207	3.1178	3401	2.9403	3597	2.7801	13
48	2830	3.5339	3019	3.3122	3211	3.1146	3404	2.9375	3600	2.7776	12
49	2833	3.5300	3022	3.3087	3214	3.1115	3408	2.9347	3604	2.7751	11
50	2836	3.5261	3026	3.3052	3217	3.1084	3411	2.9319	3607	2.7725	**10**
51	2839	3.5222	3029	3.3017	3220	3.1053	3414	2.9291	3610	2.7700	9
52	2842	3.5183	3032	3.2983	3223	3.1022	3417	2.9263	3613	2.7675	8
53	2845	3.5144	3035	3.2948	3227	3.0991	3421	2.9235	3617	2.7650	7
54	2849	3.5105	3038	3.2914	3230	3.0961	3424	2.9208	3620	2.7625	6
55	2852	3.5067	3041	3.2880	3233	3.0930	3427	2.9180	3623	2.7600	**5**
56	2855	3.5028	3045	3.2845	3236	3.0899	3430	2.9152	3627	2.7575	4
57	2858	3.4989	3048	3.2811	3240	3.0868	3434	2.9125	3630	2.7550	3
58	2861	3.4951	3051	3.2777	3243	3.0838	3437	2.9097	3633	2.7525	2
59	2864	3.4912	3054	3.2743	3246	3.0807	3440	2.9070	3636	2.7500	1
60	2867	3.4874	3057	3.2709	3249	3.0777	3443	2.9042	3640	2.7475	**0**
	cot	tan	cot	tan	cot	tan	cot	tan	cot	tan	
′	74°		73°		72°		71°		70°		′

′	20° tan	cot	21° tan	cot	22° tan	cot	23° tan	cot	24° tan	cot	′
0	3640	2.7475	3839	2.6051	4040	2.4751	4245	2.3559	4452	2.2460	60
1	3643	2.7450	3842	2.6028	4044	2.4730	4248	2.3539	4456	2.2443	59
2	3646	2.7425	3845	2.6006	4047	2.4709	4252	2.3520	4459	2.2425	58
3	3650	2.7400	3849	2.5983	4050	2.4689	4255	2.3501	4463	2.2408	57
4	3653	2.7376	3852	2.5961	4054	2.4668	4258	2.3483	4466	2.2390	56
5	3656	2.7351	3855	2.5938	4057	2.4648	4262	2.3464	4470	2.2373	55
6	3659	2.7326	3859	2.5916	4061	2.4627	4265	2.3445	4473	2.2355	54
7	3663	2.7302	3862	2.5893	4064	2.4606	4269	2.3426	4477	2.2338	53
8	3666	2.7277	3865	2.5871	4067	2.4586	4272	2.3407	4480	2.2320	52
9	3669	2.7253	3869	2.5848	4071	2.4566	4276	2.3388	4484	2.2303	51
10	3673	2.7228	3872	2.5826	4074	2.4545	4279	2.3369	4487	2.2286	50
11	3676	2.7204	3875	2.5804	4078	2.4525	4283	2.3351	4491	2.2268	49
12	3679	2.7179	3879	2.5782	4081	2.4504	4286	2.3332	4494	2.2251	48
13	3683	2.7155	3882	2.5759	4084	2.4484	4289	2.3313	4498	2.2234	47
14	3686	2.7130	3885	2.5737	4088	2.4464	4293	2.3294	4501	2.2216	46
15	3689	2.7106	3889	2.5715	4091	2.4443	4296	2.3276	4505	2.2199	45
16	3693	2.7082	3892	2.5693	4095	2.4423	4300	2.3257	4508	2.2182	44
17	3696	2.7058	3895	2.5671	4098	2.4403	4303	2.3238	4512	2.2165	43
18	3699	2.7034	3899	2.5649	4101	2.4383	4307	2.3220	4515	2.2148	42
19	3702	2.7009	3902	2.5627	4105	2.4362	4310	2.3201	4519	2.2130	41
20	3706	2.6985	3906	2.5605	4108	2.4342	4314	2.3183	4522	2.2113	40
21	3709	2.6961	3909	2.5533	4111	2.4322	4317	2.3164	4526	2.2096	39
22	3712	2.6937	3912	2.5561	4115	2.4302	4320	2.3146	4529	2.2079	38
23	3716	2.6913	3916	2.5539	4118	2.4282	4324	2.3127	4533	2.2062	37
24	3719	2.6889	3919	2.5517	4122	2.4262	4327	2.3109	4536	2.2045	36
25	3722	2.6865	3922	2.5495	4125	2.4242	4331	2.3090	4540	2.2028	35
26	3726	2.6841	3926	2.5473	4129	2.4222	4334	2.3072	4543	2.2011	34
27	3729	2.6818	3929	2.5452	4132	2.4202	4338	2.3053	4547	2.1994	33
28	3732	2.6794	3932	2.5430	4135	2.4182	4341	2.3035	4550	2.1977	32
29	3736	2.6770	3936	2.5408	4139	2.4162	4345	2.3017	4554	2.1960	31
30	3739	2.6746	3939	2.5386	4142	2.4142	4348	2.2998	4557	2.1943	30
31	3742	2.6723	3942	2.5365	4146	2.4122	4352	2.2980	4561	2.1926	29
32	3745	2.6699	3946	2.5343	4149	2.4102	4355	2.2962	4564	2.1909	28
33	3749	2.6675	3949	2.5322	4152	2.4083	4359	2.2944	4568	2.1892	27
34	3752	2.6652	3953	2.5300	4156	2.4063	4362	2.2925	4571	2.1876	26
35	3755	2.6628	3956	2.5279	4159	2.4043	4365	2.2907	4575	2.1859	25
36	3759	2.6605	3959	2.5257	4163	2.4023	4369	2.2889	4578	2.1842	24
37	3762	2.6581	3963	2.5236	4166	2.4004	4372	2.2871	4582	2.1825	23
38	3765	2.6558	3966	2.5214	4169	2.3984	4376	2.2853	4585	2.1808	22
39	3769	2.6534	3969	2.5193	4173	2.3964	4379	2.2835	4589	2.1792	21
40	3772	2.6511	3973	2.5172	4176	2.3945	4383	2.2817	4592	2.1775	20
41	3775	2.6488	3976	2.5150	4180	2.3925	4386	2.2799	4596	2.1758	19
42	3779	2.6464	3979	2.5129	4183	2.3906	4390	2.2781	4599	2.1742	18
43	3782	2.6441	3983	2.5108	4187	2.3886	4393	2.2763	4603	2.1725	17
44	3785	2.6418	3986	2.5086	4190	2.3867	4397	2.2745	4607	2.1708	16
45	3789	2.6395	3990	2.5065	4193	2.3847	4400	2.2727	4610	2.1692	15
46	3792	2.6371	3993	2.5044	4197	2.3828	4404	2.2709	4614	2.1675	14
47	3795	2.6348	3996	2.5023	4200	2.3808	4407	2.2691	4617	2.1659	13
48	3799	2.6325	4000	2.5002	4204	2.3789	4411	2.2673	4621	2.1642	12
49	3802	2.6302	4003	2.4981	4207	2.3770	4414	2.2655	4624	2.1625	11
50	3805	2.6279	4006	2.4960	4210	2.3750	4417	2.2637	4628	2.1609	10
51	3809	2.6256	4010	2.4939	4214	2.3731	4421	2.2620	4631	2.1592	9
52	3812	2.6233	4013	2.4918	4217	2.3712	4424	2.2602	4635	2.1576	8
53	3815	2.6210	4017	2.4897	4221	2.3693	4428	2.2584	4638	2.1560	7
54	3819	2.6187	4020	2.4876	4224	2.3673	4431	2.2566	4642	2.1543	6
55	3822	2.6165	4023	2.4855	4228	2.3654	4435	2.2549	4645	2.1527	5
56	3825	2.6142	4027	2.4834	4231	2.3635	4438	2.2531	4649	2.1510	4
57	3829	2.6119	4030	2.4813	4234	2.3616	4442	2.2513	4652	2.1494	3
58	3832	2.6096	4033	2.4792	4238	2.3597	4445	2.2496	4656	2.1478	2
59	3835	2.6074	4037	2.4772	4241	2.3578	4449	2.2478	4660	2.1461	1
60	3839	2.6051	4040	2.4751	4245	2.3559	4452	2.2460	4663	2.1445	0
	cot	tan	cot	tan	cot	tan	cot	tan	cot	tan	
′	69°		68°		67°		66°		65°		′

′	25° tan	cot	26° tan	cot	27° tan	cot	28° tan	cot	29° tan	cot	′
0	4663	2.1445	4877	2.0503	5095	1.9626	5317	1.8807	5543	1.8040	60
1	4667	2.1429	4881	2.0488	5099	1.9612	5321	1.8794	5547	1.8028	59
2	4670	2.1413	4885	2.0473	5103	1.9598	5325	1.8781	5551	1.8016	58
3	4674	2.1396	4888	2.0458	5106	1.9584	5328	1.8768	5555	1.8003	57
4	4677	2.1380	4892	2.0443	5110	1.9570	5332	1.8755	5558	1.7991	56
5	4681	2.1364	4895	2.0428	5114	1.9556	5336	1.8741	5562	1.7979	55
6	4684	2.1348	4899	2.0413	5117	1.9542	5340	1.8728	5566	1.7966	54
7	4688	2.1332	4903	2.0398	5121	1.9528	5343	1.8715	5570	1.7954	53
8	4691	2.1315	4906	2.0383	5125	1.9514	5347	1.8702	5574	1.7942	52
9	4695	2.1299	4910	2.0368	5128	1.9500	5351	1.8689	5577	1.7930	51
10	4699	2.1283	4913	2.0353	5132	1.9486	5354	1.8676	5581	1.7917	50
11	4702	2.1267	4917	2.0338	5136	1.9472	5358	1.8663	5585	1.7905	49
12	4706	2.1251	4921	2.0323	5139	1.9458	5362	1.8650	5589	1.7893	48
13	4709	2.1235	4924	2.0308	5143	1.9444	5366	1.8637	5593	1.7881	47
14	4713	2.1219	4928	2.0293	5147	1.9430	5369	1.8624	5596	1.7868	46
15	4716	2.1203	4931	2.0278	5150	1.9416	5373	1.8611	5600	1.7856	45
16	4720	2.1187	4935	2.0263	5154	1.9402	5377	1.8598	5604	1.7844	44
17	4723	2.1171	4939	2.0248	5158	1.9388	5381	1.8585	5608	1.7832	43
18	4727	2.1155	4942	2.0233	5161	1.9375	5384	1.8572	5612	1.7820	42
19	4731	2.1139	4946	2.0219	5165	1.9361	5388	1.8559	5616	1.7808	41
20	4734	2.1123	4950	2.0204	5169	1.9347	5392	1.8546	5619	1.7796	40
21	4738	2.1107	4953	2.0189	5172	1.9333	5396	1.8533	5623	1.7783	39
22	4741	2.1092	4957	2.0174	5176	1.9319	5399	1.8520	5627	1.7771	38
23	4745	2.1076	4960	2.0160	5180	1.9306	5403	1.8507	5631	1.7759	37
24	4748	2.1060	4964	2.0145	5184	1.9292	5407	1.8495	5635	1.7747	36
25	4752	2.1044	4968	2.0130	5187	1.9278	5411	1.8482	5639	1.7735	35
26	4755	2.1028	4971	2.0115	5191	1.9265	5415	1.8469	5642	1.7723	34
27	4759	2.1013	4975	2.0101	5195	1.9251	5418	1.8456	5646	1.7711	33
28	4763	2.0997	4979	2.0086	5198	1.9237	5422	1.8443	5650	1.7699	32
29	4766	2.0981	4982	2.0072	5202	1.9223	5426	1.8430	5654	1.7687	31
30	4770	2.0965	4986	2.0057	5206	1.9210	5430	1.8418	5658	1.7675	30
31	4773	2.0950	4989	2.0042	5209	1.9196	5433	1.8405	5662	1.7663	29
32	4777	2.0934	4993	2.0028	5213	1.9183	5437	1.8392	5665	1.7651	28
33	4780	2.0918	4997	2.0013	5217	1.9169	5441	1.8379	5669	1.7639	27
34	4784	2.0903	5000	1.9999	5220	1.9155	5445	1.8367	5673	1.7627	26
35	4788	2.0887	5004	1.9984	5224	1.9142	5448	1.8354	5677	1.7615	25
36	4791	2.0872	5008	1.9970	5228	1.9128	5452	1.8341	5681	1.7603	24
37	4795	2.0856	5011	1.9955	5232	1.9115	5456	1.8329	5685	1.7591	23
38	4798	2.0840	5015	1.9941	5235	1.9101	5460	1.8316	5688	1.7579	22
39	4802	2.0825	5019	1.9926	5239	1.9088	5464	1.8303	5692	1.7567	21
40	4806	2.0809	5022	1.9912	5243	1.9074	5467	1.8291	5696	1.7556	20
41	4809	2.0794	5026	1.9897	5246	1.9061	5471	1.8278	5700	1.7544	19
42	4813	2.0778	5029	1.9883	5250	1.9047	5475	1.8265	5704	1.7532	18
43	4816	2.0763	5033	1.9868	5254	1.9034	5479	1.8253	5708	1.7520	17
44	4820	2.0748	5037	1.9854	5258	1.9020	5482	1.8240	5712	1.7508	16
45	4823	2.0732	5040	1.9840	5261	1.9007	5486	1.8228	5715	1.7496	15
46	4827	2.0717	5044	1.9825	5265	1.8993	5490	1.8215	5719	1.7485	14
47	4831	2.0701	5048	1.9811	5269	1.8980	5494	1.8202	5723	1.7473	13
48	4834	2.0686	5051	1.9797	5272	1.8967	5498	1.8190	5727	1.7461	12
49	4838	2.0671	5055	1.9782	5276	1.8953	5501	1.8177	5731	1.7449	11
50	4841	2.0655	5059	1.9768	5280	1.8940	5505	1.8165	5735	1.7437	10
51	4845	2.0640	5062	1.9754	5284	1.8927	5509	1.8152	5739	1.7426	9
52	4849	2.0625	5066	1.9740	5287	1.8913	5513	1.8140	5743	1.7414	8
53	4852	2.0609	5070	1.9725	5291	1.8900	5517	1.8127	5746	1.7402	7
54	4856	2.0594	5073	1.9711	5295	1.8887	5520	1.8115	5750	1.7391	6
55	4859	2.0579	5077	1.9697	5298	1.8873	5524	1.8103	5754	1.7379	5
56	4863	2.0564	5081	1.9683	5302	1.8860	5528	1.8090	5758	1.7367	4
57	4867	2.0549	5084	1.9669	5306	1.8847	5532	1.8078	5762	1.7355	3
58	4870	2.0533	5088	1.9654	5310	1.8834	5535	1.8065	5766	1.7344	2
59	4874	2.0518	5092	1.9640	5313	1.8820	5539	1.8053	5770	1.7332	1
60	4877	2.0503	5095	1.9626	5317	1.8807	5543	1.8040	5774	1.7321	0
	cot	tan	cot	tan	cot	tan	cot	tan	cot	tan	′
	64°		63°		62°		61°		60°		

′	30°		31°		32°		33°		34°		′
	tan	cot	tan	cot	tan	cot	tan	cot	tan	cot	
0	5774	1.7321	6009	1.6643	6249	1.6003	6494	1.5399	6745	1.4826	60
1	5777	1.7309	6013	1.6632	6253	1.5993	6498	1.5389	6749	1.4816	59
2	5781	1.7297	6017	1.6621	6257	1.5983	6502	1.5379	6754	1.4807	58
3	5785	1.7286	6020	1.6610	6261	1.5972	6506	1.5369	6758	1.4798	57
4	5789	1.7274	6024	1.6599	6265	1.5962	6511	1.5359	6762	1.4788	56
5	5793	1.7262	6028	1.6588	6269	1.5952	6515	1.5350	6766	1.4779	55
6	5797	1.7251	6032	1.6577	6273	1.5941	6519	1.5340	6771	1.4770	54
7	5801	1.7239	6036	1.6566	6277	1.5931	6523	1.5330	6775	1.4761	53
8	5805	1.7228	6040	1.6555	6281	1.5921	6527	1.5320	6779	1.4751	52
9	5808	1.7216	6044	1.6545	6285	1.5911	6531	1.5311	6783	1.4742	51
10	5812	1.7205	6048	1.6534	6289	1.5900	6536	1.5301	6787	1.4733	50
11	5816	1.7193	6052	1.6523	6293	1.5890	6540	1.5291	6792	1.4724	49
12	5820	1.7182	6056	1.6512	6297	1.5880	6544	1.5282	6796	1.4715	48
13	5824	1.7170	6060	1.6501	6301	1.5869	6548	1.5272	6800	1.4705	47
14	5828	1.7159	6064	1.6490	6305	1.5859	6552	1.5262	6805	1.4696	46
15	5832	1.7147	6068	1.6479	6310	1.5849	6556	1.5253	6809	1.4687	45
16	5836	1.7136	6072	1.6469	6314	1.5839	6560	1.5243	6813	1.4678	44
17	5840	1.7124	6076	1.6458	6318	1.5829	6565	1.5233	6817	1.4669	43
18	5844	1.7113	6080	1.6447	6322	1.5818	6569	1.5224	6822	1.4659	42
19	5847	1.7102	6084	1.6436	6326	1.5808	6573	1.5214	6826	1.4650	41
20	5851	1.7090	6088	1.6426	6330	1.5798	6577	1.5204	6830	1.4641	40
21	5855	1.7079	6092	1.6415	6334	1.5788	6581	1.5195	6834	1.4632	39
22	5859	1.7067	6096	1.6404	6338	1.5778	6585	1.5185	6839	1.4623	38
23	5863	1.7056	6100	1.6393	6342	1.5768	6590	1.5175	6843	1.4614	37
24	5867	1.7045	6104	1.6383	6346	1.5757	6594	1.5166	6847	1.4605	36
25	5871	1.7033	6108	1.6372	6350	1.5747	6598	1.5156	6851	1.4596	35
26	5875	1.7022	6112	1.6361	6354	1.5737	6602	1.5147	6856	1.4586	34
27	5879	1.7011	6116	1.6351	6358	1.5727	6606	1.5137	6860	1.4577	33
28	5883	1.6999	6120	1.6340	6363	1.5717	6610	1.5127	6864	1.4568	32
29	5887	1.6988	6124	1.6329	6367	1.5707	6615	1.5118	6869	1.4559	31
30	5890	1.6977	6128	1.6319	6371	1.5697	6619	1.5108	6873	1.4550	30
31	5894	1.6965	6132	1.6308	6375	1.5687	6623	1.5099	6877	1.4541	29
32	5898	1.6954	6136	1.6297	6379	1.5677	6627	1.5089	6881	1.4532	28
33	5902	1.6943	6140	1.6287	6383	1.5667	6631	1.5080	6886	1.4523	27
34	5906	1.6932	6144	1.6276	6387	1.5657	6636	1.5070	6890	1.4514	26
35	5910	1.6920	6148	1.6265	6391	1.5647	6640	1.5061	6894	1.4505	25
36	5914	1.6909	6152	1.6255	6395	1.5637	6644	1.5051	6899	1.4496	24
37	5918	1.6898	6156	1.6244	6399	1.5627	6648	1.5042	6903	1.4487	23
38	5922	1.6887	6160	1.6234	6403	1.5617	6652	1.5032	6907	1.4478	22
39	5926	1.6875	6164	1.6223	6408	1.5607	6657	1.5023	6911	1.4469	21
40	5930	1.6864	6168	1.6212	6412	1.5597	6661	1.5013	6916	1.4460	20
41	5934	1.6853	6172	1.6202	6416	1.5587	6665	1.5004	6920	1.4451	19
42	5938	1.6842	6176	1.6191	6420	1.5577	6669	1.4994	6924	1.4442	18
43	5942	1.6831	6180	1.6181	6424	1.5567	6673	1.4985	6929	1.4433	17
44	5945	1.6820	6184	1.6170	6428	1.5557	6678	1.4975	6933	1.4424	16
45	5949	1.6808	6188	1.6160	6432	1.5547	6682	1.4966	6937	1.4415	15
46	5953	1.6797	6192	1.6149	6436	1.5537	6686	1.4957	6942	1.4406	14
47	5957	1.6786	6196	1.6139	6440	1.5527	6690	1.4947	6946	1.4397	13
48	5961	1.6775	6200	1.6128	6445	1.5517	6694	1.4938	6950	1.4388	12
49	5965	1.6764	6204	1.6118	6449	1.5507	6699	1.4928	6954	1.4379	11
50	5969	1.6753	6208	1.6107	6453	1.5497	6703	1.4919	6959	1.4370	10
51	5973	1.6742	6212	1.6097	6457	1.5487	6707	1.4910	6963	1.4361	9
52	5977	1.6731	6216	1.6087	6461	1.5477	6711	1.4900	6967	1.4352	8
53	5981	1.6720	6220	1.6076	6465	1.5468	6716	1.4891	6972	1.4344	7
54	5985	1.6709	6224	1.6066	6469	1.5458	6720	1.4882	6976	1.4335	6
55	5989	1.6698	6228	1.6055	6473	1.5448	6724	1.4872	6980	1.4326	5
56	5993	1.6687	6233	1.6045	6478	1.5438	6728	1.4863	6985	1.4317	4
57	5997	1.6676	6237	1.6034	6482	1.5428	6732	1.4854	6989	1.4308	3
58	6001	1.6665	6241	1.6024	6486	1.5418	6737	1.4844	6993	1.4299	2
59	6005	1.6654	6245	1.6014	6490	1.5408	6741	1.4835	6998	1.4290	1
60	6009	1.6643	6249	1.6003	6494	1.5399	6745	1.4826	7002	1.4281	0
	cot	tan	cot	tan	cot	tan	cot	tan	cot	tan	
′	59°		58°		57°		56°		55°		′

′	35°		36°		37°		38°		39°		′
	tan	cot	tan	cot	tan	cot	tan	cot	tan	cot	
0	7002	1.4281	7265	1.3764	7536	1.3270	7813	1.2799	8098	1.2349	60
1	7006	1.4273	7270	1.3755	7540	1.3262	7818	1.2792	8103	1.2342	59
2	7011	1.4264	7274	1.3747	7545	1.3254	7822	1.2784	8107	1.2334	58
3	7015	1.4255	7279	1.3739	7549	1.3246	7827	1.2776	8112	1.2327	57
4	7019	1.4246	7283	1.3730	7554	1.3238	7832	1.2769	8117	1.2320	56
5	7024	1.4237	7288	1.3722	7558	1.3230	7836	1.2761	8122	1.2312	55
6	7028	1.4229	7292	1.3713	7563	1.3222	7841	1.2753	8127	1.2305	54
7	7032	1.4220	7297	1.3705	7568	1.3214	7846	1.2746	8132	1.2298	53
8	7037	1.4211	7301	1.3697	7572	1.3206	7850	1.2738	8136	1.2290	52
9	7041	1.4202	7306	1.3688	7577	1.3198	7855	1.2731	8141	1.2283	51
10	7046	1.4193	7310	1.3680	7581	1.3190	7860	1.2723	8146	1.2276	50
11	7050	1.4185	7314	1.3672	7586	1.3182	7865	1.2715	8151	1.2268	49
12	7054	1.4176	7319	1.3663	7590	1.3175	7869	1.2708	8156	1.2261	48
13	7059	1.4167	7323	1.3655	7595	1.3167	7874	1.2700	8161	1.2254	47
14	7063	1.4158	7328	1.3647	7600	1.3159	7879	1.2693	8165	1.2247	46
15	7067	1.4150	7332	1.3638	7604	1.3151	7883	1.2685	8170	1.2239	45
16	7072	1.4141	7337	1.3630	7609	1.3143	7888	1.2677	8175	1.2232	44
17	7076	1.4132	7341	1.3622	7613	1.3135	7893	1.2670	8180	1.2225	43
18	7080	1.4124	7346	1.3613	7618	1.3127	7898	1.2662	8185	1.2218	42
19	7085	1.4115	7350	1.3605	7623	1.3119	7902	1.2655	8190	1.2210	41
20	7089	1.4106	7355	1.3597	7627	1.3111	7907	1.2647	8195	1.2203	40
21	7094	1.4097	7359	1.3588	7632	1.3103	7912	1.2640	8199	1.2196	39
22	7098	1.4089	7364	1.3580	7636	1.3095	7916	1.2632	8204	1.2189	38
23	7102	1.4080	7368	1.3572	7641	1.3087	7921	1.2624	8209	1.2181	37
24	7107	1.4071	7373	1.3564	7646	1.3079	7926	1.2617	8214	1.2174	36
25	7111	1.4063	7377	1.3555	7650	1.3072	7931	1.2609	8219	1.2167	35
26	7115	1.4054	7382	1.3547	7655	1.3064	7935	1.2602	8224	1.2160	34
27	7120	1.4045	7386	1.3539	7659	1.3056	7940	1.2594	8229	1.2153	33
28	7124	1.4037	7391	1.3531	7664	1.3048	7945	1.2587	8234	1.2145	32
29	7129	1.4028	7395	1.3522	7669	1.3040	7950	1.2579	8238	1.2138	31
30	7133	1.4019	7400	1.3514	7673	1.3032	7954	1.2572	8243	1.2131	30
31	7137	1.4011	7404	1.3506	7678	1.3024	7959	1.2564	8248	1.2124	29
32	7142	1.4002	7409	1.3498	7683	1.3017	7964	1.2557	8253	1.2117	28
33	7146	1.3994	7413	1.3490	7687	1.3009	7969	1.2549	8258	1.2109	27
34	7151	1.3985	7418	1.3481	7692	1.3001	7973	1.2542	8263	1.2102	26
35	7155	1.3976	7422	1.3473	7696	1.2993	7978	1.2534	8268	1.2095	25
36	7159	1.3968	7427	1.3465	7701	1.2985	7983	1.2527	8273	1.2088	24
37	7164	1.3959	7431	1.3457	7706	1.2977	7988	1.2519	8278	1.2081	23
38	7168	1.3951	7436	1.3449	7710	1.2970	7992	1.2512	8283	1.2074	22
39	7173	1.3942	7440	1.3440	7715	1.2962	7997	1.2504	8287	1.2066	21
40	7177	1.3934	7445	1.3432	7720	1.2954	8002	1.2497	8292	1.2059	20
41	7181	1.3925	7449	1.3424	7724	1.2946	8007	1.2489	8297	1.2052	19
42	7186	1.3916	7454	1.3416	7729	1.2938	8012	1.2482	8302	1.2045	18
43	7190	1.3908	7458	1.3408	7734	1.2931	8016	1.2475	8307	1.2038	17
44	7195	1.3899	7463	1.3400	7738	1.2923	8021	1.2467	8312	1.2031	16
45	7199	1.3891	7467	1.3392	7743	1.2915	8026	1.2460	8317	1.2024	15
46	7203	1.3882	7472	1.3384	7747	1.2907	8031	1.2452	8322	1.2017	14
47	7208	1.3874	7476	1.3375	7752	1.2900	8035	1.2445	8327	1.2009	13
48	7212	1.3865	7481	1.3367	7757	1.2892	8040	1.2437	8332	1.2002	12
49	7217	1.3857	7485	1.3359	7761	1.2884	8045	1.2430	8337	1.1995	11
50	7221	1.3848	7490	1.3351	7766	1.2876	8050	1.2423	8342	1.1988	10
51	7226	1.3840	7495	1.3343	7771	1.2869	8055	1.2415	8346	1.1981	9
52	7230	1.3831	7499	1.3335	7775	1.2861	8059	1.2408	8351	1.1974	8
53	7234	1.3823	7504	1.3327	7780	1.2853	8064	1.2401	8356	1.1967	7
54	7239	1.3814	7508	1.3319	7785	1.2846	8069	1.2393	8361	1.1960	6
55	7243	1.3806	7513	1.3311	7789	1.2838	8074	1.2386	8366	1.1953	5
56	7248	1.3798	7517	1.3303	7794	1.2830	8079	1.2378	8371	1.1946	4
57	7252	1.3789	7522	1.3295	7799	1.2822	8083	1.2371	8376	1.1939	3
58	7257	1.3781	7526	1.3287	7803	1.2815	8088	1.2364	8381	1.1932	2
59	7261	1.3772	7531	1.3278	7808	1.2807	8093	1.2356	8386	1.1925	1
60	7265	1.3764	7536	1.3270	7813	1.2799	8098	1.2349	8391	1.1918	0
	cot	tan	cot	tan	cot	tan	cot	tan	cot	tan	
′	54°		53°		52°		51°		50°		′

′	40° tan	cot	41° tan	cot	42° tan	cot	43° tan	cot	44° tan	cot	′
0	8391	1.1918	8693	1.1504	9004	1.1106	9325	1.0724	9657	1.0355	60
1	8396	1.1910	8698	1.1497	9009	1.1100	9331	1.0717	9663	1.0349	59
2	8401	1.1903	8703	1.1490	9015	1.1093	9336	1.0711	9668	1.0343	58
3	8406	1.1896	8708	1.1483	9020	1.1087	9341	1.0705	9674	1.0337	57
4	8411	1.1889	8713	1.1477	9025	1.1080	9347	1.0699	9679	1.0331	56
5	8416	1.1882	8718	1.1470	9030	1.1074	9352	1.0692	9685	1.0325	55
6	8421	1.1875	8724	1.1463	9036	1.1067	9358	1.0686	9691	1.0319	54
7	8426	1.1868	8729	1.1456	9041	1.1061	9363	1.0680	9696	1.0313	53
8	8431	1.1861	8734	1.1450	9046	1.1054	9369	1.0674	9702	1.0307	52
9	8436	1.1854	8739	1.1443	9052	1.1048	9374	1.0668	9708	1.0301	51
10	8441	1.1847	8744	1.1436	9057	1.1041	9380	1.0661	9713	1.0295	50
11	8446	1.1840	8749	1.1430	9062	1.1035	9385	1.0655	9719	1.0289	49
12	8451	1.1833	8754	1.1423	9067	1.1028	9391	1.0649	9725	1.0283	48
13	8456	1.1826	8759	1.1416	9073	1.1022	9396	1.0643	9730	1.0277	47
14	8461	1.1819	8765	1.1410	9078	1.1016	9402	1.0637	9736	1.0271	46
15	8466	1.1812	8770	1.1403	9083	1.1009	9407	1.0630	9742	1.0265	45
16	8471	1.1806	8775	1.1396	9089	1.1003	9413	1.0624	9747	1.0259	44
17	8476	1.1799	8780	1.1389	9094	1.0996	9418	1.0618	9753	1.0253	43
18	8481	1.1792	8785	1.1383	9099	1.0990	9424	1.0612	9759	1.0247	42
19	8486	1.1785	8790	1.1376	9105	1.0983	9429	1.0606	9764	1.0241	41
20	8491	1.1778	8796	1.1369	9110	1.0977	9435	1.0599	9770	1.0235	40
21	8496	1.1771	8801	1.1363	9115	1.0971	9440	1.0593	9776	1.0230	39
22	8501	1.1764	8806	1.1356	9121	1.0964	9446	1.0587	9781	1.0224	38
23	8506	1.1757	8811	1.1349	9126	1.0958	9451	1.0581	9787	1.0218	37
24	8511	1.1750	8816	1.1343	9131	1.0951	9457	1.0575	9793	1.0212	36
25	8516	1.1743	8821	1.1336	9137	1.0945	9462	1.0569	9798	1.0206	35
26	8521	1.1736	8827	1.1329	9142	1.0939	9468	1.0562	9804	1.0200	34
27	8526	1.1729	8832	1.1323	9147	1.0932	9473	1.0556	9810	1.0194	33
28	8531	1.1722	8837	1.1316	9153	1.0926	9479	1.0550	9816	1.0188	32
29	8536	1.1715	8842	1.1310	9158	1.0919	9484	1.0544	9821	1.0182	31
30	8541	1.1708	8847	1.1303	9163	1.0913	9490	1.0538	9827	1.0176	30
31	8546	1.1702	8852	1.1296	9169	1.0907	9495	1.0532	9833	1.0170	29
32	8551	1.1695	8858	1.1290	9174	1.0900	9501	1.0526	9838	1.0164	28
33	8556	1.1688	8863	1.1283	9179	1.0894	9506	1.0519	9844	1.0158	27
34	8561	1.1681	8868	1.1276	9185	1.0888	9512	1.0513	9850	1.0152	26
35	8566	1.1674	8873	1.1270	9190	1.0881	9517	1.0507	9856	1.0147	25
36	8571	1.1667	8878	1.1263	9195	1.0875	9523	1.0501	9861	1.0141	24
37	8576	1.1660	8884	1.1257	9201	1.0869	9528	1.0495	9867	1.0135	23
38	8581	1.1653	8889	1.1250	9206	1.0862	9534	1.0489	9873	1.0129	22
39	8586	1.1647	8894	1.1243	9212	1.0856	9540	1.0483	9879	1.0123	21
40	8591	1.1640	8899	1.1237	9217	1.0850	9545	1.0477	9884	1.0117	20
41	8596	1.1633	8904	1.1230	9222	1.0843	9551	1.0470	9890	1.0111	19
42	8601	1.1626	8910	1.1224	9228	1.0837	9556	1.0464	9896	1.0105	18
43	8606	1.1619	8915	1.1217	9233	1.0831	9562	1.0458	9902	1.0099	17
44	8611	1.1612	8920	1.1211	9239	1.0824	9567	1.0452	9907	1.0094	16
45	8617	1.1606	8925	1.1204	9244	1.0818	9573	1.0446	9913	1.0088	15
46	8622	1.1599	8931	1.1197	9249	1.0812	9578	1.0440	9919	1.0082	14
47	8627	1.1592	8936	1.1191	9255	1.0805	9584	1.0434	9925	1.0076	13
48	8632	1.1585	8941	1.1184	9260	1.0799	9590	1.0428	9930	1.0070	12
49	8637	1.1578	8946	1.1178	9266	1.0793	9595	1.0422	9936	1.0064	11
50	8642	1.1571	8952	1.1171	9271	1.0786	9601	1.0416	9942	1.0058	10
51	8647	1.1565	8957	1.1165	9276	1.0780	9606	1.0410	9948	1.0052	9
52	8652	1.1558	8962	1.1158	9282	1.0774	9612	1.0404	9954	1.0047	8
53	8657	1.1551	8967	1.1152	9287	1.0768	9618	1.0398	9959	1.0041	7
54	8662	1.1544	8972	1.1145	9293	1.0761	9623	1.0392	9965	1.0035	6
55	8667	1.1538	8978	1.1139	9298	1.0755	9629	1.0385	9971	1.0029	5
56	8672	1.1531	8983	1.1132	9303	1.0749	9634	1.0379	9977	1.0023	4
57	8678	1.1524	8988	1.1126	9309	1.0742	9640	1.0373	9983	1.0017	3
58	8683	1.1517	8994	1.1119	9314	1.0736	9646	1.0367	9988	1.0012	2
59	8688	1.1510	8999	1.1113	9320	1.0730	9651	1.0361	9994	1.0006	1
60	8693	1.1504	9004	1.1106	9325	1.0724	9657	1.0355	1.000	1.0000	0
′	cot	tan 49°	cot	tan 48°	cot	tan 47°	cot	tan 46°	cot	tan 45°	′

Bearing.	Distance 1.		Distance 2.		Distance 3.		Distance 4.		Distance 5.		Bearing.
° ′	Lat.	Dep.	Lat.	Dep.	Lat.	Dep.	Lat.	Dep.	Lat.	Dep.	° ′
0 15	1.000	0.004	2.000	0.009	3.000	0.013	4.000	0.017	5.000	0.022	**89** 45
30	1.000	0.009	2.000	0.017	3.000	0.026	4.000	0.035	5.000	0.044	30
45	1.000	0.013	2.000	0.026	3.000	0.039	4.000	0.052	5.000	0.065	15
1 0	1.000	0.017	2.000	0.035	3.000	0.052	3.999	0.070	4.999	0.087	**89** 0
15	1.000	0.022	2.000	0.044	2.999	0.065	3.999	0.087	4.999	0.109	45
30	1.000	0.026	1.999	0.052	2.999	0.079	3.999	0.105	4.998	0.131	30
45	1.000	0.031	1.999	0.061	2.999	0.092	3.998	0.122	4.998	0.153	15
2 0	0.999	0.035	1.999	0.070	2.998	0.105	3.998	0.140	4.997	0.174	**88** 0
15	0.999	0.039	1.998	0.079	2.998	0.118	3.997	0.157	4.996	0.196	45
30	0.999	0.044	1.998	0.087	2.997	0.131	3.996	0.174	4.995	0.218	30
45	0.999	0.048	1.998	0.096	2.997	0.144	3.995	0.192	4.994	0.240	15
3 0	0.999	0.052	1.997	0.105	2.996	0.157	3.995	0.209	4.993	0.262	**87** 0
15	0.998	0.057	1.997	0.113	2.995	0.170	3.994	0.227	4.992	0.283	45
30	0.998	0.061	1.996	0.122	2.994	0.183	3.993	0.244	4.991	0.305	30
45	0.998	0.065	1.996	0.131	2.994	0.196	3.991	0.262	4.989	0.327	15
4 0	0.998	0.070	1.995	0.140	2.993	0.209	3.990	0.279	4.988	0.349	**86** 0
15	0.997	0.074	1.995	0.148	2.992	0.222	3.989	0.296	4.986	0.371	45
30	0.997	0.078	1.994	0.157	2.991	0.235	3.988	0.314	4.985	0.392	30
45	0.997	0.083	1.993	0.166	2.990	0.248	3.986	0.331	4.983	0.414	15
5 0	0.996	0.087	1.992	0.174	2.989	0.261	3.985	0.349	4.981	0.436	**85** 0
15	0.996	0.092	1.992	0.183	2.987	0.275	3.983	0.366	4.979	0.458	45
30	0.995	0.096	1.991	0.192	2.986	0.288	3.982	0.383	4.977	0.479	30
45	0.995	0.100	1.990	0.200	2.985	0.301	3.980	0.401	4.975	0.501	15
6 0	0.995	0.105	1.989	0.209	2.984	0.314	3.978	0.418	4.973	0.523	**84** 0
15	0.994	0.109	1.988	0.218	2.982	0.327	3.976	0.435	4.970	0.544	45
30	0.994	0.113	1.987	0.226	2.981	0.340	3.974	0.453	4.968	0.566	30
45	0.993	0.118	1.986	0.235	2.979	0.353	3.972	0.470	4.965	0.588	15
7 0	0.993	0.122	1.985	0.244	2.978	0.366	3.970	0.487	4.963	0.609	**83** 0
15	0.992	0.126	1.984	0.252	2.976	0.379	3.968	0.505	4.960	0.631	45
30	0.991	0.131	1.983	0.261	2.974	0.392	3.966	0.522	4.957	0.653	30
45	0.991	0.135	1.982	0.270	2.973	0.405	3.963	0.539	4.954	0.674	15
8 0	0.990	0.139	1.981	0.278	2.971	0.418	3.961	0.557	4.951	0.696	**82** 0
15	0.990	0.143	1.979	0.287	2.969	0.430	3.959	0.574	4.948	0.717	45
30	0.989	0.148	1.978	0.296	2.967	0.443	3.956	0.591	4.945	0.739	30
45	0.988	0.152	1.977	0.304	2.965	0.456	3.953	0.608	4.942	0.761	15
9 0	0.988	0.156	1.975	0.313	2.963	0.469	3.951	0.626	4.938	0.782	**81** 0
15	0.987	0.161	1.974	0.321	2.961	0.482	3.948	0.643	4.935	0.804	45
30	0.986	0.165	1.973	0.330	2.959	0.495	3.945	0.660	4.931	0.825	30
45	0.986	0.169	1.971	0.339	2.957	0.508	3.942	0.677	4.928	0.847	15
10 0	0.985	0.174	1.970	0.347	2.954	0.521	3.939	0.695	4.924	0.868	**80** 0
15	0.984	0.178	1.968	0.356	2.952	0.534	3.936	0.712	4.920	0.890	45
30	0.983	0.182	1.967	0.364	2.950	0.547	3.933	0.729	4.916	0.911	30
45	0.982	0.187	1.965	0.373	2.947	0.560	3.930	0.746	4.912	0.933	15
11 0	0.982	0.191	1.963	0.382	2.945	0.572	3.927	0.763	4.908	0.954	**79** 0
15	0.981	0.195	1.962	0.390	2.942	0.585	3.923	0.780	4.904	0.975	45
30	0.980	0.199	1.960	0.399	2.940	0.598	3.920	0.797	4.900	0.997	30
45	0.979	0.204	1.958	0.407	2.937	0.611	3.916	0.815	4.895	1.018	15
12 0	0.978	0.208	1.956	0.416	2.934	0.624	3.913	0.832	4.891	1.040	**78** 0
15	0.977	0.212	1.954	0.424	2.932	0.637	3.909	0.849	4.886	1.061	45
30	0.976	0.216	1.953	0.433	2.929	0.649	3.905	0.866	4.881	1.082	30
45	0.975	0.221	1.951	0.441	2.926	0.662	3.901	0.883	4.877	1.103	15
13 0	0.974	0.225	1.949	0.450	2.923	0.675	3.897	0.900	4.872	1.125	**77** 0
15	0.973	0.229	1.947	0.458	2.920	0.688	3.894	0.917	4.867	1.146	45
30	0.972	0.233	1.945	0.467	2.917	0.700	3.889	0.934	4.862	1.167	30
45	0.971	0.238	1.943	0.475	2.914	0.713	3.885	0.951	4.857	1.188	15
14 0	0.970	0.242	1.941	0.484	2.911	0.726	3.881	0.968	4.851	1.210	**76** 0
15	0.969	0.246	1.938	0.492	2.908	0.738	3.877	0.985	4.846	1.231	45
30	0.968	0.250	1.936	0.501	2.904	0.751	3.873	1.002	4.841	1.252	30
45	0.967	0.255	1.934	0.509	2.901	0.764	3.868	1.018	4.835	1.273	15
15 0	0.966	0.259	1.932	0.518	2.898	0.776	3.864	1.035	4.830	1.294	**75** 0
° ′	Dep.	Lat.	Dep.	Lat.	Dep.	Lat.	Dep.	Lat.	Dep.	Lat.	° ′
Bearing	Distance 1.		Distance 2.		Distance 3.		Distance 4.		Distance 5.		Bearing.

75°—90°

Bearing.	Distance 6.		Distance 7.		Distance 8.		Distance 9.		Distance 10.		Bearing.
° '	Lat.	Dep.	Lat.	Dep.	Lat.	Dep.	Lat.	Dep.	Lat.	Dep.	° '
0 15	6.000	0.026	7.000	0.031	8.000	0.035	9.000	0.039	10.000	0.044	**89** 45
30	6.000	0.052	7.000	0.061	8.000	0.070	9.000	0.079	10.000	0.087	30
45	5.999	0.079	6.999	0.092	7.999	0.105	8.999	0.118	9.999	0.131	15
1 0	5.999	0.105	6.999	0.122	7.999	0.140	8.999	0.157	9.999	0.175	**89** 0
15	5.999	0.131	6.998	0.153	7.998	0.175	8.998	0.196	9.998	0.218	45
30	5.998	0.157	6.998	0.183	7.997	0.209	8.997	0.236	9.997	0.262	30
45	5.997	0.183	6.997	0.214	7.996	0.244	8.996	0.275	9.995	0.305	15
2 0	5.996	0.209	6.996	0.244	7.995	0.279	8.995	0.314	9.994	0.349	**88** 0
15	5.995	0.236	6.995	0.275	7.994	0.314	8.993	0.353	9.992	0.393	45
30	5.994	0.262	6.993	0.305	7.992	0.349	8.991	0.393	9.991	0.436	30
45	5.993	0.288	6.992	0.336	7.991	0.384	8.990	0.432	9.989	0.480	15
3 0	5.992	0.314	6.990	0.366	7.989	0.419	8.988	0.471	9.986	0.523	**87** 0
15	5.990	0.340	6.989	0.397	7.987	0.454	8.986	0.510	9.984	0.567	45
30	5.989	0.366	6.987	0.427	7.985	0.488	8.983	0.549	9.981	0.611	30
45	5.987	0.392	6.985	0.458	7.983	0.523	8.981	0.589	9.979	0.654	15
4 0	5.985	0.419	6.983	0.488	7.981	0.558	8.978	0.628	9.976	0 698	**86** 0
15	5.984	0.445	6.981	0.519	7.978	0.593	8.975	0.667	9.973	0.741	45
30	5.982	0.471	6.978	0.549	7.975	0.628	8.972	0.706	9.969	0.785	30
45	5.979	0.497	6.976	0.580	7.973	0.662	8.969	0.745	9.966	0.828	15
5 0	5.977	0.523	6.973	0.610	7.970	0.697	8.966	0.784	9.962	0.872	**85** 0
15	5.975	0.549	6.971	0.641	7.966	0.732	8.962	0.824	9.958	0.915	45
30	5.972	0 575	6.968	0.671	7.963	0.767	8.959	0.863	9.954	0.959	30
45	5.970	0 601	6.965	0.701	7.960	0.802	8.955	0.902	9.950	1.002	15
6 0	5.967	0.627	6.962	0.732	7.956	0.836	8.951	0.941	9.945	1.045	**84** 0
15	5.964	0.653	6.958	0.762	7.952	0.871	8.947	0.980	9.941	1.089	45
30	5.961	0.679	6.955	0.792	7.949	0.906	8.942	1.019	9.936	1.132	30
45	5.958	0.705	6.951	0.823	7.945	0.940	8.938	1.058	9.931	1.175	15
7 0	5.955	0.731	6.948	0.853	7.940	0.975	8.933	1.097	9.926	1.219	**83** 0
15	5.952	0.757	6.944	0.883	7.936	1.010	8.928	1.136	9.920	1.262	45
30	5.949	0.783	6.940	0.914	7.932	1.044	8.923	1.175	9.914	1.305	30
45	5.945	0.809	6.936	0.944	7.927	1.079	8.918	1.214	9.909	1.349	15
8 0	5.942	0.835	6.932	0.974	7.922	1.113	8.912	1.253	9.903	1.392	**82** 0
15	5.938	0.861	6.928	1.004	7.917	1.148	8.907	1.291	9.897	1.435	45
30	5.934	0.887	6.923	1.035	7.912	1.182	8.901	1.330	9.890	1.478	30
45	5.930	0 913	6.919	1.065	7.907	1.217	8.895	1.369	9.884	1.521	15
9 0	5.926	0.939	6.914	1.095	7.902	1.251	8.889	1.408	9.877	1.564	**81** 0
15	5.922	0.964	6.909	1.125	7.896	1.286	8.883	1.447	9.870	1.607	45
30	5.918	0.990	6.904	1.155	7.890	1.320	8.877	1.485	9.863	1.651	30
45	5.913	1.016	6.899	1.185	7.884	1.355	8.870	1.524	9.856	1.694	15
10 0	5.909	1.042	6.894	1.216	7.878	1.389	8.863	1.563	9.848	1.737	**80** 0
15	5.904	1.068	6.888	1.246	7.872	1.424	8.856	1.601	9.840	1.779	45
30	5.900	1.093	6.883	1.276	7.866	1.458	8.849	1.640	9.833	1.822	30
45	5.895	1.119	6.877	1.306	7.860	1.492	8.842	1.679	9.825	1.865	15
11 0	5.890	1.145	6.871	1.336	7.853	1.526	8.835	1.717	9.816	1.908	**79** 0
15	5.885	1.171	6.866	1.366	7.846	1.561	8.827	1.756	9.808	1.951	45
30	5.880	1.196	6.859	1.396	7.839	1.595	8.819	1.794	9.799	1.994	30
45	5.874	1.222	6.853	1 425	7.832	1.629	8.811	1.833	9.791	2.036	15
12 0	5.869	1.247	6.847	1.455	7.825	1.663	8.803	1.871	9 782	2.079	**78** 0
15	5.863	1.273	6.841	1.485	7.818	1.697	8.795	1.910	9.772	2.122	45
30	5.858	1.299	6.834	1.515	7.810	1.732	8.787	1.948	9.763	2.164	30
45	5.852	1.324	6.827	1.545	7.803	1.766	8.778	1.986	9.753	2.207	15
13 0	5.846	1.350	6.821	1.575	7.795	1.800	8.769	2.025	9.744	2.250	**77** 0
15	5.840	1.375	6.814	1.604	7.787	1.834	8.760	2.063	9.734	2.292	45
30	5.834	1.401	6.807	1.634	7.779	1.868	8.751	2.101	9.724	2.335	30
45	5.828	1.426	6.799	1.664	7.771	1.902	8.742	2.139	9.713	2.377	15
14 0	5.822	1.452	6.792	1.693	7.762	1.935	8.733	2.177	9.703	2.419	**76** 0
15	5.815	1.477	6.785	1.723	7.754	1.969	8.723	2.215	9.692	2.462	45
30	5.809	1.502	6.777	1.753	7.745	2.003	8.713	2.253	9.682	2.504	30
45	5.802	1.528	6.769	1.782	7.736	2.037	8.703	2.291	9.671	2.546	15
15 0	5.796	1.553	6.761	1.812	7.727	2.071	8.693	2.329	9.659	2.588	**75** 0
° '	Dep.	Lat.	Dep.	Lat.	Dep.	Lat.	Dep.	Lat.	Dep.	Lat.	° '
Bearing.	Distance 6.		Distance 7.		Distance 8.		Distance 9.		Distance 10.		Bearing.

75° — 90°

Bearing.	Distance 1.		Distance 2.		Distance 3.		Distance 4.		Distance 5.		Bearing.
° ′	Lat.	Dep.	Lat.	Dep.	Lat.	Dep.	Lat.	Dep.	Lat.	Dep.	° ′
15 15	0.965	0.263	1.930	0.526	2.894	0.789	3.859	1.052	4.824	1.315	**74** 45
30	0.964	0.267	1.927	0.534	2.891	0.802	3.855	1.069	4.818	1.336	30
45	0.962	0.271	1.925	0.543	2.887	0.814	3.850	1.086	4.812	1.357	15
16 0	0.961	0.276	1.923	0.551	2.884	0.827	3.845	1.103	4.806	1.378	**74** 0
15	0.960	0.280	1.920	0.560	2.880	0.839	3.840	1.119	4.800	1.399	45
30	0.959	0.284	1.918	0.568	2.876	0.852	3.835	1.136	4.794	1.420	30
45	0.958	0.288	1.915	0.576	2.873	0.865	3.830	1.153	4.788	1.441	15
17 0	0.956	0.292	1.913	0.585	2.869	0.877	3.825	1.169	4.782	1.462	**73** 0
15	0.955	0.297	1.910	0.593	2.865	0.890	3.820	1.186	4.775	1.483	45
30	0.954	0.301	1.907	0.601	2.861	0.902	3.815	1.203	4.769	1.504	30
45	0.952	0.305	1.905	0.610	2.857	0.915	3.810	1.220	4.762	1.524	15
18 0	0.951	0.309	1.902	0.618	2.853	0.927	3.804	1.236	4.755	1.545	**72** 0
15	0.950	0.313	1.899	0.626	2.849	0.939	3.799	1.253	4.748	1.566	45
30	0.948	0.317	1.897	0.635	2.845	0.952	3.793	1.269	4.742	1.587	30
45	0.947	0.321	1.894	0.643	2.841	0.964	3.788	1.286	4.735	1.607	15
19 0	0.946	0.326	1.891	0.651	2.837	0.977	3.782	1.302	4.728	1.628	**71** 0
15	0.944	0.330	1.888	0.659	2.832	0.989	3.776	1.319	4.720	1.648	45
30	0.943	0.334	1.885	0.668	2.828	1.001	3.771	1.335	4.713	1.669	30
· 45	0.941	0.338	1.882	0.676	2.824	1.014	3.765	1.352	4.706	1.690	15
20 0	0.940	0.342	1.879	0.684	2.819	1.026	3.759	1.368	4.698	1.710	**70** 0
15	0.938	0.346	1.876	0.692	2.815	1.038	3.753	1.384	4.691	1.731	45
30	0.937	0.350	1.873	0.700	2.810	1.051	3.747	1.401	4.683	1.751	30
45	0.935	0.354	1.870	0.709	2.805	1.063	3.741	1.417	4.676	1.771	15
21 0	0.934	0.358	1.867	0.717	2.801	1.075	3.734	1.433	4.668	1.792	**69** 0
15	0.932	0.362	1.864	0.725	2.796	1.087	3.728	1.450	4.660	1.812	45
30	0.930	0.367	1.861	0.733	2.791	1.100	3.722	1.466	4.652	1.833	30
45	0.929	0.371	1.858	0.741	2.786	1.112	3.715	1.482	4.644	1.853	15
22 0	0.927	0.375	1.854	0.749	2.782	1.124	3.709	1.498	4.636	1.873	**68** 0
15	0.926	0.379	1.851	0.757	2.777	1.136	3.702	1.515	4.628	1.893	45
30	0.924	0.383	1.848	0.765	2.772	1.148	3.696	1.531	4.619	1.913	30
45	0.922	0.387	1.844	0.773	2.767	1.160	3.689	1.547	4.611	1.934	15
23 0	0.921	0.391	1.841	0.781	2.762	1.172	3.682	1.563	4.603	1.954	**67** 0
15	0.919	0.395	1.838	0.789	2.756	1.184	3.675	1.579	4.594	1.974	45
30	0.917	0.399	1.834	0.797	2.751	1.196	3.668	1.595	4.585	1.994	30
45	0.915	0.403	1.831	0.805	2.746	1.208	3.661	1.611	4.577	2.014	15
24 0	0.914	0.407	1.827	0.813	2.741	1.220	3.654	1.627	4.568	2.034	**66** 0
15	0.912	0.411	1.824	0.821	2.735	1.232	3.647	1.643	4.559	2.054	45
30	0.910	0.415	1.820	0.829	2.730	1.244	3.640	1.659	4.550	2.073	30
45	0.908	0.419	1.816	0.837	2.724	1.256	3.633	1.675	4.541	2.093	15
25 0	0.906	0.423	1.813	0.845	2.719	1.268	3.625	1.690	4.532	2.113	**65** 0
15	0.904	0.427	1.809	0.853	2.713	1.280	3.618	1.706	4.522	2.133	45
30	0.903	0.431	1.805	0.861	2.708	1.292	3.610	1.722	4.513	2.153	30
45	0.901	0.434	1.801	0.869	2.702	1.303	3.603	1.738	4.503	2.172	15
26 0	0.899	0.438	1.798	0.877	2.696	1.315	3.595	1.753	4.494	2.192	**64** 0
15	0.897	0.442	1.794	0.885	2.691	1.327	3.587	1.769	4.484	2.211	45
30	0.895	0.446	1.790	0.892	2.685	1.339	3.580	1.785	4.475	2.231	30
45	0.893	0.450	1.786	0.900	2.679	1.350	3.572	1.800	4.465	2.250	15
27 0	0.891	0.454	1.782	0.908	2.673	1.362	3.564	1.816	4.455	2.270	**63** 0
15	0.889	0.458	1.778	0.916	2.667	1.374	3.556	1.831	4.445	2.289	45
30	0.887	0.462	1.774	0.923	2.661	1.385	3.548	1.847	4.435	2.309	30
45	0.885	0.466	1.770	0.931	2.655	1.397	3.540	1.862	4.425	2.328	15
28 0	0.883	0.469	1.766	0.939	2.649	1.408	3.532	1.878	4.415	2.347	**62** 0
15	0.881	0.473	1.762	0.947	2.643	1.420	3.524	1.893	4.404	2.367	45
30	0.879	0.477	1.758	0.954	2.636	1.431	3.515	1.909	4.394	2.386	30
45	0.877	0.481	1.753	0.962	2.630	1.443	3.507	1.924	4.384	2.405	15
29 0	0.875	0.485	1.749	0.970	2.624	1.454	3.498	1.939	4.373	2.424	**61** 0
15	0.872	0.489	1.745	0.977	2.617	1.466	3.490	1.954	4.362	2.443	45
30	0.870	0.492	1.741	0.985	2.611	1.477	3.481	1.970	4.352	2.462	30
45	0.868	0.496	1.736	0.992	2.605	1.489	3.473	1.985	4.341	2.481	15
30 0	0.866	0.500	1.732	1.000	2.598	1.500	3.464	2.000	4.330	2.500	**60** 0
° ′	Dep.	Lat.	Dep.	Lat.	Dep.	Lat.	Dep.	Lat.	Dep.	Lat.	° ′
Bearing.	Distance 1.		Distance 2.		Distance 3.		Distance 4.		Distance 5.		Bearing.

Bearing.		Distance 6.		Distance 7.		Distance 8.		Distance 9.		Distance 10.		Bearing.	
°	'	Lat.	Dep.	Lat.	Dep.	Lat.	Dep.	Lat.	Dep.	Lat.	Dep.	°	'
15	15	5.789	1.578	6.754	1.841	7.718	2.104	8.683	2.367	9.648	2.630	74	45
	30	5.782	1.603	6.745	1.871	7.709	2.138	8.673	2.405	9.636	2.672		30
	45	5.775	1.629	6.737	1.900	7.700	2.172	8.662	2.443	9.625	2.714		15
16	0	5.768	1.654	6.729	1.929	7.690	2.205	8.651	2.481	9.613	2.756	74	0
	15	5.760	1.679	6.720	1.959	7.680	2.239	8.640	2.518	9.601	2.798		45
	30	5.753	1.704	6.712	1.988	7.671	2.272	8.629	2.556	9.588	2.840		30
	45	5.745	1.729	6.703	2.017	7.661	2.306	8.618	2.594	9.576	2.882		15
17	0	5.738	1.754	6.694	2.047	7.650	2.339	8.607	2.631	9.563	2.924	73	0
	15	5.730	1.779	6.685	2.076	7.640	2.372	8.595	2.669	9.550	2.965		45
	30	5.722	1.804	6.676	2.105	7.630	2.406	8.583	2.706	9.537	3.007		30
	45	5.714	1.829	6.667	2.134	7.619	2.439	8.572	2.744	9.524	3.049		15
18	0	5.706	1.854	6.657	2.163	7.608	2.472	8.560	2.781	9.511	3.090	72	0
	15	5.698	1.879	6.648	2.192	7.598	2.505	8.547	2.818	9.497	3.132		45
	30	5.690	1.904	6.638	2.221	7.587	2.538	8.535	2.856	9.483	3.173		30
	45	5.682	1.929	6.629	2.250	7.575	2.572	8.522	2.893	9.469	3.214		15
19	0	5.673	1.953	6.619	2.279	7.564	2.605	8.510	2.930	9.455	3.256	71	0
	15	5.665	1.978	6.609	2.308	7.553	2.638	8.497	2.967	9.441	3.297		45
	30	5.656	2.003	6.598	2.337	7.541	2.670	8.484	3.004	9.426	3.338		30
	45	5.647	2.028	6.588	2.365	7.529	2.703	8.471	3.041	9.412	3.379		15
20	0	5.638	2.052	6.578	2.394	7.518	2.736	8.457	3.078	9.397	3.420	70	0
	15	5.629	2.077	6.567	2.423	7.506	2.769	8.444	3.115	9.382	3.461		45
	30	5.620	2.101	6.557	2.451	7.493	2.802	8.430	3.152	9.367	3.502		30
	45	5.611	2.126	6.546	2.480	7.481	2.834	8.416	3.189	9.351	3.543		15
21	0	5.601	2.150	6.535	2.509	7.469	2.867	8.102	3.225	9.336	3.584	69	0
	15	5.592	2.175	6.524	2.537	7.456	2.900	8.388	3.262	9.320	3.624		45
	30	5.582	2.199	6.513	2.566	7.443	2.932	8.374	3.299	9.304	3.665		30
	45	5.573	2.223	6.502	2.594	7.430	2.964	8.359	3.335	9.288	3.706		15
22	0	5.563	2.248	6.490	2.622	7.417	2.997	8.345	3.371	9.272	3.746	68	0
	15	5.553	2.272	6.479	2.651	7.404	3.029	8.330	3.408	9.255	3.787		45
	30	5.543	2.296	6.467	2.679	7.391	3.061	8.315	3.444	9.239	3.827		30
	45	5.533	2.320	6.455	2.707	7.378	3.094	8.300	3.480	9.222	3.867		15
23	0	5.523	2.344	6.444	2.735	7.364	3.126	8.285	3.517	9.205	3.907	67	0
	15	5.513	2.368	6.432	2.763	7.350	3.158	8.269	3.553	9.188	3.947		45
	30	5.502	2.392	6.419	2.791	7.336	3.190	8.254	3.589	9.171	3.988		30
	45	5.492	2.416	6.407	2.819	7.322	3.222	8.238	3.625	9.153	4.028		15
24	0	5.481	2.440	6.395	2.847	7.308	3.254	8.222	3.661	9.136	4.067	66	0
	15	5.471	2.464	6.382	2.875	7.294	3.286	8.206	3.696	9.118	4.107		45
	30	5.460	2.488	6.370	2.903	7.280	3.318	8.190	3.732	9.100	4.147		30
	45	5.449	2.512	6.357	2.931	7.265	3.349	8.173	3.768	9.081	4.187		15
25	0	5.438	2.536	6.344	2.958	7.250	3.381	8.157	3.804	9.063	4.226	65	0
	15	5.427	2.559	6.331	2.986	7.236	3.413	8.140	3.839	9.045	4.266		45
	30	5.416	2.583	6.318	3.014	7.221	3.444	8.123	3.875	9.026	4.305		30
	45	5.404	2.607	6.305	3.041	7.206	3.476	8.106	3.910	9.007	4.345		15
26	0	5.393	2.630	6.292	3.069	7.190	3.507	8.089	3.945	8.988	4.384	64	0
	15	5.381	2.654	6.278	3.096	7.175	3.538	8.072	3.981	8.969	4.423		45
	30	5.370	2.677	6.265	3.123	7.160	3.570	8.054	4.016	8.949	4.462		30
	45	5.358	2.701	6.251	3.151	7.144	3.601	8.037	4.051	8.930	4.501		15
27	0	5.346	2.724	6.237	3.178	7.128	3.632	8.019	4.086	8.910	4.540	63	0
	15	5.334	2.747	6.223	3.205	7.112	3.663	8.001	4.121	8.890	4.579		45
	30	5.322	2.770	6.209	3.232	7.096	3.694	7.983	4.156	8.870	4.618		30
	45	5.310	2.794	6.195	3.259	7.080	3.725	7.965	4.190	8.850	4.656		15
28	0	5.298	2.817	6.181	3.286	7.064	3.756	7.947	4.225	8.829	4.695	62	0
	15	5.285	2.840	6.166	3.313	7.047	3.787	7.928	4.260	8.809	4.733		45
	30	5.273	2.863	6.152	3.340	7.031	3.817	7.909	4.294	8.788	4.772		30
	45	5.260	2.886	6.137	3.367	7.014	3.848	7.891	4.329	8.767	4.810		15
29	0	5.248	2.909	6.122	3.394	6.997	3.878	7.872	4.363	8.746	4.848	61	0
	15	5.235	2.932	6.107	3.420	6.980	3.909	7.852	4.398	8.725	4.886		45
	30	5.222	2.955	6.093	3.447	6.963	3.939	7.833	4.432	8.704	4.924		30
	45	5.209	2.977	6.077	3.474	6.946	3.970	7.814	4.466	8.682	4.962		15
30	0	5.196	3.000	6.062	3.500	6.928	4.000	7.794	4.500	8.660	5.000	60	0
°	'	Dep.	Lat.	Dep.	Lat.	Dep.	Lat.	Dep.	Lat.	Dep.	Lat.	°	'
Bearing.		Distance 6.		Distance 7.		Distance 8.		Distance 9.		Distance 10.		Bearing.	

60° — 75°

Bearing.	Distance 1.		Distance 2.		Distance 3.		Distance 4.		Distance 5.		Bearing.
° '	Lat.	Dep.	Lat.	Dep.	Lat.	Dep.	Lat.	Dep.	Lat.	Dep.	° '
30 15	0.864	0.504	1.728	1.008	2.592	1.511	3.455	2.015	4.319	2.519	**59** 45
30	0.862	0.508	1.723	1.015	2.585	1.523	3.447	2.030	4.308	2.538	30
45	0.859	0.511	1.719	1.023	2.578	1.534	3.438	2.045	4.297	2.556	15
31 0	0.857	0.515	1.714	1.030	2.572	1.545	3.429	2.060	4.286	2.575	**59** 0
15	0.855	0.519	1.710	1.038	2.565	1.556	3.420	2.075	4.275	2.594	45
30	0.853	0.522	1.705	1.045	2.558	1.567	3.411	2.090	4.263	2.612	30
45	0.850	0.526	1.701	1.052	2.551	1.579	3.401	2.105	4.252	2.631	15
32 0	0.848	0.530	1.696	1.060	2.544	1.590	3.392	2.120	4.240	2.650	**58** 0
15	0.846	0.534	1.691	1.067	2.537	1.601	3.383	2.134	4.229	2.668	45
30	0.843	0.537	1.687	1.075	2.530	1.612	3.374	2.149	4.217	2.686	30
45	0.841	0.541	1.682	1.082	2.523	1.623	3.364	2.164	4.205	2.705	15
33 0	0.839	0.545	1.677	1.089	2.516	1.634	3.355	2.179	4.193	2.723	**57** 0
15	0.836	0.548	1.673	1.097	2.509	1.645	3.345	2.193	4.181	2.741	45
30	0.834	0.552	1.668	1.104	2.502	1.656	3.336	2.208	4.169	2.760	30
45	0.831	0.556	1.663	1.111	2.494	1.667	3.326	2.222	4.157	2.778	15
34 0	0.829	0.559	1.658	1.118	2.487	1.678	3.316	2.237	4.145	2.796	**56** 0
15	0.827	0.563	1.653	1.126	2.480	1.688	3.306	2.251	4.133	2.814	45
30	0.824	0.566	1.648	1.133	2.472	1.699	3.297	2.266	4.121	2.832	30
45	0.822	0.570	1.643	1.140	2.465	1.710	3.287	2.280	4.108	2.850	15
35 0	0.819	0.574	1.638	1.147	2.457	1.721	3.277	2.294	4.096	2.868	**55** 0
15	0.817	0.577	1.633	1.154	2.450	1.731	3.267	2.309	4.083	2.886	45
30	0.814	0.581	1.628	1.161	2.442	1.742	3.257	2.323	4.071	2.904	30
45	0.812	0.584	1.623	1.168	2.435	1.753	3.246	2.337	4.058	2.921	15
36 0	0.809	0.588	1.618	1.176	2.427	1.763	3.236	2.351	4.045	2.939	**54** 0
15	0.806	0.591	1.613	1.183	2.419	1.774	3.226	2.365	4.032	2.957	45
30	0.804	0.595	1.608	1.190	2.412	1.784	3.215	2.379	4.019	2.974	30
45	0.801	0.598	1.603	1.197	2.404	1.795	3.205	2.393	4.006	2.992	15
37 0	0.799	0.602	1.597	1.204	2.396	1.805	3.195	2.407	3.993	3.009	**53** 0
15	0.796	0.605	1.592	1.211	2.388	1.816	3.184	2.421	3.980	3.026	45
30	0.793	0.609	1.587	1.218	2.380	1.826	3.173	2.435	3.967	3.044	30
45	0.791	0.612	1.581	1.224	2.372	1.837	3.163	2.449	3.953	3.061	15
38 0	0.788	0.616	1.576	1.231	2.364	1.847	3.152	2.463	3.940	3.078	**52** 0
15	0.785	0.619	1.571	1.238	2.356	1.857	3.141	2.476	3.927	3.095	45
30	0.783	0.623	1.565	1.245	2.348	1.868	3.130	2.490	3.913	3.113	30
45	0.780	0.626	1.560	1.252	2.340	1.878	3.120	2.504	3.899	3.130	15
39 0	0.777	0.629	1.554	1.259	2.331	1.888	3.109	2.517	3.886	3.147	**51** 0
15	0.774	0.633	1.549	1.265	2.323	1.898	3.098	2.531	3.872	3.164	45
30	0.772	0.636	1.543	1.272	2.315	1.908	3.086	2.544	3.858	3.180	30
45	0.769	0.639	1.538	1.279	2.307	1.918	3.075	2.558	3.844	3.197	15
40 0	0.766	0.643	1.532	1.286	2.298	1.928	3.064	2.571	3.830	3.214	**50** 0
15	0.763	0.646	1.526	1.292	2.290	1.938	3.053	2.584	3.816	3.231	45
30	0.760	0.649	1.521	1.299	2.281	1.948	3.042	2.598	3.802	3.247	30
45	0.758	0.653	1.515	1.306	2.273	1.958	3.030	2.611	3.788	3.264	15
41 0	0.755	0.656	1.509	1.312	2.264	1.968	3.019	2.624	3.774	3.280	**49** 0
15	0.752	0.659	1.504	1.319	2.256	1.978	3.007	2.637	3.759	3.297	45
30	0.749	0.663	1.498	1.325	2.247	1.988	2.996	2.650	3.745	3.313	30
45	0.746	0.666	1.492	1.332	2.238	1.998	2.984	2.664	3.730	3.329	15
42 0	0.743	0.669	1.486	1.338	2.229	2.007	2.973	2.677	3.716	3.346	**48** 0
15	0.740	0.672	1.480	1.345	2.221	2.017	2.961	2.689	3.701	3.362	45
30	0.737	0.676	1.475	1.351	2.212	2.027	2.949	2.702	3.686	3.378	30
45	0.734	0.679	1.469	1.358	2.203	2.036	2.937	2.715	3.672	3.394	15
43 0	0.731	0.682	1.463	1.364	2.194	2.046	2.925	2.728	3.657	3.410	**47** 0
15	0.728	0.685	1.457	1.370	2.185	2.056	2.913	2.741	3.642	3.426	45
30	0.725	0.688	1.451	1.377	2.176	2.065	2.901	2.753	3.627	3.442	30
45	0.722	0.692	1.445	1.383	2.167	2.075	2.889	2.766	3.612	3.458	15
44 0	0.719	0.695	1.439	1.389	2.158	2.084	2.877	2.779	3.597	3.473	**46** 0
15	0.716	0.698	1.433	1.396	2.149	2.093	2.865	2.791	3.582	3.489	45
30	0.713	0.701	1.427	1.402	2.140	2.103	2.853	2.804	3.566	3.505	30
45	0.710	0.704	1.420	1.408	2.131	2.112	2.841	2.816	3.551	3.520	15
45 0	0.707	0.707	1.414	1.414	2.121	2.121	2.828	2.828	3.536	3.536	**45** 0
° '	Dep.	Lat.	Dep.	Lat.	Dep.	Lat.	Dep.	Lat.	Dep.	Lat.	° '
Bearing.	Distance 1.		Distance 2.		Distance 3.		Distance 4.		Distance 5.		Bearing.

45° — 60°

° ′	Lat.	Dep.	Lat.	Dep.	Lat.	Dep.	Lat.	Dep.	Lat.	Dep.	° ′
30 15	5.183	3.023	6.047	3.526	6.911	4.030	7.775	4.534	8.638	5.038	**59** 45
30	5.170	3.045	6.031	3.553	6.893	4.060	7.755	4.568	8.616	5.075	30
45	5.156	3.068	6.016	3.579	6.875	4.090	7.735	4.602	8.594	5.113	15
31 0	5.143	3.090	6.000	3.605	6.857	4.120	7.715	4.635	8.572	5.150	**59** 0
15	5.129	3.113	5.984	3.631	6.839	4.150	7.694	4.669	8.549	5.188	45
30	5.116	3.135	5.968	3.657	6.821	4.180	7.674	4.702	8.526	5.225	30
45	5.102	3.157	5.952	3.683	6.803	4.210	7.653	4.736	8.504	5.262	15
32 0	5.088	3.180	5.936	3.709	6.784	4.239	7.632	4.769	8.481	5.299	**58** 0
15	5.074	3.202	5.920	3.735	6.766	4.269	7.612	4.802	8.457	5.336	45
30	5.060	3.224	5.904	3.761	6.747	4.298	7.591	4.836	8.434	5.373	30
45	5.046	3.246	5.887	3.787	6.728	4.328	7.569	4.869	8.410	5.410	15
33 0	5.032	3.268	5.871	3.812	6.709	4.357	7.548	4.902	8.387	5.446	**57** 0
15	5.018	3.290	5.854	3.838	6.690	4.386	7.527	4.935	8.363	5.483	45
30	5.003	3.312	5.837	3.864	6.671	4.416	7.505	4.967	8.339	5.519	30
45	4.989	3.333	5.820	3.889	6.652	4.445	7.483	5.000	8.315	5.556	15
34 0	4.974	3.355	5.803	3.914	6.632	4.474	7.461	5.033	8.290	5.592	**56** 0
15	4.960	3.377	5.786	3.940	6.613	4.502	7.439	5.065	8.266	5.628	45
30	4.945	3.398	5.769	3.965	6.593	4.531	7.417	5.098	8.241	5.664	30
45	4.930	3.420	5.752	3.990	6.573	4.560	7.395	5.130	8.217	5.700	15
35 0	4.915	3.441	5.734	4.015	6.553	4.589	7.372	5.162	8.192	5.736	**55** 0
15	4.900	3.463	5.716	4.040	6.533	4.617	7.350	5.194	8.166	5.772	45
30	4.885	3.484	5.699	4.065	6.513	4.646	7.327	5.226	8.141	5.807	30
45	4.869	3.505	5.681	4.090	6.493	4.674	7.304	5.258	8.116	5.843	15
36 0	4.854	3.527	5.663	4.115	6.472	4.702	7.281	5.290	8.090	5.878	**54** 0
15	4.839	3.548	5.645	4.139	6.452	4.730	7.258	5.322	8.064	5.913	45
30	4.823	3.569	5.627	4.164	6.431	4.759	7.235	5.353	8.039	5.948	30
45	4.808	3.590	5.609	4.188	6.410	4.787	7.211	5.385	8.013	5.983	15
37 0	4.792	3.611	5.590	4.213	6.389	4.815	7.188	5.416	7.986	6.018	**53** 0
15	4.776	3.632	5.572	4.237	6.368	4.842	7.164	5.448	7.960	6.053	45
30	4.760	3.653	5.554	4.261	6.347	4.870	7.140	5.479	7.934	6.088	30
45	4.744	3.673	5.535	4.286	6.326	4.898	7.116	5.510	7.907	6.122	15
38 0	4.728	3.694	5.516	4.310	6.304	4.925	7.092	5.541	7.880	6.157	**52** 0
15	4.712	3.715	5.497	4.334	6.283	4.953	7.068	5.572	7.853	6.191	45
30	4.696	3.735	5.478	4.358	6.261	4.980	7.043	5.603	7.826	6.225	30
45	4.679	3.756	5.459	4.381	6.239	5.007	7.019	5.633	7.799	6.259	15
39 0	4.663	3.776	5.440	4.405	6.217	5.035	6.994	5.664	7.772	6.293	**51** 0
15	4.646	3.796	5.421	4.429	6.195	5.062	6.970	5.694	7.744	6.327	45
30	4.630	3.816	5.401	4.453	6.173	5.089	6.945	5.725	7.716	6.361	30
45	4.613	3.837	5.382	4.476	6.151	5.116	6.920	5.755	7.688	6.394	15
40 0	4.596	3.857	5.362	4.500	6.128	5.142	6.894	5.785	7.660	6.428	**50** 0
15	4.579	3.877	5.343	4.523	6.106	5.169	6.869	5.815	7.632	6.461	45
30	4.562	3.897	5.323	4.546	6.083	5.196	6.844	5.845	7.604	6.495	30
45	4.545	3.917	5.303	4.569	6.061	5.222	6.818	5.875	7.576	6.528	15
41 0	4.528	3.936	5.283	4.592	6.038	5.248	6.792	5.905	7.547	6.561	**49** 0
15	4.511	3.956	5.263	4.615	6.015	5.275	6.767	5.934	7.518	6.594	45
30	4.494	3.976	5.243	4.638	5.992	5.301	6.741	5.964	7.490	6.626	30
45	4.476	3.995	5.222	4.661	5.968	5.327	6.715	5.993	7.461	6.659	15
42 0	4.459	4.015	5.202	4.684	5.945	5.353	6.688	6.022	7.431	6.691	**48** 0
15	4.441	4.034	5.182	4.707	5.922	5.379	6.662	6.051	7.402	6.724	45
30	4.424	4.054	5.161	4.729	5.898	5.405	6.635	6.080	7.373	6.756	30
45	4.406	4.073	5.140	4.752	5.875	5.430	6.609	6.109	7.343	6.788	15
43 0	4.388	4.092	5.119	4.774	5.851	5.456	6.582	6.138	7.314	6.820	**47** 0
15	4.370	4.111	5.099	4.796	5.827	5.481	6.555	6.167	7.284	6.852	45
30	4.352	4.130	5.078	4.818	5.803	5.507	6.528	6.195	7.254	6.884	30
45	4.334	4.149	5.057	4.841	5.779	5.532	6.501	6.224	7.224	6.915	15
44 0	4.316	4.168	5.035	4.863	5.755	5.557	6.474	6.252	7.193	6.947	**46** 0
15	4.298	4.187	5.014	4.885	5.730	5.582	6.447	6.280	7.163	6.978	45
30	4.280	4.206	4.993	4.906	5.706	5.607	6.419	6.308	7.133	7.009	30
45	4.261	4.224	4.971	4.928	5.681	5.632	6.392	6.336	7.102	7.040	15
45 0	4.243	4.243	4.950	4.950	5.657	5.657	6.364	6.364	7.071	7.071	**45** 0
° ′	Dep.	Lat.	Dep.	Lat.	Dep.	Lat.	Dep.	Lat.	Dep.	Lat.	° ′

45° — 60°

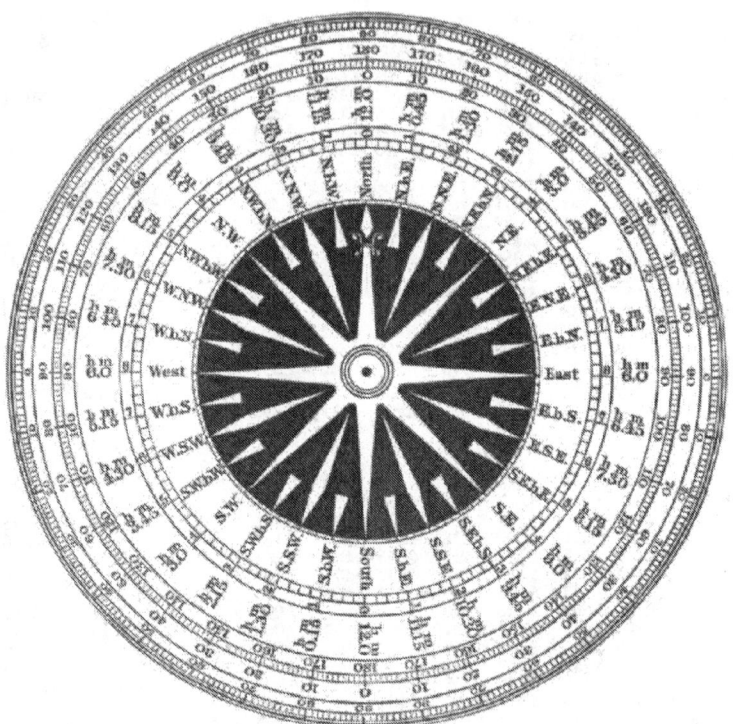

A TABLE OF THE ANGLES

Which every Point and Quarter Point of the Compass makes with the Meridian.

North.		Points.	° ′ ″	Points.	South.	
		0 – ¼	2 48 45	0 – ¼		
		0 – ½	5 37 30	0 – ½		
		0 – ¾	8 26 15	0 – ¾		
N. by E.	N. by W.	1	11 15 0	1	S. by E.	S. by W.
		1 – ¼	14 3 45	1 – ¼		
		1 – ½	16 52 30	1 – ½		
		1 – ¾	19 41 15	1 – ¾		
N.N.E.	N.N.W.	2	22 30 0	2	S.S.E.	S.S.W.
		2 – ¼	25 18 45	2 – ¼		
		2 – ½	28 7 30	2 – ½		
		2 – ¾	30 56 15	2 – ¾		
N.E. by N.	N.W. by N.	3	33 45 0	3	S.E. by S.	S.W. by S.
		3 – ¼	36 33 45	3 – ¼		
		3 – ½	39 22 30	3 – ½		
		3 – ¾	42 11 15	3 – ¾		
N.E.	N.W.	4	45 0 0	4	S.E.	S.W.
		4 – ¼	47 48 45	4 – ¼		
		4 – ½	50 37 30	4 – ½		
		4 – ¾	53 26 15	4 – ¾		
N.E. by E	N.W. by W.	5	56 15 0	5	S.E. by E.	S.W. by W.
		5 – ¼	59 3 45	5 – ¼		
		5 – ½	61 52 30	5 – ½		
		5 – ¾	64 41 15	5 – ¾		
E.N.E.	W.N.W.	6	67 30 0	6	E.S.E.	W.S.W.
		6 – ¼	70 18 45	6 – ¼		
		6 – ½	73 7 30	6 – ½		
		6 – ¾	75 56 15	6 – ¾		
E. by N.	W. by N.	7	78 45 0	7	E. by S.	W. by S.
		7 – ¼	81 33 45	7 – ¼		
		7 – ½	84 22 30	7 – ½		
		7 – ¾	87 11 15	7 – ¾		
East.	West.	8	90 0 0	8	East.	West.